Study Guide and
Student Solutions Manual

for use with

Statistics
A First Course

Sixth Edition

Donald H. Sanders

Robert K. Smidt
California Polytechnic State University

Prepared by
Robert K. Smidt
California Polytechnic State University

John A. Banks
San Jose City College and Evergreen Valley College

Boston Burr Ridge, IL Dubuque, IA Madison, WI New York San Francisco St. Louis
Bangkok Bogotá Caracas Lisbon London Madrid
Mexico City Milan New Delhi Seoul Singapore Sydney Taipei Toronto

McGraw-Hill Higher Education

A Division of The McGraw-Hill Companies

Study Guide and Student Solutions Manual for use with
STATISTICS: A FIRST COURSE, SIXTH EDITION

1 2 3 4 5 6 7 8 9 0 CUS CUS 9 0 3 2 1 0 9

ISBN 0-07-229554-6

www.mhhe.com

Contents

Preface

This Study Guide and Student Solutions Manual is intended to be used with the sixth edition of *Statistics: A First Course* by Don Sanders and Robert Smidt. Its purpose is to help you learn statistics, and succeed in your course.

The Study Aids and Practice Exercises is not meant to be a replacement for the textbook. It is intended to complement the text, distilling and illustrating the many ideas and applications of statistics. Each section of the Study Aids and Practice Exercises corresponds to a section of the text and contains the following elements:

Study Objectives The expected outcomes for each section are presented here. You should read these before working on a section and reread them afterwards to verify that the goals have been accomplished.

Section Overview Sometimes in reading a technical book, it is possible to find yourself lost in details. This overview should help to summarize and bring to the forefront the important concepts and techniques of each section.

Key Terms & Formulas This provides a brief summary of the important terms and formulas from a section. This is a helpful reference after you have digested the text and study guide material, but have forgotten some detail.

Worked Examples Statistics is a problem solving science. There are many data-analytic techniques the text explains. You will find examples of all the important methods here. Each has a detailed solution that you can use as a model in solving problems.

Practice Exercises You learn statistics by doing statistics. For each example in the *Study Guide*, you will find corresponding exercises. To truly understand the details of the methods, you will need to do a substantial portion or even all these problems.

Solutions to Practice Exercises For every exercise in the *Study Guide*, a detailed solution is provided. These allow you to do more than just see if you did some calculation correctly--you can verify that your assumptions, interpretations and conclusions are all correct.

Hints & Cautions Throughout the *Study Guide* you will find Hints and Cautions. These provide details, insights, tricks, and warnings that should make your studying go a little smoother and help your understanding go a little deeper.

The Solutions to Odd-Numbered Exercises provides detailed solutions to all odd-numbered exercises from *Statistics: A First Course*. In almost all cases, solutions are written with all of the steps shown. Where appropriate, output from MINITAB is placed after the written solution. This allows MINITAB users to compare the two results and check their own MINITAB output. Differences between the hand-worked solutions and the MINITAB solutions are due to rounding, and can be ignored.

With all the applications of statistics to various fields and interests, you are bound to see uses for statistics in your own life. Use the *Study Guide* to help your transition from the statistically unaware (sometimes just referred to as "those poor, poor people"), to the statistically enlightened (or "the few, the brave . . ."). Enjoy yourself.

A few words of thanks. The MINITAB program was provided by MINITAB Inc., 3081 Enterprise Drive, State College, PA 16801-3008. David Dietz and Susan Brusch of McGraw-Hill were supportive, helpful, and efficient. They made the work more pleasant than it otherwise would have been.

Robert Smidt
John Banks

Chapter 1 Let's Get Started

Study Aids and Practice Exercises

1-1 What to Expect

Study Objective
You should be able to:
1. Recognize the objective of this text.
2. Relax.

Section Overview
   Statistics is a science that we can use. As opposed to esoteric fields that only the expert can practice, many people use statistics on their job or in their everyday life. You do not need to be an expert mathematician to be a user of statistics. More important than any mathematical skill is the ability to think logically, examine critically, and explain clearly.

1-2 Purpose and Organization of the Text

Study Objective
You should be able to:
1. Understand the definition of Statistics.
2. Distinguish between a population and a sample.
3. Differentiate between parameters and statistics.
4. Understand the functions of descriptive and inferential statistics.

Section Overview
   A statistician tries to make sense of data. Primarily, a statistician is interested in understanding a data set called a population. Examples of populations are the percentages of body fat of overweight males, heights of thoroughbred horses, incomes of residents of Wyoming (other than the horses), amounts of oil in shale found in Alberta, stock market prices over a year, cap sizes of "Ma'am on Motorcycle" mushrooms (that's really their name), or speeds of cars receiving tickets on Highway 1 over a one week period. A complete examination of a population is called a census. Rather than taking a census, most people prefer to take a sample from the population. Usually a sample will provide enough information to make reasonable judgments about the population. Often these judgments are about a characteristic of a population, called a parameter. A parameter is important because it gives us information about the nature of the population. Examples of parameters include the average percentage of body fat of overweight males, the percent of thoroughbred horses that are taller than their sire, the percent of residents of Wyoming with incomes below the poverty level, the mean cap sizes of "Ma'am on Motorcycle"

mushrooms. When a parameter describing a population is unknown, we try to provide a reasonable guess for it by calculating a similar quantity from a sample. We might sample 100 Wyoming residents and determine what percent have incomes below the poverty level or we could measure the cap sizes of 15 "Ma'am on Motorcycle" mushrooms and calculate their average size. Such sample based quantities are called statistics.

Descriptive statistics includes all the methods used to organize, present, and summarize a data set. We might graph the body fat of overweight males, prepare a table of the incomes for Wyoming residents, or obtain the minimum and maximum cap size of mushrooms. Inferential statistics includes techniques we use to reach decisions about a population based on the information contained in a sample. These inferences usually take the form of a statement about a population parameter based on the value of a sample statistic. For example, we might use our sample of 100 Wyoming residents and estimate that the percent below the poverty level is somewhere between 8% and 14%. Or we might use a sample of 15 mushrooms found in Washington to see if their average size is greater than that of Oregon mushrooms.

Key Terms & Formulas

Statistics The science of collecting, presenting, and analyzing data to see what information the data contains and to see what conclusions the data supports.

Population The complete collection under study. This collection could be people, animals, objects, transactions, measurements, or anything else of interest.

Census An examination of every member of a population.

Sample A part or subset of a population. A sample is taken because we want to use it to gain information about the population.

Parameter A numerical characteristic of a population.

Statistic A numerical characteristic of a sample.

Descriptive Statistics The procedures for collecting, classifying, summarizing and presenting data.

Inferential Statistics The procedures for using sample statistics to reach conclusions about population parameters.

Worked Examples

Power

A power consortium is considering the construction of a coal generated power plant near Kissimmee Florida. They want to know the proportion of Kissimmee residents that oppose the plant. They plan to ask the opinions of 200 residents selected from a phone book. Identify the population, parameter, sample and statistic for this situation.

Solution

The population is the opinions of the Kissimmee residents. The parameter is the proportion of residents that oppose the plant. The sample is the opinions of the 200 residents phoned. The statistic is the proportion in the sample that oppose the plant.

> Hint: Notice that a population and its characteristic, a parameter, both begin with the letter p. Also notice that a sample and its characteristic, a statistic, both start with the letter s. It makes it easier to keep the definitions straight.

Prison Testosterone

A prison psychologist suspects that male prison inmates, on the average, have elevated amounts of testosterone. The psychologist samples 20 inmates from a prison in Lompoc, California and measures the amounts of testosterone in each. Identify the population, parameter, sample, and statistic for this situation. Also give two examples of inferences that might be made.

Solution

The population is the testosterone levels of male prison inmates. The parameter is the mean testosterone level of all such inmates. The sample is the testosterone levels of the 20 sampled inmates and the statistic is their mean. One inference would be to use the sample mean to estimate the population mean. Alternatively, we could use the sample mean as the basis of a test to decide if the male inmates have an elevated mean level of testosterone.

> Hint: There is a problem with the way this sample was taken. While the target population might be the testosterone amounts in all prison inmates, the sample came exclusively from the Lompoc prison. So the population actually sampled is those inmates incarcerated at Lompoc. Any extension to a larger group of prison inmates is of dubious validity.

Migrating Monarchs

An ecologist collected 120 Monarch butterflies that recently migrated from Mexico and for each found the number of damaged wings. Give two examples of descriptive statistics the ecologist might use and two examples of statistical inferences that might be made.

Solution

To describe the data, the ecologist could create a table giving the number of butterflies with no wings damaged, the number with one wing damaged, etc. Alternatively, the ecologist could calculate the average number of damaged wings. Inferentially, the ecologist could use the sample mean as an estimate of the mean number of damaged wings for all recently migrated Monarch butterflies. The ecologist could also test to see if the mean number of damaged wings is greater than before the migration.

Practice Exercises 1-2

1-4. A mortgage broker is interested in the mean appraisal time for new construction projects. He samples 44 new projects and learns their mean appraisal time. For this situation:

1. Identify the population.
2. Define the parameter.
3. Describe the sample.
4. Determine the statistic involved.

5-8. A politician wants to know the proportion of her constituents who favor a balanced national budget. The politician asks 330 constituents for their opinions on a balanced budget. For this situation:

5. Identify the population.
6. Define the parameter.
7. Describe the sample.
8. Determine the statistic involved.

Solutions to Practice Exercises 1-2

1. The population is the appraisal times for new construction projects.
2. The parameter is the mean appraisal time for new construction projects.
3. The sample is the appraisal times for the 44 projects.
4. The statistic is the mean appraisal time of the 44 projects.
5. The population is the opinions of the politician's constituents on a balanced budget.
6. The parameter is the proportion of the constituents in favor of a balanced budget.
7. The sample is the opinions of the 330 constituents.
8. The statistic is the proportion of the 330 constituents who favor a balanced budget.

1-3 Need for Statistics

Study Objective

You should be able to:
1. See the need to describe relationships between variables.
2. Learn how statistics can be used to make better decisions.

Section Overview

This section gives many examples of two of the more important functions of statistics, describing relationships between variables and making decisions.

Worked Examples

Insurance Risks

An insurance company assigns people to different risk groups based on personal characteristics and the types of insurance carried. Why?

Solution

The insurance company recognizes that there is a relation between characteristics of a person and the chance that the insurance company will have to pay on an insurance policy. For example, a 20-year old male will probably pay a large premium on auto insurance, but a smaller amount on life insurance.

Fabrics

A designer is considering a switch from cotton to a new fabric for a line of clothing. The designer will switch to this new fabric if he believes that more than 50% of his potential clients will prefer this fabric. What suggestion would you make to help him decide?

Solution

The designer could take a survey (i.e., sample) of his potential clients and learn which fabric they prefer. The percent in the sample that prefer the new fabric would suggest if the switch would be a good idea.

Practice Exercises 1-3

1. A criminologist suggests to the mayor of a city that the mayor increase the budget of the neighborhood watch program. Why?
2. High school juniors and seniors often take career preference tests. These tests are multiple choice and do not ask the student what career they would prefer. Why?
3. The EPA is investigating a plant that discharges heated water into a lake. If the water temperature increases an average of over twenty degrees, damage is done to the ecosystem of the lake. What would you suggest the EPA do?

Solutions to Practice Exercises 1-3

1. The criminologist obviously believes that there is a relationship (hopefully a negative one) between the number of neighborhood watch programs and the crime rate.
2. Often juniors and senior in high school do not know enough about what is involved in different careers to make educated choices. The career preference tests are created with the idea that people who enjoy certain activities would prefer particular careers. School counselors use these tests to help suggest to students possible careers.
3. The EPA should sample the temperature increases throughout the lake and use this sample to make a determination regarding the fate of the plant. If the sample provides strong evidence that the average temperature increase exceeds 20 degrees, it would be appropriate for the EPA to take further steps in controlling the plant's discharge.

1-4 Statistical Problem-Solving Methodology

Study Objective

You should be able to:
1. Understand the six steps of statistical problem-solving methodology.
2. Identify various sampling methods.

Section Overview

Good planning can prevent wasted time and effort. The text suggests a series of steps for statistical problem solving. These steps help our experiment produce results that are valid, accurate, and useful. These steps include identifying the problem, gathering available facts, gathering new data (sampling), classifying and summarizing the data, presenting and analyzing the data, and reaching a conclusion. An important part of this process is sampling. We want samples that have a good chance of being representative of the population. Probability samples are such samples.

Key Terms & Formulas

Statistical Problem Solving A series of steps used in an experiment to insure that the resulting conclusions will be valid and accurate. These steps are:
  1. Identify the problem or opportunity.
  2. Decide on the method of data collection.
  3. Collecting the data.
  4. Classify and summarize the data.
  5. Present and analyze the data.
  6. Reach a decision.

Judgment Sample A sample based on a person's opinion about which items are characteristic of a population.

Voluntary Sample A sample based only on the responses of voluntary participants.

Convenience Sample A sample collected in a manner convenient for the interviewer.

Probability Sample A sample in which the chance of selecting each item is known before the sample is chosen.
    Included among probability samples are simple random, systematic, stratified, and cluster samples.

Simple Random Sample A sample taken so that every group of size n has an equal chance of being selected.

Stratified Sample The population is divided into groups or strata. Then a simple random sample is taken from each strata.

Systematic Sample The population is ordered and divided into n groups. Then one item is randomly selected from the first group and the items in the corresponding position of the subsequent groups are also selected.

Cluster Sample The population is divided into clusters. Then a certain number of clusters are randomly selected and all items in those clusters are selected.

Worked Examples

Sampling a Class

Suppose a teacher wants to sample eight students from a class of 32. The class is arranged in eight rows of four chairs as diagrammed. Illustrate each of the sampling plans discussed in the text with this situation.

Solution

For a judgment sample, the teacher would select eight people that he thinks best represents the class. To take a simple random sample from the class, the teacher could number the seats 1 through 32. Then he would randomly select eight numbers from 32 to select the seats to pull into the sample. A computer could generate these random numbers, or he could pick eight pieces of paper from 32 that have numbered 1 through 32 and placed in a hat. To take a systematic sample from the class, he could number the seats in each row from 1 to 4. Then he would randomly select one number from 1 to 4 and select that seat in every row. For example, if a 3 is selected, the third person in each row would be taken in the sample. To take a stratified sample from the class, the teacher would again number the seats in each row from 1 to 4. However, then he would randomly select one number from 1 to 4 for each row and select from each row the seat selected. For example, if 3, 4, 4, 1, 2, 3, 1, 3 are selected, the third person in the first row, the fourth person in the second row, the fourth person in the third row, etc., would be taken into the sample. To take a cluster sample from the class, the teacher would number the rows from 1 to 8. Then the teacher would randomly select two numbers from 1 to 8 and select everyone in those two rows. For example, if 3 and 8 are selected, everyone in the third and eighth row would be taken into the sample.

Comment: Any valid probability sample can yield valid, analyzable results. However, in the text, only the analysis of simple random samples is considered. Luckily, few adjustments are necessary to analyze other probability samples.

Practice Exercises 1-4

1-6. The director of an economic research firm wants to take a survey at Guernsey, Wyoming. She wants the residents' opinions about a nearby strip-mining operation. The director would like to obtain the opinion of 40 of the 800 residents. For each of the following situations, identify the type of sample obtained (judgment, voluntary, convenience, simple random, stratified, systematic, cluster):

1. From the town hall, the director obtains a list of all 800 residents and numbers them 1 to 800. Then 40 numbers between 1 and 800 are randomly selected and people with those numbers are taken into the sample.
2. From the town hall, the director obtains a list of all 800 residents and numbers them 1 to 800. Then a random number between 1 and 20 is selected. That numbered person and every twentieth person from that point on is included in the sample.
3. The city is partitioned geographically into voting precincts of 20 people each. One person is randomly selected from each precinct.
4. The city is partitioned geographically into voting precincts of 20 people each. Two voting precincts are randomly selected and all people in those two precincts are included in the sample.
5. The director obtains a list of phone numbers for the residents and, starting at the top of the list, dials numbers until he obtains the opinions of 40 people who answer their phone.
6. The director places the survey in the local newspaper and uses the responses of the first 40 surveys mailed into the town hall.

Solutions to Practice Exercises 1-4

1. Simple random sample.
2. Systematic sample.
3. Stratified sample.
4. Cluster sample.
5. Convenience sample.
6. Voluntary sample.

1-5 Role of the Computer in Statistics

Study Objective

You should be able to:
    Understand the significance of the computer in statistical procedures.

Section Overview

Many statistical tasks are done by computers. Luckily, we do not need to know how to program a computer to use one. Spreadsheets are available that allow us to enter data into a computer and obtain basic statistics. Statistical analysis packages such as MINITAB have a spreadsheet-like function, but provide a greater range of statistical functionality.

Worked Examples

Speeding Tickets

Following is a listing of the speeds at which motorists received tickets. MINITAB was used to describe the speeds and create a histogram, a graphical display with which you will soon become acquainted.

Solution

```
MTB > print 'speeds'

speeds
    96    73    76    74    90    75    82    73    75    81    83
    77    79    81    82    82    87    88    69    93    70    66
    86    79    79    80    75    70    84    93    73    89    77
    78    85    77    80    81    85    77    84    75    76    70
    84    73    78    91    74    70    77    80    83    85    75
    84    82    85    88    76    78    74    79    66    81    85
    84    90    83    87    70    81    79    80    81    77    79
    70    66    88    78    67    67    88    76    71    81    80
    68    89    60    80    73    71    82    88    86    80    86
    73    83    77    77    82    67    69    78    85    72    78
    65    75   105    75    84    87    68    84    86    70    76
    67    76    75    77    86    79
```

```
MTB > desc 'speeds'
```

Descriptive Statistics

Variable	N	Mean	Median	TrMean	StDev	SE Mean
speeds	127	78.858	79.000	78.757	7.274	0.645

Variable	Minimum	Maximum	Q1	Q3
speeds	60.000	105.000	74.000	84.000

```
MTB > histogram 'speeds'
```

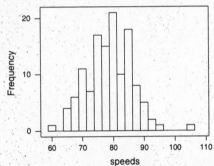

Comment: While just a few of the elements from the "describe" command might be familiar, the output from the histogram command should be clear. The greater the rectangle above any speeds, the more often those speeds were ticketed.

Practice Exercises 1-5

1-2. A fire inspector examined 77 buildings for code violations in the last month and obtained the following number of violations for each building. Use this data and MINITAB or a similar package to answer the following two questions.

Violations

```
    5     7    11     5    10     6     6     5     4     9     8    15    11
    2     9    12     6     5     9     7     5     7     8     9    12     7
   10     9     7    10     6     4     9     9    10     7     6     8    10
    8     5     7     7    11     6     9     9     8    10     5     7     9
    7     9    12     5     9     5     7     5     9     6     8     6    15
   12     6     6     4     3     9     7    13     7     5    12    10
```

1. Use MINITAB's describe command (or a similar command from a different program) to summarize the data.

2. Use MINITAB to obtain a histogram for the data.

Solutions to Practice Exercises 1-5

1.

```
MTB > desc 'Violatio'
Descriptive Statistics

Variable          N        Mean      Median      TrMean      StDev    SE Mean
Violatio         77       7.831       7.000       7.754      2.643      0.301

Variable    Minimum     Maximum          Q1          Q3
Violatio      2.000      15.000       6.000       9.000
```

2.

```
MTB > hist 'Violatio'
```

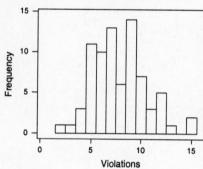

Solutions to Odd-Numbered Exercises

1. d

3. h

5. a

7. f

9. Systematic

11. Simple random

13. Cluster

15. Systematic

17. The answer depends on the procedures you've selected.

Chapter 2 Thinking Critically About Data: Liars, #$%& Liars, and a Few Statisticians

Study Aids and Practice Exercises

2-1 Unfavorable Opinions and The Bias Obstacle

Study Objective
You should be able to:
   Recognize several ways in which statistics is misused.

Section Overview
   Statistics has gotten a bad press. The reason is not that there is something wrong with statistics itself. The problem lies (no pun intended) with people who try to use statistics to support whatever they want to prove. While sometimes unintentional, it is possible to distort the results of a statistical analysis by various means. One common way is by introducing a bias into a study. This is done by poor wording, leading questions, or ignoring whatever contradicts our beliefs and emphasizing whatever strengthens our position. Whether our own work or someone else's, we should consider both how the data was collected and how it was analyzed. It is foolish to blindly accept a statement because it begins with "statistics says." We should be informed consumers of statistics, questioning any vagueness or unjustified claim.

Key Terms & Formulas
Bias A tendency either consciously or unconsciously to favor one side over another.

Worked Examples
Gum
Find any potentially misleading portion of the statement: "Of all dentists who recommend chewing gum, two out of three recommend Chewzit."

Solution
Very few dentists recommend chewing any gum. Those dentists who are among "Of all dentists who recommend chewing gum" would probably not represent the consensus opinion.

2-2 Aggravating Averages

Study Objective
You should be able to:
1.   Calculate a mean, median, and mode.
2.   Analyze a statement using the word average critically.

Section Overview

We should be careful when someone tells us "an average". While that person might be telling us the mean of some data, sometimes what they call an average might really be a median or mode. Because they can be quite different, it is important we identify exactly the measure involved.

Key Terms & Formulas

<u>Mean</u> The arithmetic mean is calculated by adding all the data and dividing by the total number of values.

<u>Median</u> The middle number of a data set after the numbers have been arranged in order of magnitude (either ascending or descending).

Hint: The median is an example of a percentile. A percentile is a dividing line between two parts of a data set. Since the median divides the data in half, i.e., "fifty-fifty," it is the fiftieth percentile.

<u>Mode</u> The most frequently occurring number in a data set. If there is a tie in the values that occur most often, there is more than one mode.

Worked Examples

<u>Texts</u>

Five students purchased 5, 6, 9, 5, and 10 textbooks. Calculate the mean, median, and mode for this data.

<u>Solution</u>

The mean is $\dfrac{5+6+9+5+10}{5}=7$. The median is obtained by reordering the sample into 5, 5, 6, 9, 10, and finding the middle value, 6. The mode is the only value that repeats, 5.

Caution: Beware. It is a common mistake to forget to order the numbers before obtaining the median. For example, if we neglect to reorder the above data, we would say that the median is 9.

<u>Texts + 1</u>

A sixth student purchased 9 texts, so now the numbers of texts purchased are 5, 6, 9, 5, 10, and 9. Recalculate the mean, median, and mode.

<u>Solution</u>

The mean is $\dfrac{5+6+9+5+10+9}{6}=7\dfrac{1}{3}\approx 7.33$. The reordered sample is 5, 5, 6, 9, 9, 10. There are two values occupying the middle most positions, 6 and 9. The median is halfway between (or the average of) 6 and 9, 7.5. This data is bimodal, as both 5 and 9 occur twice.

<u>Salaries</u>

An owner of a small company is negotiating with a labor union. The owner pays himself $250,000 a year, while his nine employees earn between $20,000 and $30,000 a year. What measure of central tendency should the owner use if he wants to show how high his company's salaries are? What should the union use if it wants to show how low the average salary is?

Hint: When there is a value in a data set that is substantially different from the rest, it is called an outlier. An outlier pulls the mean in its direction but leaves the median unaffected.

<u>Solution</u>

Because the owner's salary is so large, the mean is greater than the median. Therefore, the owner would prefer using the mean while the union would prefer using the median.

Practice Exercises 2-2

1-3. In a five-day period, a highway patrol officer wrote 11, 5, 13, 21, and 7 tickets. For this data calculate the following.

1. The mean.
2. The median.
3. The mode.

4-7. The number of classes dropped by eight recent college graduates is 0, 33, 0, 1, 2, 5, 4, 6. For this data,

    4. what is the mean?

    5. what is the median?

    6. what is the mode?

    7. If it were discovered that the 33 was an error and should have been a 13, which of the three measures would be most drastically affected?

8-9. A bowler bowls games of 222, 217, and 110. What measure of central tendency should the bowler report if

    8. she wants to impress her friends?

    9. she wants to increase her handicap by reporting a low value.

Solutions to Practice Exercises 2-2

    1. The mean is $\dfrac{11 + 5 + 13 + 21 + 7}{5} = \dfrac{57}{5} = 11.4$.

    2. The median is the middle number of 5, 7, 11, 13, 21, i.e., 11.

    3. There are no repeats, so there is no mode.

    4. The mean is $\dfrac{0 + 33 + 0 + 1 + 2 + 5 + 4 + 6}{8} = \dfrac{51}{8} = 6\dfrac{3}{8} = 6.375$.

    5. The median is the average of the two middle numbers from 0, 0, 1, 2, 4, 5, 6, 33, i.e., 3.

    6. The mode is 0.

    7. The mean changes to 3.875; the median and mode are unchanged.

    8. The median is 217 while the mean is 183, so if she wants to impress her friends, she would report the median.

    9. The mean, being smaller than the median, would be the one she would use.

2-3 Disregarded Dispersions

Study Objective

You should be able to:

  Recognize the need to examine the dispersion of a data set.

Section Overview

Knowing an average does not give us a complete picture of a data set. Suppose you know that the mean weight of gray whales is 30 tons. You purchase a gray whale, William, at your pet store. When full grown, William weighs 27 tons. Does this mean William is malnourished? If typical gray whale weights include 37, 23, 30, 25, and 34 tons, the answer is no; if typical gray whale weights are 29.9, 30.2, 30.1, 29.7, and 30.0 tons, the answer is yes. In the first case, the dispersion is great and 27 is not unusual; in the second case, the dispersion is small and 27 is unusual. To get a good picture of a data set, we should know the dispersion as well as the average of the data.

Key Terms & Formulas

Dispersion The spread or scatter of a set of numbers.

Worked Examples

Mpg

You are the purchaser of a new automobile, a Sanders Sedan. The salesperson told you that the average mileage for the Sanders Sedan is 44 mpg. You find that you get 37 mpg. Can you complain to the dealer and be sure that you have a valid claim?

<u>Solution</u>

Not necessarily. If Sanders Sedans individually get mileages such as 31, 67, 19, 83, 54, 39, etc., then there is nothing unusual about getting 37 mpg. The dispersion of the data set is too great. If the mileage were more typically 44.2, 43.9, 44.5, 43.3, 43.7, 45.1, 44.3, 43.6, etc., then 37 mpg would stick out like a sore thumb. You would be justified in complaining.

Practice Exercises 2-3

1. The average yards gained per run by Barry Sanders is 4.9. Would it be surprising if Barry Sanders made a 55-yard run?

2. The average time for a varsity runner to run a mile is 4 minutes, 20 seconds. Would it be surprising if the runner ran a mile in under 4 minutes?

3. A person's weight fluctuates during any period of time. Would you be surprised if you gained 10 pounds in a 24-hour period?

Solutions to Practice Exercises 2-3

1. No. As any football fan knows, the dispersion in the gains of a football back, particularly a fast, shifty runner, is highly variable. So a large gain on any one run is not at all surprising.

2. Yes. Most runners will vary only a few seconds in their times for a mile run, i.e., the dispersion is small. To differ from the average speed by as much as 20 seconds is startling.

3. Surprised and annoyed. Your weight, while fluctuating, does not vary that greatly over a short period. To gain ten pounds in a 24-hour period would be a biological anomaly.

2-4 The Persuasive Artist

Study Objective

You should be able to:
1. Use and understand statistical tables and graphical displays such as line charts, bar charts, and pie charts.
2. Detect when a chart is giving a misleading impression.

Section Overview

There are many ways to display data. Included are statistical tables, line charts, bar charts, and pie charts. Many of these are available from graphical or statistical packages. Whatever displays are used, we must always examine them carefully. To prevent any misunderstanding of the data, we should scan the titles and headings of any table and scrutinize the axes and legend of any chart.

Key Terms & Formulas

<u>Line Chart</u> To create a line chart, points representing quantities or percentages are plotted against time or another variable and connected by lines.

<u>Bar Chart</u> Bar charts are similar to line charts, but they use rectangles to represent the quantities, instead of connected lines.

<u>Pie Chart</u> A pie chart is created by dividing a circle is into slices, each slice proportional in size to the percentage it represents.

Worked Examples

SAT's

A university with three colleges, Arts, Humanities, and Engineering, requires its applicants to take the SAT's. The scores from 167 applicants have been taken and broken down by subtest (Verbal and Math), gender, and college to which each applied. What are some reasonable ways of presenting the data?

Solution

One way is to organize the information in statistical tables. We decided to make tables for the mean combined SAT scores broken down by gender and college. The tables that follow were generated by MINITAB. We first obtained the mean SAT scores for each gender.

```
           SATtotal
             MEAN

  Male        933.7
Female        930.8
   ALL        932.3
```

The row labeled ALL is the grand average of the males and females. We can see from this table that there is little difference between the mean scores of the males and females. Next, the mean SAT scores were broken down by college:

```
             SATtotal
               MEAN
Humanities     940.1
     Arts      863.4
  Engineer    1038.0
      ALL      932.3
```

We see that the mean SAT's are markedly different between colleges. To further examine the scores, we can create a two-way table with the means broken down both by gender and college:

	Human	Arts	Enginr	ALL
Male	951.7	879.3	1014.8	933.7
Female	928.6	843.6	1056.6	930.8
ALL	940.1	863.4	1038.0	932.3

We note that, while males who apply to the Arts or Humanities colleges have higher SAT's than females, the reverse is true for applicants to the engineering college.

We can also present the data with charts. The first is a MINITAB line chart of the mean SAT scores broken down by college. It pictures the information presented two tables ago. We can again see the differences among the three colleges.

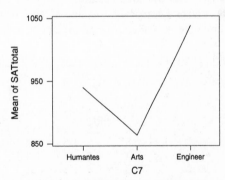

Line Chart: SAT by College
Total SAT versus College

We also subdivided the SAT scores by subtests, verbal and math, and graphed these versus colleges. The colleges are represented by positions along the horizontal axis, and the subtests are represented by different lines, solid for verbal, dashed for math. We found some interesting results. The humanities' applicants have basically the same mean score for the verbal and math subtests. However, the arts' applicants have higher verbal scores, and the engineers have higher math scores. The line chart makes this identification simple.

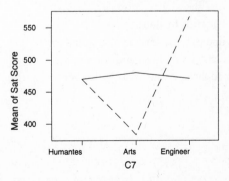

Line Chart: SAT Subtest by College
Sat Verbal & SAT Math verses College

Next we had MINITAB make a bar chart of mean SAT total scores broken down by both college and gender. From this we can see that the female engineers scored the highest, while the female arts' applicants scored the lowest. There is one problem with this and the previous graphs. If we carefully examine the vertical axis, we find that it does not extend all the way down to zero. From this, we can easily get the impression that the differences are much greater than they are. For example, in this last graph it appears that the female applicants to the arts college scored almost nothing on the SAT's. Unfortunately, this is the way most computer packages create their charts. However, adaptable programs like MINITAB allow us to specify exactly the vertical axis we want, including one that extends to zero. Redoing the last graph with such a vertical axis gives a fairer impression of the means for each group.

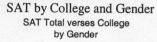

SAT by College and Gender

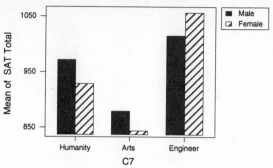

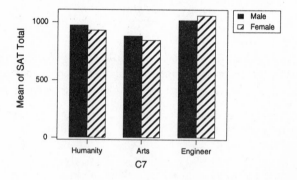

Practice Exercises 2-4

1-6. A manufacturer of communication equipment has begun a training program for its managers. After training, each manager takes a competency exam. MINITAB tabled the mean scores on the exam broken down by the managers' level (Low, Middle, and High) and their departments (Administration, Design, Programming, and Quality Control).

```
ROWS: Deprtmnt     COLUMNS: Level

          1         2         3        ALL
   1   17.105    19.105    19.632    18.614
   2   15.111    16.294    18.611    16.679
   3   14.435    15.857    16.471    15.444
   4   12.524    14.217    16.500    14.155
 ALL   14.716    16.288    17.926    16.216
```

For the mean competency scores presented in the table, obtain the following charts:

1. A line chart broken down by level of manager.
2. A line chart by department.
3. A bar chart by department.
4. A line chart by level and department.
5. A bar chart by level and department.
6. Redo the previous chart with a vertical axis that extends all the way to zero.

Solutions to Practice Exercises 2-4

The follow graphs were created using MINITAB.

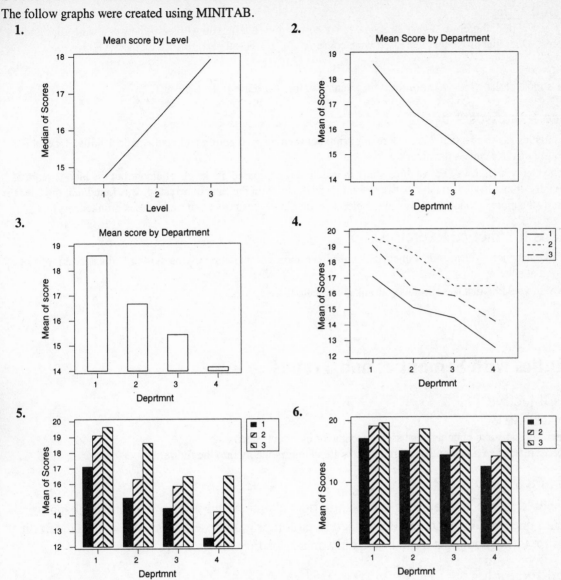

2-5 The Post Hoc Ergo Propter Hoc Trap

Study Objective

You should be able to:
   Recognize when people "go too far" with statistics and reach erroneous conclusions.

Section Overview

   Statistics can show that two things are related, it cannot show that one "causes" the other.
While a theory of causality might be supported, it is not proven by statistical results.

Worked Examples

SAT's and Wine

For a 15-year period, there was a very strong negative correlation (a term you will learn) between wine sales and SAT scores. Does this prove that increased wine consumption decreased performance on the SAT's?

Solution

Not unless most of the wine was consumed the evening before (or during) the SAT test.

Practice Exercises 2-5

1. Statistics has shown that there is a relationship between musical ability and mathematical skills. Does this imply that one causes the other?
2. The Georgia Highway Patrol did a study in which they compared the levels of alcohol in the bloodstream of legally drunk drivers arrested between 8 p.m. and 4 a.m. with the time of arrest. It was found that the later the time, the more alcohol there was in the bloodstream. Does this prove that time causes drunkenness?

Solutions to Practice Exercises 2-5

1. Obviously not. Music and Mathematics must have some common underlying skill or skills necessary to be good at either.
2. No, it would be just as valid to say drunkenness causes time.

2-6 Antics with Semantics and Trends

Study Objective

You should be able to:
1. Identify statistical conclusions based on vague or incorrect statements.
2. Recognize when a person has extrapolated a trend unreasonably into the future.

Section Overview

We should ignore any spurious cause-and-effect claims based solely on a statistical analysis. We should question any statements couched in jargon or grandiose terms. We should discredit any attempt to extrapolate a trend based only on a hope that things won't change.

Worked Examples

Sexual Harassment

According to a recent magazine article, the reported cases of sexual harassment in the workplace have increased 225%. Does this imply that sexual harassment is increasing at a dramatic rate?

Solution

Perhaps, but a key word in the above is "reported." Over time, there has been an increased willingness by victims of sexual harassment to report the incidents. So the increase may be partially due to fewer cases of sexual harassment going unreported.

Gasoline

During the mid to late 1970's, much was made of the steadily increasing price of gasoline. Projections were made that by the year 1990, the price of gasoline would be over 10 dollars per gallon. What happened?

Solution

Obviously nothing. The projections were based on the dramatic increase in the price of gasoline due to the OPEC oil embargo. To assume that such a dramatic upswing would continue indefinitely is absurd.

Practice Exercises 2-6

1. A telephone survey of Omaha residents was made during daytime in the first week of June. Of those who responded, over 70% favored floating a bond to pay for the construction of an Olympic-sized pool. Are there any problems with this survey?

2. The marketing director of a Fortune 500 company noticed that the revenues of her company increased 20% when the company increased advertising expenditures 50%. Based on this data, the director suggests to the executive board that an increase of another 50% for advertising should increase revenues another 20%. What problems should the members of the executive board have with this statement?

Solutions to Practice Exercises 2-6

1. Several things seem bothersome. First, the group that would be home during the day might have different views than other Omaha residents. For instance, it is possible that those home during the day would be the primary caregivers of children. These might be more interested in having a pool available than the general population. Second, since we only have the opinions of those who responded, we do not know the opinions of those who refused to participate. This more uncommunicative group might have different opinions. Third, it is possible that many people do not know the financial consequences of "floating a bond" and might have a different opinion if they fully understood the consequences.

2. The most basic problem is that there is no guarantee that an increase of the past will automatically imply a corresponding increase in the future. Suppose the marketplace is now tired of this company's previously effective but now becoming intrusive commercials? Suppose the people interested in this company's products have already purchased all they wanted? Suppose the advertising campaign is not as effective as the previous company's? Suppose the economy takes a downswing?

2-7 Follow the Bouncing Base (if you can)

Study Objective
You should be able to:
Recognize and employ the appropriate base when working with changes in percentages.

Section Overview
We should make sure we know the base used in the calculation of any percentage.

Worked Examples
Raises
Two workers were earning $800/week five years ago. Sam received a 20% raise after two years and another 10% raise last year. Bassam was passed over for a raise after two years, but received a 30% raise last year. Do they now have the same weekly salary?
   Hint: There is difference between percentage increase and percentage relative. For example, Bassam's salary change is a 30% increase and his current salary is 130% relative to his base salary of five years ago.

Solution
No. Bassam's salary is now 130% of his $800/week salary of five years ago, i.e., it is $1.3(\$800) = \$1040$. Sam's first pay raise changed his salary to $1.2(\$800) = \$960$. When he received his second raise, it was a 10% increase over $960 instead of the $800 salary. So Sam's latest salary is $1.1(\$960) = \$1056$.

Practice Exercises 2-7

1. Jean, Patrice, and Rick both started their jobs with identical salaries of $30,000$ per annum. Each year, Jean has received a 10% increase. Patrice received a 20% increase after two years, and received a second 20% increase two years later. Rick received a 30% increase after three years and a 10% increase a year later. Which person has the highest current salary? Who has the lowest?

Solutions to Practice Exercises 2-7

1. After one year, Jean's salary would be 1.1($30,000) = $33,000. After two years, her salary becomes 1.1($33,000) = $36,300. In the third year, her salary rises to 1.1($36,300) = $39,930. The fourth raise makes her salary 1.1($39,930) = $43,923. After two years, Patrice's salary is 1.2($30,000) = $36,000. After four years, her salary is raised to 1.2($36,000) = $43,200. For Rick, the first raise changes his salary to 1.3($30,000) = $39,000. The second raise leaves his salary at 1.1($39,000) = $42,900. So the highest paid person is Jean, and the lowest is Rick.

2-8 Avoiding Spurious Accuracy and Other Pitfalls

Study Objective

You should be able to:
1. Identify spurious significant digits in a result.
2. Avoid accepting a statistically based statement unless you understand the source of the information, believe the evidence that supports the statement, and can figure out what information is missing.

Section Overview

We should not add 2 plus 2 and get 4.00000000. We should question all sources of information and the evidence that supports conclusions. We should think.

Worked Examples

Juvenile Delinquents

A study reported that, among the juvenile delinquent arrests made in a small town in South Carolina, the percent of repeat offenders was 42.85714%. How impressive is this?

Solution

The number of digits in the percentage is silly. It turns out that this number was based on seven juvenile delinquent arrests, three of which involved repeat offenders. Carrying five decimal gives an unwarranted note of authority to the percentage.

   Caution: Computers usually spew out results with many decimal places. The fact that these magical machines provide results with a plethora of decimal places does not imply that we need that much "accuracy."

Practice Exercises 2-8

1. There was a recent newspaper article about undercover drug arrests in a California coastal village. It reported that 33.3333% of all such arrests occurred in one local bar. What questions might you ask about this report?

Solutions to Practice Exercises 2-8

1. The first question you might ask is about that 33.3333%. This is in a 1 in 3 ratio. Is it 1 out of 3, 2 out of 6, 30 out of 90, etc.? Turned out that it was 1 in three. Reporting 1 out of 3 with 6 decimal places was absurd in the extreme.

Solutions to Odd-Numbered Exercises

1. We can't really evaluate this performance without knowing how other comparable funds performed during the same period. And we can't be sure that the performance trend will continue and produce the same rate of growth in the future.

3. This is a case of aggravating averages. Both $2,000 (the mode and the median) and $6,400 (the mean) are legitimate averages.

5. Not necessarily. This issue has become more publicized and more people are willing to report it than in the past. Perhaps the number of incidents hasn't increased much, but more of the incidents that are occurring are being reported.

7. The studies were conducted by organizations that had something to gain. It's possible that an element of bias has entered the picture.

9. What was Krinkle Gum compared against? Perhaps Krinkle Gum users had 36 percent fewer cavities than those addicted to candy bars. The data may be spurious, and it is certainly curious since we don't know how it was obtained.

11. This is always true for the median of a set of values.

13. Much more contact is made with friends, acquaintances, and relatives than with strangers in lonely parks at night. Thus, rape victims are more likely to be acquainted with their assailants.

15. The second graph represents the information more honestly. The scale on the first does not tart at zero,thus making the Nutrition majors values seem several times larger than the Business majors values, when in reality there is only a 10% difference.

17. A bar graph could be used, with the before and after proportions side by side. It would be obvious very quickly that the set up time had been reduced significantly, and the running time had increased.

19. The percentages of users seems to be decreasing over time.

21. Marijuana use steadily increased for all three age levels over the years shown, with usage being consistently higher as the grade level increased. In 1996 the 10th graders' use increased at a much faster rate than ever before, to the point of almost reaching the 12th graders' level.

23. The second graph shows the rate of change is near 0, the change went neither up nor down. In the line graph, the line stays nearly horizontal during the time period with almost no change.

25. The projection for 2007 will more likely be inaccurate. The farther a prediction gets from actual data, the less accurate it will be, since it is more likely that other unforeseeable factors may affect the data as time goes on.

27. One might expect that border states might get a higher percentage of LEP individuals that have come from bordering countries. States that are entry-ways from Europe and other continents would hold more non-English speakers. If people from other countries come to one place, it is likely that they would stay put until they became more comfortable with the language, thus explaining the lack of LEP students in the more central states.

Chapter 3 Descriptive Statistics

Study Aids and Practice Exercises

3-1 Introduction to Data Collection

Study Objective
You should be able to:
1. Distinguish between attribute and numerical data.
2. Identify whether a numerical variable is discrete or continuous.

Section Overview
In statistics we analyze data. Data comes in two types. We have attribute (or qualitative or categorical) data and numerical data. Attribute data is nonnumerical, such as a list of people's eye colors or college majors. Numerical data consists of quantities such as observations on people's heights or incomes. Data is generated by observing or measuring values of a variable, a characteristic that each item in a study possesses. There are two types of numerical variables, discrete and continuous. A discrete variable is one whose possible values are finite or countably infinite. A continuous variable is one whose possible values consist of a countless number of values.

> Hint: Phrases such as countably infinite can be confusing. It is easy to judge if a variable is discrete or continuous. Determine reasonable low and high values for the variable. If every value between the low and high values is possible, then the variable is continuous. If there are gaps in the possible values, the variable is discrete.

Key Terms & Formulas
Attribute data Data whose possible values consist of a list of categories. Examples of attribute data include observations on religion, political party, gender, and marital status.

Numerical data Data that consist of counts or measurements on some phenomenon. Examples of numerical data include measurements on GPA, weight, time, and age.

Variable A characteristic that can be obtained from each member of a sample (or population).

Discrete variable A variable whose possible values consist of a finite or countably infinite set of distinct values.

Continuous variable A variable whose possible values consist of a countless number of values along a line interval.

Worked Examples
Attribute or Numerical?
Identify each of the following variables as attribute or numerical variables.
1. Whether people favor, oppose, or have no opinion about the latest actions of the president concerning the budget deficit.
2. The amount charged for a one-minute commercial on network television shows.
3. The crimes for which prisoners at a state penal institute were convicted.
4. The altitudes of each of the state capitals in the United States.

5.   The number of errors made in major league baseball games played on next Monday.

Solution
1.   Peoples' opinions would be attribute data.
2.   Money consists of quantities, which makes it numerical data.
3.   The types of crimes for which convicts have been incarcerated would fall into several categories, making this attribute data.
4.   Altitudes are quantities and are therefore numerical data.
5.   Number of errors is a count and would generate numerical data.

Discrete or Continuous?
Identify each of the following numerical variables as discrete or continuous.
1.   The number of textbooks purchased from a bookstore in a day.
2.   The speeds reached by automobiles traveling on Interstate 80 through Iowa.
3.   The number of times a person throws a dart at a dart board until they hit the bull's-eye.
4.   The ages of the professors in a statistics department.
5.   The number of points received by the students in a statistics class.

Solution
1.   While a bookstore may sell 1000 or 1001 texts, it is not possible for the bookstore to sell 1000.6. Since there are gaps in the possible values, the variable is discrete.
2.   A car could be traveling at 65 or 66 mph. The car could also be traveling at 65.5434 mph. The fact that we do not measure speed this precisely is irrelevant; speed is continuous.
3.   The number of dart throws could be 1 or 2 or 3 or ..., but it could not be 1.5 or 2.2; i.e., there are gaps in the possible values. Therefore, the number of dart throws is discrete.
>   Hint: Notice that there is no limit to the number of dart throws. Theoretically, a person might never hit the bull's-eye. This does not prevent the variable from being discrete. This is an example of a variable that is discrete because it is, as the book calls it, countably infinite.
4.   It is possible (though we wouldn't think of this way) that a person is 22.456432 years old. That is, any age within biological limits is possible, so age is continuous.
>   Caution: Do not be misled by the fact that we usually report our ages as integers. We age continuously--a year of age does not enter our body on our birthday.
5.   Even if the instructor does strange things like giving half points, it is unlikely that anyone would receive a grade of 72.257483. Because there are gaps in the possible values of the variable, it is discrete.

Practice Exercises 3-1

1-5. Identify the following as attribute or numerical variables.
1.   Brands of automobiles.
2.   The costs of various brands of cars.
3.   The colors of cars driving through a particular intersection.
4.   The number of doors on various brands of cars.
5.   Whether cars have tinted windows.

6-9. Identify each of the following numerical variables as attribute or numerical. If they are numerical, identify whether they are discrete or continuous.
6.   The number of college courses you have successfully completed.
7.   The amount of time you spent the last time you read the textbook.
8.   The majors of the people taking this class.
9.   Your final grade in this course (A, B, C, D, or F).

10-14. Identify each of the following numerical variables as attribute or numerical. If they are numerical, identify whether they are discrete or continuous.
10.   The number of defective transistors received in a shipment.
11.   The time required to prepare a report.

12. The department for which a person works.
13. The number of error messages in a compilation of a computer program.
14. The reaction of a client (favorable, unfavorable, neutral) to a new product.

Solutions to Practice Exercises 3-1

1. Attribute. The possibilities are Ford, Toyota, Volkswagen, Sanders Sedan, etc.
2. Numerical. The possible values are dollar amounts.
3. Attribute. Red, blue, magenta, teal, etc.
4. Numerical. 2, 3, 4, perhaps more for vans.
5. Attribute. Yes, no.
6. Numerical discrete. Possibilities include 0, 1, 2, but not 1.7 courses completed.
7. Numerical continuous. It is possible to spend any amount of time reading the text with no gaps in the possibilities.
8. Attribute. Possibilities include Psychology, Kinesiology, Education, etc.
9. Attribute.
10. Numerical discrete. The number of defective transistors would be a nonnegative integer.
11. Numerical continuous. Time can be any positive real number.
12. Attribute. Personnel, administration, sales, etc.
13. Numerical discrete. The number of errors could be 2 or 3 but not 2.5.
14. Attribute. The possible reactions are given as favorable, unfavorable, and neutral.

3-2 Data Organization and Frequency Distributions

Study Objective
You should be able to:
1. Form a data array.
2. Create a frequency distribution from raw data.

Section Overview
An unorganized set of data is difficult to comprehend, so we try to organize and present the data in an understandable form. We might create a data array, an arrangement of the data from smallest to largest. Or we could summarize the data in a frequency table, a table containing possible values of the data and counts of how frequently these values occur. These possible values are usually grouped into classes or class intervals that optimally have the same class width and are not open-ended. The middle of the class is called the class midpoint, while the dividing line between classes is called the class boundary.

Key Terms & Formulas
Array An arrangement of data in size order, usually from the smallest to the largest.

Range The difference between the largest and smallest value in a data set.

Frequency Distribution A frequency distribution first groups the data items into intervals called classes and then counts the number of data items in each class, called the class frequency.

Worked Examples

Grammar Errors I

An instructor of a writing class counted the number of grammatical errors in student themes. The data is presented below. Create an array for this data.

```
 9 12   8 19   9 11   6 12 11   8 12 10 15 10   5   7 10 12 14   2   5
17   4 14 13   9 16 11 16 11   8 23   5 13 20 13 13   4 11 17   6 13
13 17   9 11   9   6   8 20 14 17 16   8   9 14 10 17   9 12   9   7 15
12 14   6 14   6   9 11 13   7   6   5 12   6 10 13   7   7   7 14   8 12
20  9
```

Solution

The data array:

```
 2   4   4   5   5   5   5   6   6   6   6   6   6   6   7   7   7   7   7   7
 8   8   8   8   8   8   9   9   9   9   9   9   9   9   9   9 10 10 10 10
10 11 11 11 11 11 11 11 12 12 12 12 12 12 12 12 13 13 13 13
13 13 13 13 14 14 14 14 14 14 14 15 15 16 16 16 17 17 17 17
17 19 20 20 20 23
```

  Hint: While it is easier to read this array than the raw data, it is still difficult to absorb this information. That is one reason for also using other descriptive techniques to describe data.

Grammar Errors II

Form a frequency distribution for the error's data.

Solution

The first thing to do is form classes. We want a reasonable number of classes that are easy to understand. The first class must include 2 and the last must include 23. One choice is:

    0 to less than 5
    5 to less than 10
    10 to less than 15
    15 to less than 20
    20 to less than 25

  Hint: We have been trained in base 10 arithmetic. It is easier for us to understand classes whose widths are multiples or fractions of 10. Here we have class widths of 5, one-half of 10.

Once the classes have been selected, we count the number of data points that fall into each class.

ERRORS	FREQUENCY
0 to less than 5	3
5 to less than 10	33
10 to less than 15	35
15 to less than 20	11
20 to less than 25	4

Practice Exercises 3-2

1-2. A psychologist used a standardized test to measure the feelings of autonomy of inner-city children. Sampling 62 children, the psychologist obtained the following scores.

```
221  164  143  121  191  169  217  231   93  176  188
248  153  193  235  129  154  132  134  213  168  141
149   85  141  134  141  185  231  204  163   84  101
189  103  219  141   92  170  120  175  156  192  143
123  152  172  136  101  185  172  161  199  133  136
191  158  171  163  257  254  165
```

  1. Arrange the data in an ascending array.
  2. Create a frequency distribution with a first class of 75 up to less than 100.

3-5. Part of the real estate licensing process involves a 25-question exam on real estate law. A recent exam generated the following grades.

```
19   25   23   22   18   16   21    9   24   19   21   21   22   20   22
18   14   18   17   17   19   20   21   19   17   17   17   17   15   19
```

| 20 | 16 | 21 | 10 | 21 | 19 | 19 | 20 | 16 | 20 | 22 | 19 | 14 | 19 | 22 |
| 20 | 21 | 24 | 18 | 22 | 16 | 14 | 25 | 18 | 17 | 18 | 20 | | | |

3. Arrange the data in an ascending array.

4. Create a frequency distribution with a first class of 9 up to less than 11.

5. Create a frequency distribution with a first class of 9 up to less than 12.

Solutions to Practice Exercises 3-2

1. Array:

84	85	92	93	101	101	103	120	121	123	129
132	133	134	134	136	136	141	141	141	141	143
143	149	152	153	154	156	158	161	163	163	164
165	168	169	170	171	172	172	175	176	185	185
188	189	191	191	192	193	199	204	213	217	219
221	231	231	235	248	254	257				

2.

AUTONOMY	FREQUENCY
75 to less than 100	4
100 to less than 125	6
125 to less than 150	14
150 to less than 175	16
175 to less than 200	11
200 to less than 225	5
225 to less than 250	4
250 to less than 275	2

3. Array:

9	10	14	14	14	15	16	16	16	16	17	17	17	17	17
17	17	18	18	18	18	18	18	19	19	19	19	19	19	19
19	19	20	20	20	20	20	20	20	21	21	21	21	21	21
21	22	22	22	22	22	22	23	24	24	25	25			

4.

LICENSE EXAM	FREQUENCY
9 to less than 11	2
11 to less than 13	0
13 to less than 15	3
15 to less than 17	5
17 to less than 19	13
19 to less than 21	16
21 to less than 23	13
23 to less than 25	3
25 to less than 27	2

5.

LICENSE EXAM	FREQUENCY
9 to less than 12	2
12 to less than 15	3
15 to less than 18	12
18 to less than 21	22
21 to less than 24	14
24 to less than 27	4

3-3 Graphic Presentations of Frequency Distributions

Study Objective
You should be able to:
1. Create and read histograms, frequency polygons, and ogives.
2. Create and read exploratory data analysis graphs: stem-and-leaf displays, dotplots, and boxplots.

Section Overview
   Graphical representations help us understand a frequency distribution. Both histograms and frequency polygons are graphs of a frequency distribution. The difference is that a histogram represents the frequencies with rectangles, while a frequency polygon does so with lines. An ogive is similar to a frequency polygon except it represents cumulative instead of ordinary frequencies. Newer graphical methods have arisen that fall under the category of exploratory data analysis (EDA) techniques. These include stem-and-leaf displays, dotplots, and boxplots.

Key Terms & Formulas
Histogram A bar graph used to represent a frequency distribution.

Frequency Polygon A line chart used to represent a frequency distribution.

Ogive A frequency polygon of cumulative frequencies.

Stem-and-Leaf Display A presentation of an entire data set created by partitioning the numbers into two parts, the stem and the leaf. Each stem is put on a line with all its leaves.

Dotplot A presentation of an entire data set created by positioning dots over an interval to represent each observation.

Worked Examples
Grammar Errors III
Obtain a histogram for the grammar errors data.

Solution
The frequency distribution for the errors data was:

ERRORS	FREQUENCY
0 to less than 5	3
5 to less than 10	33
10 to less than 15	35
15 to less than 20	11
20 to less than 25	4

The histogram for this frequency distribution can be given with either vertical or horizontal bars.

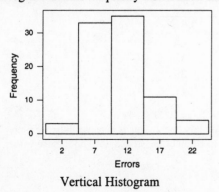

Vertical Histogram

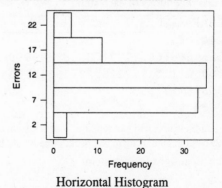

Horizontal Histogram

Grammar Errors IV

Obtain a frequency polygon and ogive for the errors data set.

Solution

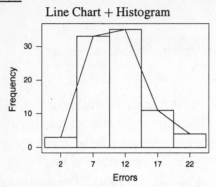

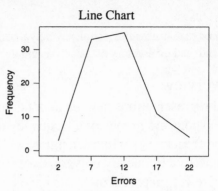

Hint: To better see how to create a line graph, we have placed a line graph on top of a histogram for the frequency distribution.

To obtain an ogive, we need the cumulative frequency distribution for the errors data:

ERRORS	FREQUENCY
0 to less than 5	3
5 to less than 10	36
10 to less than 15	71
15 to less than 20	82
20 to less than 25	86

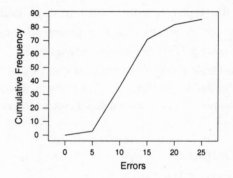

Grammar Errors V

Obtain a stem-and-leaf display and dotplot for the errors data.

Solution

Rather than obtaining a single stem-and-leaf display, we decided to create three, each with a different number of repetitions of the stem. The first has no repetitions.

STEM	LEAF VALUES	COUNT
0	2445555666666677777788888899999999999	36
1	00000111111122222222233333333444444455666777779	46
2	0003	4

The second has two repetitions of each stem.

STEM	LEAF VALUES	COUNT
0	244	3
0	555566666667777778888889999999999	33
1	00000111111122222222233333333444444	35
1	55666777779	11
2	0003	4

Hint: Notice that the above had two repetitions of the stems of 0 and 1, but only one 2. This is because there were no 25's through 29's in the data. If there had been, there would have been a repetition of the stem of 2 to hold those values.

The third has five repetitions of each stem.

STEM	LEAF VALUES	COUNT
0	2	1
0	445555	6
0	6666666777777	13
0	8888889999999999	16
1	000001111111	12
1	2222222233333333	16
1	444444455	9
1	66677777	8
1	9	1
2	000	3
2	3	1

Hint: The purpose of creating three (or more) stem-and-leaf displays is to see which presents the data in the way that is most easily understood. It is that stem-and-leaf display that we should use to represent the data.

The dotplot was generated using MINITAB, but could easily have been created by hand from an ordered array or stem-and-leaf display.

Dotplot for Errors

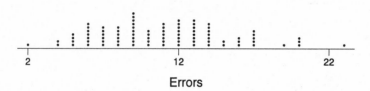

Errors

Practice Exercises 3-3

1-6 A problem from Exercises 3-2 involved a psychologist who measured the feelings of autonomy of 62 inner-city children. The data array and its frequency distribution follow.

84	85	92	93	101	101	103	120	121	123	129
132	133	134	134	136	136	141	141	141	141	143
143	149	152	153	154	156	158	161	163	163	164
165	168	169	170	171	172	172	175	176	185	185
188	189	191	191	192	193	199	204	213	217	219
221	231	231	235	248	254	257				

AUTONOMY	FREQUENCY
75 to less than 100	4
100 to less than 125	6
125 to less than 150	14
150 to less than 175	16
175 to less than 200	11
200 to less than 225	5
225 to less than 250	4
250 to less than 275	2

For this data:
1. Make a histogram.
2. Create a frequency polygon.
3. Obtain a cumulative frequency distribution.
4. Draw an ogive.
5. Create a stem-and-leaf display.
6. Make a dotplot.

7-12. Another problem from Exercises 3-2 involved an exam on real estate law. The data array and frequency distribution follow.

```
 9  10  14  14  14  15  16  16  16  16  17  17  17
17  17  17  17  18  18  18  18  18  18  19  19  19
19  19  19  19  19  19  20  20  20  20  20  20  20
21  21  21  21  21  21  21  22  22  22  22  22  22
23  24  24  25  25
```

LICENSE EXAM	FREQUENCY
9 to less than 11	2
11 to less than 13	0
13 to less than 15	3
15 to less than 17	5
17 to less than 19	13
19 to less than 21	16
21 to less than 23	13
23 to less than 25	3
25 to less than 27	2

For this data:

7. Create a histogram.
8. Draw a frequency polygon.
9. Obtain the cumulative frequency distribution.
10. Draw an ogive.
11. Create a stem-and-leaf display.
12. Make a dotplot.

Solutions to Practice Exercises 3-3

1.
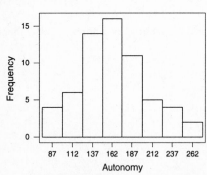

2.
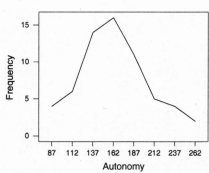

3.

AUTONOMY	CUMULATIVE FREQUENCY
75 to less than 100	4
100 to less than 125	10
125 to less than 150	24
150 to less than 175	40
175 to less than 200	51
200 to less than 225	56
225 to less than 250	60
250 to less than 275	62

4.

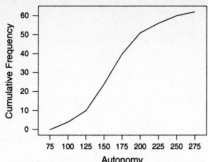

5.

STEM	LEAF VALUES	COUNT
8	45	2
9	23	2
10	113	3
11		0
12	0139	4
13	234466	6
14	1111339	7
15	23468	5
16	1334589	7
17	012256	6
18	5589	4
19	11239	5
20	4	1
21	379	3
22	1	1
23	115	3
24	8	1
25	47	2

6. Using MINITAB we get

Dotplot for Autonomy

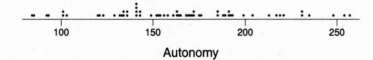

7.

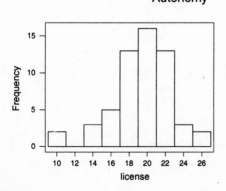

8.

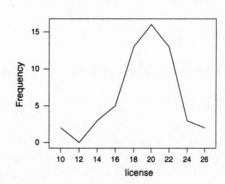

9.

LICENSE EXAM	CUMULATIVE FREQUENCY
9 to less than 11	2
11 to less than 13	2
13 to less than 15	5
15 to less than 17	10
17 to less than 19	23
19 to less than 21	39
21 to less than 23	52
23 to less than 25	55
25 to less than 27	57

10.

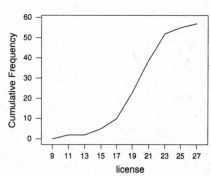

11.

STEM	LEAF VALUES	COUNT
0	9	1
1	0	1
1		0
1	4445	4
1	66667777777	11
1	888888999999999	15
2	00000001111111	14
2	2222223	7
2	4455	4

12. Using MINITAB we get

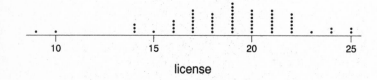

Dotplot for license

3-4 Computing Measures of Central Tendency

Study Objective

You should be able to:

1. Recognize symmetrical, positively skewed, and negatively skewed distributions.
2. Calculate the mean, median, and mode for ungrouped data.
3. Estimate the mean, median, and mode for grouped data.

Section Overview

Measures of central tendency include the mean, median and mode. These all give different information about the nature of the data. If we have a symmetric, unimodal distribution, they should all be about the same. However, if the data is skewed, then these will be different.

Hint: Instead of trying to decide which measure best describes the center of some data, many people give more than one measure.

Each of these measures may be calculated for populations and samples. Additionally, they can be calculated for ungrouped data or estimated from grouped data, i.e., data in a frequency distribution.

Key Terms & Formulas

Symmetrical distribution The distribution of a data set is symmetrical if, when you draw a line from the peak of the distribution's curve to the horizontal axis, the curves on either side of the line are mirror images of each other.

Skewed distribution The distribution of a data set is skewed if there are a few values much larger or much smaller than the rest of the data set. If the extreme values are larger than the rest, the curve has an extended tail in the positive direction and is said to be a positively skewed distribution. If the extreme values are smaller than the rest, the curve has an extended tail in the negative direction and is said to be a negatively skewed distribution.

Ungrouped data Data in its raw, untabled form.

Grouped data Data organized into a frequency distribution.

Population The complete set of data under study.

Sample A part or subset of a population.

Parameter A numerical characteristic of a population.

Statistic A numerical characteristic of a sample.

Population Mean $\mu = \dfrac{\Sigma x}{N}$, where N = population size.

Sample Mean $\bar{x} = \dfrac{\Sigma x}{n}$, where n = sample size.

Median The middle number of a set of values after the numbers have been placed in order of magnitude (either ascending or descending).

Mode The most frequently occurring number in a data set. If there are ties in the values that occur most often, there is more than one mode.

Population Mean for Grouped Data $\mu = \dfrac{\Sigma fm}{N}$, where f = frequency of a class and m = its midpoint.

Sample Mean for Grouped Data $\bar{x} = \dfrac{\Sigma fm}{n}$.

Median Class The class of a frequency distribution in which the median is found.

Modal Class The class of a frequency distribution with the largest frequency.

Worked Examples

Symmetric vs. Skewed

A manufacturing company gives its assembly line workers a test of manual dexterity. This test involves eight tasks and the number that each worker successfully completes is counted. Three different plants operated by this company yielded the following histograms for this test. How should each distribution be described?

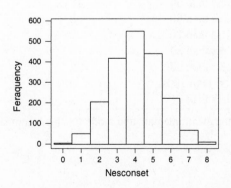

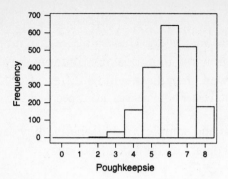

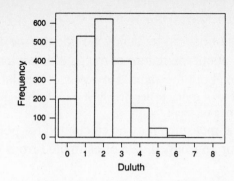

Solution

The Nesconset plant seems to have a symmetric distribution. The Duluth plant has an extended tail in the right direction and by that is a positively skewed distribution. Conversely, the Poughkeepsie plant is negatively skewed.

Mites

A botanist is interested in the number of mites found on the leaves of Eucalyptus trees. The botanist samples 10 trees and determines the number of mites on each. The data is listed below. Obtain the mean, median, and mode for this sample.

27   43   19   23   19   5   3   76   53   22

Solution

The mean is $\bar{x} = \dfrac{27 + 43 + 19 + 23 + 19 + 5 + 3 + 76 + 53 + 22}{10} = \dfrac{290}{10} = 29$.

    Hint: If this had been a population, although a very small one, the calculations would be the same. However, the symbols would change. The symbol for a population mean, μ, replaces $\bar{x}$, and the symbol for the population size, N, replaces n.

The median is the middle number of the ordered array for the sample. Here we have:

3   5   19   19   22   23   27   43   53   76.

There are 10 numbers in this sample. Because 10 is an even number, we do not have a single middle number. Rather, 22 and 23 are the middle most two numbers. Taking the average of these two, the median is 22.5. The only number repeated is 19, so the mode of this data is 19.

Kids Up in Arms

A psychologist took a sample of 258 1-year-old children and examined their ability to understand nonverbal instructions. One task was to get a child to raise both arms. The psychologist observed the times (in seconds) for the child to raise his or her hands and recorded the results in the following frequency distribution.

TIMES TO RAISE HANDS	FREQUENCY
0 to less than 10	97
10 to less than 20	65
20 to less than 30	36
30 to less than 40	21
40 to less than 50	17
50 to less than 60	8
60 to less than 70	7
70 to less than 80	3
80 to less than 90	2
90 to less than 100	0
100 to less than 110	2

Approximate the mean, median, and mode for this sample.

Solution

We can organize our calculations in a table.

TIMES TO RAISE HANDS	FREQUENCY (f)	MIDPOINTS (m)	fm
0 to less than 10	97	5	485
10 to less than 20	65	15	975
20 to less than 30	36	25	900
30 to less than 40	21	35	735
40 to less than 50	17	45	765
50 to less than 60	8	55	440
60 to less than 70	7	65	455
70 to less than 80	3	75	225
80 to less than 90	2	85	170
90 to less than 100	0	95	0
100 to less than 110	2	105	210
	$n = \Sigma f = 258$		$\Sigma fm = 5360$

Then $\bar{x} = \dfrac{\Sigma fm}{n} = \dfrac{5360}{258} = 20.8.$

Hint: We previously calculated the mean of this sample, 20.59. This differs from our approximate mean by 0.18.

The median of 258 numbers is approximately the $\dfrac{258}{2} = 129^{\text{th}}$ observation. The first class contains 97 observations. The second class has another 65, so the first two classes account for 162 of the observations. The median must be in the second class. The 129th observation would be the $129 - 97 = 32^{\text{nd}}$ observation in the second class. Since there are 65 observations in the second class, the median would be approximately $\dfrac{32}{65}$ of the way into the class. Since $\dfrac{32}{65}$ is approximately 0.492, the median would be 49.2% into the class. Because the class is 10 to less than 20, the median is approximately 14.92. The first class has the greatest frequency and is the modal class. The midpoint of this class is 5 and is the approximate mode.

Practice Exercises 3-4

1-4. A sample of the weights of 20 American made automobiles is listed below.

3450	2750	4400	3680	2550	3000	3300	3300	2900	4550
4100	3700	2650	2300	3050	3700	3500	2600	2450	2950

  1. Calculate the mean weight of this sample.

  2. Obtain the median of the weights.

  3. Find the sample mode.

  4. Would the mean of this data set be denoted by μ or $\bar{x}$? Why?

5-7. The car weights from the previous set of exercises were organized into the following frequency distribution.

WEIGHT	FREQUENCY
2000 to less than 2500	2
2500 to less than 3000	6
3000 to less than 3500	5
3500 to less than 4000	4
4000 to less than 4500	2
4500 to less than 5000	1

  5. Approximate the mean weight.

  6. Calculate the approximate median.

  7. Approximate the mode for this sample.

8-10. The data below represents a sample of the number of books borrowed from a library by 25 people.

4	10	7	9	13	12	2	6	1	13	9	12	3
2	12	3	9	1	2	7	3	4	5	5	11	

For this sample:

  8. calculate the mean.

9. obtain the median.

10. find the mode.

11. The previous data was organized into the following frequency distribution. Use this to approximate the mean.

BOOKS	FREQUENCY
1-3	8
4-6	5
7-9	5
10-12	5
13-15	2

Hint: Notice that the classes are in a slightly different form than the previous examples. This is an alternate way of creating classes for data that is always integer valued, as here.

Solutions to Practice Exercises 3-4

1. $\bar{x} = \dfrac{\Sigma x}{n} = \dfrac{3450 + 2750 + 4400 + \cdots + 2950}{20} = \dfrac{64880}{20} = 3244.$

2. The median is the middle number of the ordered array for the weights:

    2300 2450 2550 2600 2650 2750 2900 2950 3000 3050
    3300 3300 3450 3500 3680 3700 3700 4100 4400 4550.

    There are 20 numbers in this sample. Because 20 is an even number, we take the average of the two meddle most observations. These are 3050 and 3300, so the median is 3175.

3. Both 3300 and 3700 are repeated twice, so this is a bimodal distribution with these two values as the modes.

4. Because this is a sample of weights, the mean is represented by $\bar{x}$.

5. We organize our calculations in a table.

WEIGHT	FREQUENCY (f)	MIDPOINTS (m)	fm
2000 to less than 2500	2	2250	4500
2500 to less than 3000	6	2750	16500
3000 to less than 3500	5	3250	16250
3500 to less than 4000	4	3750	15000
4000 to less than 4500	2	4250	8500
4500 to less than 5000	1	4750	4750
	$n = \Sigma f = 20$		$\Sigma fm = 65500$

Then $\bar{x} = \dfrac{\Sigma fm}{n} = \dfrac{65500}{20} = 3275.$

6. The median of 20 numbers would be approximately the 10th observation. The first two classes account for 8 observations. The third class has another 5, so the first three classes account for 13 of the observations. The median must be in the third class. The 10th observation would be the $10 - 8 = 2^{nd}$ observation in the third class. Since there are 5 observations in the third class, the median would be approximately $\dfrac{2}{5}$ of the way into the class. Because the class is 3000 to less than 3500, the median is approximately 3200.

7. The second class has the greatest frequency and is the modal class. The midpoint of this class is 2750 and therefore is the approximate mode.

8. $\bar{x} = \dfrac{\Sigma x}{n} = \dfrac{4 + 10 + 7 + \cdots + 11}{25} = \dfrac{165}{25} = 6.60.$

9. The median is the middle number of the ordered array for the borrowed books:

    1   1   2   2   2   3   3   3   4   4   5   5   6
    7   7   9   9   9   10   11   12   12   12   13   13

    There are 25 numbers in this sample. Because 25 is an odd number, the median is the $\dfrac{n + 1}{2} = \dfrac{25 + 1}{2} = 13^{th}$ number in the array, 6.

10. Each of 2, 3, 9, and 12 occurs three times, so there are four modes.

11. We organize our calculations in a table.

BOOKS	FREQUENCY (f)	MIDPOINTS (m)	fm
1-3	8	2	16
4-6	5	5	25
7-9	5	8	40
10-12	5	11	55
13-15	2	14	28
$n = \Sigma f = 25$			$\Sigma fm = 164$

Then $\overline{x} = \dfrac{\Sigma fm}{n} = \dfrac{165}{25} = 6.56$.

3-5 Computing Measures of Dispersion and Relative Position

Study Objective

You should be able to:
1. Calculate the range, mean absolute deviation, and standard deviation of a data set.
2. Calculate the standard deviation for grouped data.
3. Interpret the standard deviation by using Chebyshev's theorem and the empirical rule.
4. Compute standard (z) scores and percentiles.
5. Work with a box-and-whiskers display and a five-number summary of a data set.

Section Overview

In the previous chapter, we discussed the importance of considering the dispersion of a data set. Here we examine a few ways to measure this dispersion. One measure is the range, which is the difference between the largest and smallest number in a data set. Another is the mean absolute deviation, the average of the absolute value of the deviations. The most important is the standard deviation, which is the square root of the average of the squared deviations.

Once we know its mean and standard deviation, we can use Chebyshev's theorem and the empirical rule to gain insight into a data set. Chebyshev's theorem tells us the minimum amount of data that is within k standard deviations of the mean. The empirical rule tells us approximately how much data is within k standard deviations of the mean if the distribution is bell shaped.

In order to identify how high or how low an individual value is, we have measures of relative position. One is a standard (z) score, which indicates how many standard deviations a value is above or below the mean. Another is a percentile which specifies the percentage of the data below a data point.

A useful summary of a data set is presented by a box-and-whiskers display (or boxplot). To create a boxplot, we first find the first and third quartiles. The first quartile (Q_1) is the dividing line between the smallest 25% of the data and the rest; the third quartile (Q_3) is the dividing line between the largest 25% of the data and the rest. The boxplot places a box around the middle 50% of the data, i.e., between Q_1 and Q_3. Then lines (whiskers) are drawn extending from the box in either direction until they touch the smallest and largest observation.

Key Terms & Formulas

Range The difference between the largest and smallest number in a data set.

Deviations The differences between the observations of a data set and their mean.

Mean Absolute Deviation The average of the absolute value of the deviations in a data set. It is $\text{MAD} = \dfrac{\Sigma |x - \overline{x}|}{n}$.

Population Standard Deviation The square root of the average of the squared deviations. It is $\sigma = \sqrt{\dfrac{\Sigma(x-\mu)^2}{N}}$.

Sample Standard Deviation A statistic used to estimate σ. It can be calculated two ways:

$$s = \sqrt{\frac{\Sigma(x-\bar{x})^2}{n-1}} = \sqrt{\frac{n(\Sigma x^2)-(\Sigma x)^2}{n(n-1)}}.$$

Variance The square of a standard deviation. The variance is symbolized by σ^2 and s^2 for populations and samples, respectively.

Standard Deviation for Grouped Data An approximation of a standard deviation based on grouped data. The formula for a population is $\sigma = \sqrt{\dfrac{\Sigma f(m-\mu)^2}{N}}$. For a sample, two formulae are available:

$$s = \sqrt{\frac{\Sigma f(m-\bar{x})^2}{n-1}} = \sqrt{\frac{n(\Sigma f(m^2))-(\Sigma fm)^2}{n(n-1)}}.$$

Chebyshev's Theorem For any data set, at least $1 - \dfrac{1}{k^2}$ of the data will be within k standard deviations of the mean.

Empirical Rule For a "bell-shaped" distribution (aka a normal curve) about 68% of the data will be within 1 standard deviation of the mean, about 95% of the data will be within 2 standard deviations of the mean, about 99.7% of the data will be within 3 standard deviations of the mean.

Standard Score $z = \dfrac{x-\mu}{\sigma}$ (population); $z = \dfrac{x-\bar{x}}{s}$ (sample).

P^{th} Percentile A number that has $p\%$ of the data no more than its value and $(100-p)\%$ of the data at least its value. For example, the 40th percentile has 40% of the observations at or below it and 60% of the observations at or above it.

First Quartile (Q_1) Same as 25th percentile.

Third Quartile (Q_3) Same as 75th percentile.

Interquartile Range The range that contains the middle 50% of the data. $IQR = Q_3 - Q_1$.

Quartile Deviation One-half the interquartile range.

Box-and-Whiskers Display Also known as a boxplot, it is a graphical display involving the use of the median, both quartiles, and the smallest and largest values (and therefore is described as a five-number summary of a data set).

Worked Examples

Mites Revisited

The botanist who counted the number of mites on the leaves of Eucalyptus trees now wants to calculate the range, MAD, and s for the sample. Its array is reproduced below.

$$3 \quad 5 \quad 19 \quad 19 \quad 22 \quad 23 \quad 27 \quad 43 \quad 53 \quad 76$$

Solution

The range is the difference between the largest and smallest observation. Here that is $76 - 3 = 73$. To calculate the mean absolute deviation and s, we need $\bar{x}$. We found that the mean was 29.0. Then the calculations are done in the following table.

x	$x-\bar{x}$	$|x-\bar{x}|$
3	-26	26
5	-24	24
19	-10	10
19	-10	10
22	-7	7
23	-6	6
27	-2	2
43	14	14
53	24	24
76	47	47
		170

Then MAD $= \dfrac{170}{10} = 17.0$. A similar table is used to calculate the sample standard deviation.

x	$x - \bar{x}$	$(x - \bar{x})^2$
3	−26	676
5	−24	576
19	−10	100
19	−10	100
22	−7	49
23	−6	36
27	−2	4
43	14	196
53	24	576
76	47	2209
		4522

So $s = \sqrt{\dfrac{\Sigma(x - \bar{x})^2}{n - 1}} = \sqrt{\dfrac{4522}{9}} = \sqrt{502.4444} = 22.4153 = 22.4$. Alternatively, we can use the shortcut formula. We would have a different table for the calculations (and we would not need to calculate the mean before we form the table).

x	x^2
3	9
5	25
19	361
19	361
22	484
23	529
27	729
43	1849
53	2809
76	5776
290	12932

Then the shortcut formula yields:

$$s = \sqrt{\dfrac{n(\Sigma x^2) - (\Sigma x)^2}{n(n - 1)}} = \sqrt{\dfrac{10(12932) - (290)^2}{10(9)}} = \sqrt{\dfrac{45220}{90}} = \sqrt{502.4444} = 22.4153 = 22.4.$$

Exam Grades
A professor of History is teaching a section of 100 students. Her first exam's grade distribution follows. Calculate the standard deviation for this grouped data.

EXAM GRADES	FREQUENCY
45 to less than 50	1
50 to less than 55	2
55 to less than 60	6
60 to less than 65	19
65 to less than 70	12
70 to less than 75	22
75 to less than 80	12
80 to less than 85	13
85 to less than 90	11
90 to less than 95	0
95 to less than 100	2

Solution

To calculate s, we need the mean for grouped data. The arithmetic necessary to calculate it and the standard deviation is in the following table.

EXAM GRADES	m	f	fm	$m - \bar{x}$	$(m - \bar{x})^2$	$f(m - \bar{x})^2$
45 to less than 50	47	1	47	-25	625	625
50 to less than 55	52	2	104	-20	400	800
55 to less than 60	57	6	342	-15	225	1350
60 to less than 65	62	19	1178	-10	100	1900
65 to less than 70	67	12	804	-5	25	300
70 to less than 75	72	22	1584	0	0	0
75 to less than 80	77	12	924	5	25	300
80 to less than 85	82	13	1066	10	100	1300
85 to less than 90	87	11	957	15	225	2475
90 to less than 95	92	0	0	20	400	0
95 to less than 100	97	2	194	25	625	1250
			$\Sigma fm = 7200$			$\Sigma f(m - \bar{x})^2 = 10300$

$$\bar{x} = \frac{\Sigma fm}{n} = \frac{7200}{100} = 72.0. \text{ Then } s = \sqrt{\frac{\Sigma f(m - \bar{x})^2}{n - 1}} = \sqrt{\frac{10300}{99}} = \sqrt{104.04} = 10.2.$$

We can calculate s without calculating the sample mean. Again the work is made easier by placing the calculations in a table.

EXAM GRADES	m	f	fm	m^2	$f(m^2)$
45 to less than 50	47	1	47	2209	2209
50 to less than 55	52	2	104	2704	5408
55 to less than 60	57	6	342	3249	19494
60 to less than 65	62	19	1178	3844	73036
65 to less than 70	67	12	804	4489	53868
70 to less than 75	72	22	1584	5184	114048
75 to less than 80	77	12	924	5929	71148
80 to less than 85	82	13	1066	6724	87412
85 to less than 90	87	11	957	7569	83259
90 to less than 95	92	0	0	8464	0
95 to less than 100	97	2	194	9409	18818
			$\Sigma fm = 7200$		$\Sigma f(m^2) = 10300$

Here $s = \sqrt{\dfrac{n(\Sigma f(m^2)) - (\Sigma fm)^2}{n(n - 1)}} = \sqrt{\dfrac{100(528700) - (7200)^2}{100(99)}} = \sqrt{\dfrac{1030000}{9900}} = \sqrt{104.04} = 10.2.$

Scuba Dive I

The times scuba divers can stay underwater at a depth of 40 feet has a mean of 50 minutes with a standard deviation of 4 minutes. Use Chebyshev's theorem to decide what times at least 75% of all divers can stay underwater. Do the same for 89% of the divers.

Solution

Chebyshev's theorem says that at least 75% of the divers can remain underwater within two standard deviations of the mean time. One standard deviation is 4, so two standard deviations are 8. If we add and subtract 8 from the mean of 50, we get an interval from 42 to 58. Therefore, at least 75% of scuba divers can remain underwater at 40 feet for between 42 and 58 minutes. Working similarly, at least 89% of scuba divers can stay underwater for times within three standard deviations of the mean. Three standard deviations are 12. Adding and subtracting 12 from 50, we find that at least 89% of scuba divers can stay underwater between 38 and 62 minutes.

   Hint: It is important to remember that Chebyshev's theorem includes the phrase "at least." Suppose in the last example we had a boat of 60 scuba divers all diving at 40 feet. According to Chebyshev's theorem at least 75% of the divers, i.e., 45 divers, can remain underwater between 42 and 58 minutes. It is possible that 48 or 52 or 59 or all 60 of them stay underwater within these times. So to say that "45 can remain underwater" rather than "at least 45 ... " is a misstatement and should be avoided.

Scuba Diving: Dive II

Again suppose that the time scuba divers can stay underwater at a depth of 40 feet has a mean of 50 minutes with a standard deviation of 4 minutes. Using Chebyshev's theorem, what can you say about the proportion of divers that can stay underwater between 30 and 70 minutes?

Solution

We first notice that 30 is 20 minutes below the mean while 70 is 20 minutes above the mean. Because one standard deviation is 4 minutes, 20 minutes is $\frac{20}{4} = 5$ standard deviations. Chebyshev's theorem says that at least $1 - \frac{1}{k^2} = 1 - \frac{1}{5^2} = 96\%$ will be within this interval.

Scuba Diving: Last Dive

Now suppose that the dive times are symmetrical and bell-shaped (normal). Use the empirical rule to determine the times that would include 68% and 95% of the dives.

Solution

For normal data, approximately 68% will be within one standard deviation. Therefore, 68% of the divers can stay underwater between 46 and 54 minutes. The empirical rule says that 95% of the times will be within two standard deviations of the mean. From our first dive, we know that this would be from 42 to 58 minutes.

Mites Revisited

The botanist who counted the number of mites on the leaves of Eucalyptus trees now wants to calculate the IQR, QD, and create a box-and-whiskers display. Its array is reproduced below.

   3   5   19   19   22   23   27   43   53   76 .

Solution

We first calculate the first and third quartile positions. Using MINITAB's method, we calculate the position of Q_1 as $\frac{n+1}{4} = \frac{10+1}{4} = 2.75$, and the position of Q_3 as $\frac{3(n+1)}{4} = \frac{3(10+1)}{4} = 8.25$. So Q_1 = three quarters of the way from the second to the third number, i.e., three quarters of the way from 5 to 19. The difference is $19 - 5 = 14$. So $Q_1 = 5 + 0.75(14) = 15.5$. Similarly, Q_3 = one quarter of the way from the eighth to the ninth number, i.e., one quarter of the way from 43 to 53. The difference is $53 - 43 = 10$. This gives $Q_3 = 43 + 0.25(10) = 45.5$. From this we can calculate the interquartile range and the quartile, deriation

$IQR = 45.5 - 15.5 = 30.0$, and $QD = \frac{30.0}{2} = 15.0$.

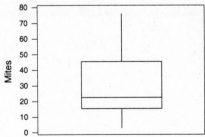

To form a box-and-whiskers display (boxplot), we need five numbers. Besides the quartiles above, we need the smallest, largest, and median observation. The smallest is 3, the largest is 76, and the median is the mean of 22 and 23, or 22.5. These are the numbers MINITAB used to produce the boxplot.

Practice Exercises 3-5

1-4. A marine biologist is interested in the width of abalones found in extreme cold water off the Oregon coast. She obtained the measurements (in inches) from a sample of such abalones. An ordered array of this data is presented below.

   3.4   3.8   3.9   4.0   4.0   4.1   4.1   4.2   4.2   4.3

1. Obtain the range of the abalone widths.
2. Calculate the mean absolute deviation.
3. Calculate the standard deviation of the sample using the original definition of s.

4. Calculate the standard deviation using the shortcut formula.

5-8. A realtor wants to examine the asking price for two-bedroom houses in Bismarck, North Dakota. She takes a listing of houses for sale and takes a sample of 8 two-bedroom houses. An ordered array of this data is presented below.

$$97 \quad 109 \quad 110 \quad 119 \quad 121 \quad 125 \quad 129 \quad 138$$

5. Obtain the range of the asking prices.

6. Calculate the mean absolute deviation.

7. Calculate the standard deviation of the sample using the original definition of s.

8. Calculate the standard deviation using the shortcut formula.

9-12. The time necessary to fly on a commercial airline from NYC to LA has a mean of 330 minutes with a standard deviation of 5 minutes. What can be said about the proportion of times that a flight will take between:

9. 320 and 340 minutes?

10. 315 and 345 minutes?

11. 320 and 340 minutes if flight times have a normal distribution?

12. 315 and 345 minutes if flight times have a normal distribution?

13-14. The lengths of "8 by" boards from Timbuktu Lumber have a mean of 8 feet, 3 inches, with a standard deviation of 2 inches. Using Chebyshev's theorem,

13. How long are at least 75% of the boards?

14. How long are at least 89% of the boards?

15-16. Timbuktu Lumber sells boxes of nails labeled 50 lbs. The actual weight of such boxes is normal with a mean of 52.4 lbs. and a standard deviation of 1.2 lbs. Using the empirical rule,

15. How much will approximately 68% of the boxes weigh?

16. How much will approximately 95% of the boxes weigh?

17-18. We previously used the table below to calculate an approximate mean for the weights of automobiles. The mean was 3275 lbs.

WEIGHT	FREQUENCY (f)	MIDPOINTS (m)
2000 to less than 2500	2	2250
2500 to less than 3000	6	2750
3000 to less than 3500	5	3250
3500 to less than 4000	4	3750
4000 to less than 4500	2	4250
4500 to less than 5000	1	4750

17. Calculate the standard deviation for grouped data using the original definition.

18. Calculate the standard deviation for grouped data using the shortcut formula.

19-20. We previously used the table below to approximate the mean number of books borrowed from a library. The mean was 6.56 books.

BOOKS	FREQUENCY (f)	MIDPOINTS (m)
1-3	8	2
4-6	5	5
7-9	5	8
10-12	5	11
13-15	2	14
$n = \Sigma f = 25$		

19. Approximate the standard deviation for grouped data using the original definition.

20. Approximate the standard deviation for grouped data using the shortcut formula.

21-25. The array of the widths of abalones from Exercises 1-4 of this section is reproduced below.

$$3.4 \quad 3.8 \quad 3.9 \quad 4.0 \quad 4.0 \quad 4.1 \quad 4.1 \quad 4.2 \quad 4.2 \quad 4.3$$

21. What is the z score of 4.2?

22. What percentile is 3.95?

23. Obtain the interquartile range of the abalone widths.

24. Calculate the quartile deviation.

25. Draw a box-and-whiskers display.

26-30. An array of the asking prices for two-bedroom houses in Bismarck, N.D., from Exercises 5-8 of this section is reproduced below.

$$97 \quad 109 \quad 110 \quad 119 \quad 121 \quad 125 \quad 129 \quad 138$$

26. What is the z score of 109?

27. What percentile is 127?

28. Obtain the interquartile range of the asking prices.

29. Calculate the quartile deviation.

30. Draw a box-and-whiskers display.

Solutions to Practice Exercises 3-5

1. The range is $4.3 - 3.4 = 0.9$.

2. To calculate the mean absolute deviation, we first need $\bar{x}$.

$$\bar{x} = \frac{3.4 + 3.8 + 3.9 + 4.0 + 4.0 + 4.1 + 4.2 + 4.2 + 4.3}{10} = \frac{40.0}{10} = 4.0.$$ Then the calculations can be summarized in tabular form.

x	$x - \bar{x}$	$|x - \bar{x}|$
4.0	0.0	0.0
3.4	−0.6	0.6
4.1	0.1	0.1
4.3	0.3	0.3
4.2	0.2	0.2
4.0	0.0	0.0
3.9	−0.1	0.1
4.1	0.1	0.1
4.2	0.2	0.2
3.8	−0.2	0.2
		1.8

$$\text{MAD} = \frac{1.8}{10} = 0.18.$$

3. A table can be used to help calculate the sample standard deviation.

x	$x - \bar{x}$	$(x - \bar{x})^2$
4.0	0.0	0.00
3.4	−0.6	0.36
4.1	0.1	0.01
4.3	0.3	0.09
4.2	0.2	0.04
4.0	0.0	0.00
3.9	−0.1	0.01
4.1	0.1	0.01
4.2	0.2	0.04
3.8	−0.2	0.04
		0.60

Then $s = \sqrt{\dfrac{\Sigma(x - \bar{x})^2}{n - 1}} = \sqrt{\dfrac{0.60}{9}} = \sqrt{0.066666} = 0.258199 \approx 0.26.$

4. Using the shortcut formula we have a different table to do the calculations.

x	x^2
4.0	16.00
3.4	11.56
4.1	16.81
4.3	18.49
4.2	17.64
4.0	16.00
3.9	15.21
4.1	16.81
4.2	17.64
40.0	160.60

Then using the shortcut formula yields:

$$s = \sqrt{\frac{n(\Sigma x^2) - (\Sigma x)^2}{n(n-1)}} = \sqrt{\frac{10(160.60) - (40.0)^2}{10(9)}} = \sqrt{0.066666} = 0.258199 \approx 0.26.$$

5. The range is $138 - 97 = 41$.

6. To calculate the mean absolute deviation, we first need $\overline{x}$.

$$\overline{x} = \frac{129 + 97 + 119 + 109 + 121 + 110 + 138 + 125}{8} = \frac{948}{8} = 118.5.$$ Then the calculations can be summarized in tabular form.

x	$x - \overline{x}$	$|x - \overline{x}|$
129	10.5	10.5
97	−21.5	21.5
119	0.5	0.5
109	−9.5	9.5
121	2.5	2.5
110	−8.5	8.5
138	19.5	19.5
125	6.5	6.5
		79.0

So MAD $= \dfrac{79.0}{8} = 9.875$.

7. A table will be used to help calculate the sample standard deviation.

x	$x - \overline{x}$	$(x - \overline{x})^2$
129	10.5	110.25
97	21.5	462.25
119	0.5	0.25
109	9.5	90.25
121	2.5	6.25
110	8.5	72.25
138	19.5	380.25
125	6.5	42.25
		1164.00

Then $s = \sqrt{\dfrac{\Sigma(x - \overline{x})^2}{n-1}} = \sqrt{\dfrac{1164.00}{7}} = \sqrt{166.286} = 12.8952 \approx 12.9$.

8. To calculate s using the shortcut formula, we work with the following table.

x	x^2
129	16641
97	9409
119	14161
109	11881
121	14641
110	12100
138	19044
125	15625
948	113502

Then using the shortcut formula yields:

$$s = \sqrt{\frac{n(\Sigma x^2) - (\Sigma x)^2}{n(n-1)}} = \sqrt{\frac{8(113502) - (948)^2}{8(7)}} = \sqrt{166.286} = 12.8952 \approx 12.9.$$

9. Because 340 and 320 are both 10 minutes from the mean and the standard deviation is 5, this interval consists of values that are within $\frac{10}{5} = 2$ standard deviations of the mean. Chebyshev's theorem says that at least $1 - \frac{1}{2^2} = \frac{3}{4}$ of the times will be within this interval.

10. Because 345 and 315 are both 15 minutes from the mean and the standard deviation is 5, this interval consists of values that are within $\frac{15}{5} = 3$ standard deviations of the mean. Chebyshev's theorem says that at least $1 - \frac{1}{3^2} = \frac{8}{9}$ of the times will be within this interval.

11. From problem 9 we know that this interval consists of data points that are within 2 standard deviations of the mean. The empirical rule says that approximately 95% of the times will be in such an interval.

12. From problem 10 we know that this interval consists of data points that are within 3 standard deviations of the mean. The empirical rule says that approximately 99.7% of the times will be in such an interval.

13. Chebyshev's theorem says that at least 75% will be within 2 standard deviations of the mean. Two standard deviations here are 2(2 inches) = 4 inches. Adding and subtracting this amount from 8 feet, 3 inches, gives an interval from 7 feet, 11 inches, to 8 feet, 7 inches.

14. Chebyshev's theorem says that at least 89% will be within 3 standard deviations of the mean. Three standard deviations here are 3(2 inches) = 6 inches. Adding and subtracting this amount from 8 feet, 3 inches, gives an interval from 7 feet, 9 inches, to 8 feet, 9 inches.

15. The empirical rule says that approximately 68% will be within 1 standard deviation of the mean. One standard deviation is 1.2 lbs. Adding and subtracting this amount from 52.4 lbs. gives an interval from 51.2 lbs. to 53.6 lbs.

16. The empirical rule says that approximately 95% will be within 2 standard deviations of the mean. Two standard deviations are 2(1.2 lbs.) = 2.4 lbs. Adding and subtracting this amount from 52.4 lbs. gives an interval from 50.0 lbs. to 54.8 lbs.

17. We can organize the necessary calculations for the standard deviation in a table.

WEIGHT	f	m	$m - \bar{x}$	$(m - \bar{x})^2$	$f(m - \bar{x})^2$
2000 to less than 2500	2	2250	−1025	1050625	2101250
2500 to less than 3000	6	2750	−525	275625	1653750
3000 to less than 3500	5	3250	−25	625	3125
3500 to less than 4000	4	3750	475	225625	902500
4000 to less than 4500	2	4250	975	950625	1901250
4500 to less than 5000	1	4750	1475	2175625	2175625
	$n = 20$				$\Sigma f(m - \bar{x})^2 = 8737500$

$$.s = \sqrt{\frac{\Sigma f(m - \bar{x})^2}{n - 1}} = \sqrt{\frac{8737500}{19}} = \sqrt{459868} = 678.136 \approx 678.$$

18. The calculations are based on the table below.

WEIGHT	f	m	fm	fm^2
2000 to less than 2500	2	2250	4500	10125000
2500 to less than 3000	6	2750	16500	45375000
3000 to less than 3500	5	3250	16250	52812500
3500 to less than 4000	4	3750	15000	56250000
4000 to less than 4500	2	4250	8500	36125000
4500 to less than 5000	1	4750	4750	22562500
	$n = 20$		$\Sigma fm = 65500$	$\Sigma fm^2 = 223250000$

$$s = \sqrt{\frac{n(\Sigma f(m^2)) - (\Sigma fm)^2}{n(n-1)}} = \sqrt{\frac{20(223250000) - (65500)^2}{20(19)}} = \sqrt{459868.42} = 678.136 \approx 678.$$

19. We organize the necessary calculations for the standard deviation in a table.

BOOKS	f	m	$m - \bar{x}$	$(m - \bar{x})^2$	$f(m - \bar{x})^2$
1-3	8	2	−4.56	20.7936	166.349
4-6	5	5	−1.56	2.4336	12.168
7-9	5	8	1.44	2.0736	10.368
10-12	5	11	4.44	19.7136	98.568
13-15	2	14	7.44	55.3536	110.707
	$n = 25$				$\Sigma f(m - \bar{x})^2 = 398.160$

$$s = \sqrt{\frac{\Sigma f(m - \bar{x})^2}{n - 1}} = \sqrt{\frac{398.16}{24}} = \sqrt{16.59} = 4.073 \approx 4.1.$$

20. The calculations are based on the table below.

BOOKS	f	m	fm	fm^2
1-3	8	2	16	32
4-6	5	5	25	125
7-9	5	8	40	320
10-12	5	11	55	605
13-15	2	14	28	392
	$n = 25$		$\Sigma fm = 164$	$\Sigma fm^2 = 1474$

$$s = \sqrt{\frac{n(\Sigma f(m^2)) - (\Sigma fm)^2}{n(n-1)}} = \sqrt{\frac{25(1474) - (164)^2}{25(24)}} = \sqrt{16.59} = 4.073 \approx 4.1.$$

21. $z = \dfrac{4.2 - 4.0}{0.26} \approx 0.77.$

22. 3 of 10 values are below 3.95, so it is the 30th percentile.

23. The position of Q_1 is $\dfrac{n+1}{4} = \dfrac{10+1}{4} = 2.75$. This means Q_1 is 75% of the way between 3.8 and 3.9 and is therefore 3.75. Q_3 has position $\dfrac{3(n+1)}{4} = \dfrac{33}{4} = 8.25$. But the eighth and ninth observations are both 4.2, so Q_3 is 4.2. Then IQR $= 4.2 - 3.75 = 0.45$.

24. QD $= \dfrac{\text{IQR}}{2} = \dfrac{0.45}{2} = 0.225.$

25. Boxplot

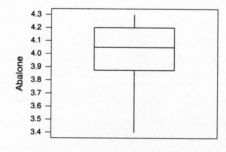

26. $z = \dfrac{109 - 118.5}{12.9} \approx -0.74.$

27. 6 of 8 values are below 127, so it is the 75th percentile.

28. The position of Q_1 is $\dfrac{n+1}{4} = \dfrac{8+1}{4} = 2.25.$ This means Q_1 is 25% of the way between 109 and 110 and is therefore 109.25. Q_3 has position $\dfrac{3(n+1)}{4} = \dfrac{27}{4} = 6.75.$ So it is 75% of the way between 125 and 129 and is 128. Then IQR $= 128 - 109.25 = 18.75.$

29. $QD = \dfrac{IQR}{2} = \dfrac{18.75}{2} = 9.375.$

30. Boxplot

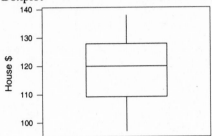

3-6 Summarizing Qualitative Data

Study Objective
You should be able to:
    Work with percentages when dealing with attribute data.

Section Overview
    It does not make sense to work with averages or measures of dispersion with attribute data. Instead, we examine the proportion or percentage of the time that events occur. For example, we might want to know the percentage of Washington, D.C., residents who favor statehood or the proportion of Americans who are bilingual. We will find that attribute data can be just as informative as numerical data.

Worked Examples
Unions
In a recent survey, 1011 of 2427 people surveyed agreed with the statement "Unions have become less effective in the last 10 years." What percent agreed with the statement?

Solution
Percent agreed $= \dfrac{1011}{2427} \times 100 = 41.65\% \approx 42\%.$

Practice Exercises 3-6
1. A recent survey at an Indiana college revealed that 234 of 515 students who entered as first-year students five years earlier had graduated. What percent graduated?
2. In the state of California, a recent poll of adults revealed that 587 of 1693 had been born in the state. What percent were born in California?

Solutions to Practice Exercises 3-6

1. Percent graduated $= \dfrac{234}{515} \times 100 = 45.44 \approx 45\%.$

2. Percent California born $= \dfrac{587}{1693} \times 100 = 34.67 \approx 35\%.$

Solutions to Odd-Numbered Exercises

1.

10	10	12	12	14	15	17	20	20	20	22	22	23	24	24	25
25	25	28	30	30	30	32	35	35	35	40	48	50	50	50	50
50	50	60	65	65	79	120									

3. The histogram of the scale price data is prepared using the frequency distribution we found in Exercise 2.

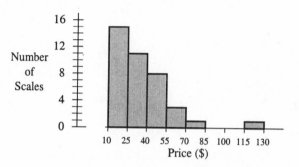

5. Since there are 39 scales, the position of the median price is $\dfrac{n+1}{2} = \dfrac{39+1}{2} = 20.$ Using the array we found in Exercises 1, the median is \$30. The position of Q_1 in the array is found using the formula $\dfrac{n+1}{4} = \dfrac{39+1}{4} = 10.$ The value of the 10th position is \$20. Similarly, the position of Q_3 is $\dfrac{3(n+1)}{4} = \dfrac{3(39+1)}{4} = 30$ and the value of the 30th position is \$50.

7.

Price

20 40 60 80 100 120

9. We expand the frequency distribution we found in Exercise 2.

Price of Scale	Number of Scales (f)	Class midpoint (m)	fm
\$10 and less than \$25	15	17.5	262.5
\$25 and less than \$40	11	32.5	357.5
\$40 and less than \$55	8	47.5	380
\$55 and less than \$70	3	62.5	187.5
\$70 and less than \$85	1	77.5	77.5
\$85 and less than \$100	0	92.5	0
\$100 and less than \$115	0	107.5	0
\$115 and less than \$130	1	122.5	122.5
	$\Sigma f = 39$		$\Sigma fm = 1387.5$

Sample mean, $xbar = \dfrac{\Sigma fm}{\Sigma f} = \dfrac{1387.5}{39} = \$35.58.$

11.

$4,792	6,642	7,440	10,418	10,696	13,129	15,020	16,000	16,020
16,229	16,300	16,950	17,300	17,300	17,500	17,600	17,655	17,700
17,745	17,750	17,757	17,802	18,000	18,570	18,700		

13. The histogram of the tuition data is prepared from the frequency distribution we found in Exercise 12.

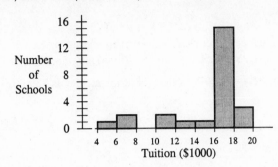

15. Since there are 25 schools, the median tuition is found in the $\dfrac{n+1}{2} = \dfrac{25+1}{2} = 13^{\text{th}}$ position of the array; thus the median tuition is $17,300. The position of Q_1 in the array is found using the formula $\dfrac{n+1}{4} = \dfrac{25+1}{4} = 6.5$. The value of Q_1 is found by averaging the tuition in the 6th and 7th position; hence $Q_1 = \dfrac{13129 + 15020}{2} = \$14,074.50$. The position of Q_3 is $\dfrac{3(n+1)}{4} = \dfrac{3(25+1)}{4} = 19.5$ and the value of Q_3 is the average of the 19th and 20th position which is $\dfrac{17745 + 17750}{2} = \$17,747.50$.

17.

Tuition ($1000)

```
 o   o    *    [===|=]
 4   6   8  10  12  14  16  18  20
```

19. We expand the frequency distribution we found in Exercise 12.

Tuition	Number of Schools (f)	Class midpoint (m)	fm
$4,000 and less than $6,000	1	5000	5000
$6,000 and less than $8,000	2	7000	14000
$8,000 and less than $10,000	0	9000	0
$10,000 and less than $12,000	2	11000	22000
$12,000 and less than $14,000	1	13000	13000
$14,000 and less than $16,000	1	15000	15000
$16,000 and less than $18,000	15	17000	255000
$18,000 and less than $20,000	3	19000	57000
	$\Sigma f = 25$		$\Sigma fm = 381000$

Population mean, $\mu = \dfrac{\Sigma fm}{\Sigma f} = \dfrac{381000}{25} = \$15240.$

21.

Stem	Leaf
1	7
2	479
3	45
4	01146778
5	1
6	00357
7	3

23. The median is the 11th score which is 46.

25. Range = (high score) − (low score) = 73 − 17 = 56.

27. Since there are 21 values and 40.5 is between the 7th score and the 8th score, the score 40.5 is approximately the 33rd percentile.

29. First arrange the 9 price per share values in ascending order.
25.50 30.75 31.25 33.50 35.00 53.25 57.25 58.50 104.75
The median is the $\dfrac{9+1}{2} = 5$th value. Thus the median is \$35.00.

31. Using $\mu = \$47.75$ from Exercise 50 and formula 3.6, $\text{MAD} = \dfrac{\Sigma|x - \mu|}{n}$ and $\mu = \$47.75$ (from Exercise 28).

Price per share, x	$\lvert x - \mu \rvert$
25.50	22.25
30.75	17.00
31.25	16.50
33.50	14.25
35.00	12.75
53.25	5.50
57.25	9.50
58.50	10.75
104.75	57.00
	165.50

$$\text{MAD} = \frac{\Sigma|x - \mu|}{n} = \frac{165.50}{9} = 18.389.$$

33. First arrange the 12 oxygen consumption values in ascending order.
7.0 8.3 8.6 9.7 10.9 12.5 12.8 14.3 15.2 19.1 19.5 21.0
The median is the $\dfrac{12+1}{2} = 6.5$ term in the array. Thus the median is $\dfrac{12.5 + 12.8}{2} = 12.65$. Q_1 is the $\dfrac{12+1}{4} = 3.25$ term in the array. So $Q_1 = 8.6 + 0.25(9.7 - 8.6) = 8.875$. Likewise, Q_3 is the $\dfrac{3(12+1)}{4} = 9.75$ term in the array. So $Q_3 = 15.2 + 0.75(19.1 - 15.2) = 18.125$.

35. $\mu = \dfrac{8.12 + 6.67 + 6.64 + 6.46 + 6.40 + 6.19 + 6.18 + 5.96 + 5.64 + 5.60}{10} = \dfrac{63.86}{10} = 6.386.$

37. Using formula 3.7, $\sigma = \sqrt{\dfrac{\Sigma(x - \mu)^2}{N}}$ with $\mu = 13.24$

Rating Score	$(x - \mu)^2$
8.12	3.006756
6.67	0.080656
6.64	0.064516
6.46	0.005476
6.40	0.000196
6.19	0.038416
6.18	0.042436
5.96	0.181476
5.64	0.556516
5.60	0.617796
Total	4.594240

$$\sigma = \sqrt{\frac{\Sigma(x-\mu)^2}{N}} = \sqrt{\frac{4.594240}{10}} = \sqrt{0.459424} = 0.6778.$$

39. Use the classes given to produce the table:

Grades	Frequency
40 < 50	4
50 < 60	6
60 < 70	10
70 < 80	4
80 < 90	4
90 < 100	2
	30

41. We expand the frequency distribution we found in Exercise 39 and use $\mu = 66.33$ that was found in Exercise 40.

Grades	Frequency	Class midpoint (m)	Deviation $(m-\mu)$	$(m-\mu)^2$	$f(m-\mu)^2$
40 < 50	4	45	−21.33	454.9689	1819.8756
50 < 60	6	55	−11.33	128.3689	770.2134
60 < 70	10	65	−1.33	1.7689	17.6890
70 < 80	4	75	8.67	75.1689	300.6756
80 < 90	4	85	18.67	348.5689	1394.2756
90 < 100	2	95	28.67	821.9689	1643.9378
	30				5946.6670

$$\sigma = \sqrt{\frac{\Sigma f(m-\mu)^2}{N}} = \sqrt{\frac{5946.6670}{30}} = 14.08 \text{ points.}$$

43. Q_1 is the $\dfrac{12+1}{4} = 3.25$ term in the array. So $Q_1 = 2{,}673 + 0.25(2{,}882 - 2{,}673) = 2725.25$. Likewise, Q_3 is the $\dfrac{3(12+1)}{4} = 9.75$ term in the array. So $Q_3 = 3{,}254 + 0.75(3{,}259 - 3{,}254) = 3{,}257.75$.

Interquartile range $\text{IQR} = Q_3 - Q_1 = 3{,}257.75 - 2{,}725.25 = \$532.50$. $\text{QD} = \dfrac{\text{IQR}}{2} = \dfrac{532.50}{2} = \$266.25$.

45. Using formula 3.8, $s = \sqrt{\dfrac{\Sigma(x-\bar{x})^2}{n-1}}$ and $\bar{x} = 2{,}996.70$.

Cost ($)	$(x-\bar{x})$	$(x-\bar{x})^2$
2,448	−548.7	301071.69
2,624	−372.7	138905.29
2,673	−323.7	104781.69
2,882	−114.7	13156.09
2,939	−57.7	3329.29
2,963	−33.7	1135.69
3,025	28.3	800.89
3,133	136.3	18577.69
3,254	257.3	66203.29
3,259	262.3	68801.29
3,295	298.3	88982.89
3,465	468.3	219304.89
		1025050.68

$$s = \sqrt{\frac{\Sigma(x-\bar{x})^2}{n-1}} = \sqrt{\frac{1025050.68}{11}} = \sqrt{93186.425} = \$305.26$$

47. The data is given to us in a descending array (except for the last two points). Thus the median $= \dfrac{1{,}517 + 1{,}301}{2} = 1{,}409$ employees.

49. Using formula 3.8 and $\overline{x} = 1{,}490.4$.

Employees	$x - \overline{x}$	$(x - \overline{x})^2$
2,829	1338.6	1791850
2,359	868.6	754466
1,784	293.6	86201
1,569	78.6	6178
1,517	26.6	708
1,301	−189.4	35872
1,267	−223.4	49908
873	−617.4	381183
750	−740.4	548192
655	−835.4	697893
		4352451

$$s = \sqrt{\frac{\Sigma(x - \overline{x})^2}{n - 1}} = \sqrt{\frac{4352451}{9}} = \sqrt{483605.67} = 695.4 \text{ employees.}$$

51. The z score is $z = \dfrac{x - \overline{x}}{s} = \dfrac{2829 - 1490.4}{695.4} = 1.93$.

53. Median $= \dfrac{440 + 567}{2} = \dfrac{1007}{2} = 503.5$ calories.

55. Range $= 615 - 387 = 228$ calories.

57. Since 78 degrees is between the 1st and 2nd term, the value of 78 degrees is the 10th percentile.

59. Since 90 degrees is between the 9th and 10th term, the value of 90 degrees is the 90th percentile.

61.. Using Chebyshev's Theorem with $k = 2$, we know that at least 75% of functional families should score between $\mu - 2\sigma$ and $\mu + 2\sigma$ scale points. $\mu - 2\sigma = 7.17 - 2(1.49) = 4.19$ and $\mu + 2\sigma = 7.17 + 2(1.49) = 10.15$.

63. Using Chebyshev's Theorem with $k = 2$, we know that at least 75% of functional families should score between $\mu - 2\sigma$ and $\mu + 2\sigma$ scale points. $\mu - 2\sigma = 5.57 - 2(2.49) = 0.59$ and $\mu + 2\sigma = 5.57 + 2(2.49) = 10.55$.

65. College A $= \dfrac{32}{148 + 32} = \dfrac{32}{180} = 17.78\%$; College B $= \dfrac{42}{64 + 42} = \dfrac{42}{106} = 39.62\%$;

College C $= \dfrac{26}{12 + 26} = \dfrac{26}{38} = 68.42\%$; College D $= \dfrac{48}{102 + 48} = \dfrac{48}{150} = 32\%$.

67. Use $\mu = 73.7$ which was found in Exercise 66.

Test scores	Students f	Midpoint m	Deviation $m - \mu$	$(m - \mu)^2$	$f(m - \mu)^2$
$40 < 50$	10	45	-28.7	823.69	8236.90
$50 < 60$	5	55	-18.7	349.69	1748.45
$60 < 70$	7	65	-8.7	75.69	529.83
$70 < 80$	3	75	1.3	1.69	5.07
$80 < 90$	16	85	11.3	127.69	2043.04
$90 < 100$	12	95	21.3	453.69	5444.28
	53				18007.57

Population standard deviation, $\sigma = \sqrt{\dfrac{\Sigma f(m - \mu)^2}{N}} = \sqrt{\dfrac{18007.57}{53}} = \sqrt{339.76} = 18.43$ points.

69. The median is the 3rd term in the ascending array: 5 8 9 10 16. So median $= 9$ tickets.

71. *New* population mean $= \dfrac{100 \cdot 33,000 + 30,000}{100} = \dfrac{3,330,000}{100} = \$33,300.$

73. The column labeled x^2 is not required to find the mean. However, it is included here to help with Exercise 75.

Exam Scores, x	x^2
269	72361
345	119025
381	145161
428	183184
455	207025
582	338724
760	577600
763	582169
898	806404
990	980100
999	998001
1059	1121481
1141	1301881
1181	1394761
1213	1471369
1264	1597696
1352	1827904
1403	1968409
15483	15693255

$\bar{x} = \dfrac{15483}{18} = 860.2$ points.

75. Using the formula $s = \sqrt{\dfrac{n(\Sigma x^2) - (\Sigma x)^2}{n(n-1)}}$ and the value found in the table above we have

$s = \sqrt{\dfrac{18 \cdot 15693255 - (15483)^2}{18 \cdot 17}} = \sqrt{139,723.22} = 373.80$ points.

77. The median is the 4th term in the ascending array:
11,740   13,316   13,733   14,193   15,077   15,536   15,704
So median $= 14,193.$

79. Using the formulas $\bar{x} = \dfrac{\Sigma x}{n}$ and $s = \sqrt{\dfrac{n(\Sigma x^2) - (\Sigma x)^2}{n(n-1)}}$, compute the table:

Weight, x	x^2
7.98	63.6804
9.74	94.8676
8.98	80.6404
11.26	126.7876
4.41	19.4481
1.87	3.4969
44.24	388.921

Mean $\bar{x} = \dfrac{44.24}{6} = 7.37$ grams; Standard deviation $s = \sqrt{\dfrac{6 \cdot 388.921 - (44.24)^2}{6 \cdot 5}} = \sqrt{12.545} = 3.54$ grams.

81. Using the formulas $\bar{x} = \dfrac{\Sigma x}{n}$ and $s = \sqrt{\dfrac{n(\Sigma x^2) - (\Sigma x)^2}{n(n-1)}}$, compute the table:

Density, x	x^2
43	1849
31	961
25	625
40	1600
10	100
11	121
160	5256

Mean $\bar{x} = \dfrac{160}{6} = 26.67$ gm/m^2; Standard deviation $s = \sqrt{\dfrac{6 \cdot 5256 - (160)^2}{6 \cdot 5}} = \sqrt{197.867} = 14.07$ gm/m^2.

Chapter 4 Probability Concepts

Study Aids and Practice Exercises

4-1 Some Basic Considerations

Study Objective

You should be able to:
1. Understand terms such as probability experiment, sample space, and event.
2. Recognize the various interpretations of the word probability.
3. Work with the two properties of probability.

Section Overview

Everyone has a "feel" for the word probability. If we say that the probability of flipping a head with a coin is one-half, no one would look askance at us. However, to calculate probabilities in nontrivial situations, we need a basic framework of definitions and rules about probability. In this section we examine some fundamental terminology and meanings of probability.

The basic framework of probability begins with probability experiments and their sample spaces, the complete set of simple outcomes to probability experiments. Events are parts or subsets of such a sample space. If we cannot decompose an event into smaller events, we call it a simple event. For example, rolling an even number with a die is an event that is not a simple event. It occurs if we roll a 2 or a 4 or a 6. However, rolling a 2 is a simple event. It cannot be broken into any smaller parts.

The probability of an event E, $P(E)$, is the relative likelihood that the event occurs. We can assign probabilities three ways. If we assign a probability *a priori*, it means we can figure out the probability from some means before any experiments. For example, we would *a priori* assign $\frac{1}{2}$ for the probability that the gender of an unborn child will be female. If we assign a probability using relative frequency, we do so by counting the number of outcomes in an event and dividing that by the total number of all outcomes. For example, suppose there is a class with 15 females and 20 males and we decide to randomly select one person from the class. Based on the relative frequency of females, we would assign a probability of $\frac{15}{35} = \frac{3}{7}$ for the probability of the selected person being a female. Finally, we can assign probabilities through intuition, i.e., using subjective judgment. For example, a person might subjectively decide that the probability that he will use statistics in the future is 0.90.

Probability must satisfy two conditions. First, all probabilities are between 0 and 1. An event that has a 0 probability should never occur, while an event with a probability of 1 must occur. For example, when rolling two dice, the probability of rolling a total of 27 is 0, while the probability of rolling a total between 2 and 12, inclusive, is 1.

Key Terms & Formulas

Probability Experiment Any action for which the outcome, response, or measurement is in doubt.

Sample Space The set of all possible simple outcomes, responses, or measurements from an experiment.

> Hint: The key word in the above definition is simple. If we throw a die (the singular of dice) it is possible to roll an even number. This is not a simple outcome because it can happen three ways. It happens if we roll a 2 or if we roll a 4 or if we roll a 6. An example of a simple outcome is rolling a 6. It can happen only one way--we get a 6.

Event A subset of a sample space.

> Hint: Do not be thrown by a word such as "subset." An event is just a collection of some possible outcomes of an experiment. For example, rolling an even number in the previous hint is an event. It consists of {2, 4, 6}, all part of or a subset of the sample space (which happens to consist of {1, 2, 3, 4, 5, 6}).

Simple Event An event that cannot be broken down any further.

> Hint: Looking at the previous term, it is possible that the sample space can be described as a collection of simple events. So in throwing a single die, the simple events consist of {1, 2, 3, 4, 5, 6}.

Probability The probability of an event E, $P(E)$, is the relative likelihood that the event occurs.

> Hint: It is hard to define probability without relying on synonyms like chance or likely or such. It is probably sufficient to recognize that an event that has a small probability, while possible, should not occur very often. So, while it is unlikely that in a game of Monopoly or Backgammon that we would roll doubles three times in a row, it does happen. But if someone rolls ten doubles in a row, we would consider checking the dice--such a streak of luck is just too unlikely to be believed.

Properties of Probabilities

1. $0 \leq P(E) \leq 1$
2. The probabilities of all the simple events in a sample space sum to 1.

Worked Examples

Possible Probabilities

Suppose a marketing firm is considering four different background colors (teal, azure, ocher, and magenta) for the package of a new cereal. If a randomly selected person is asked their preference for background color, which of the following assignments of probabilities for the selection would be valid?

1. $P(\text{teal}) = 0.2$, $P(\text{azure}) = 0.2$, $P(\text{ocher}) = 0.2$, $P(\text{magenta}) = 0.2$
2. $P(\text{teal}) = 0.2$, $P(\text{azure}) = 0.2$, $P(\text{ocher}) = 0.4$, $P(\text{magenta}) = 0.4$
3. $P(\text{teal}) = 0.6$, $P(\text{azure}) = -0.2$, $P(\text{ocher}) = 0.2$, $P(\text{magenta}) = 0.4$
4. $P(\text{teal}) = 0.2$, $P(\text{azure}) = 0$, $P(\text{ocher}) = 0.4$, $P(\text{magenta}) = 0.4$

Solution

Set 4. is the only valid set of probabilities--all the probabilities are at least zero and they sum to 1. The probabilities of set 1 sum to 0.8, less than 1. The probabilities of set 2 sum to 1.2, more than 1. Set 3 has a negative probability.

Ages

In a Statistics class, there are eight 19-year-olds, seven 20-year-olds, six 21-year-olds, and four 22-year-olds. If you randomly select one person from the class, what is the probability that a 22-year-old is selected? What is the probability that someone at least 20 years old is selected?

Solution

There is a total of 25 people in the class. If you select a person at random, the chance of getting a 22-year-old is $\frac{4}{25} = 0.16$. Because all but eight of the students are at least 20 years old, the probability of such a selection is $\frac{17}{25} = 0.68$.

Practice Exercises 4-1

1-4. The operator of a mountaineering shop sells three types of fill for his sleeping bags: down, cotton, and polyester. He is interested in predicting the fill type of the next sleeping bag purchase in his store. He assigns probabilities for each fill type. Identify which of the following assignments are valid. For those that are not, explain why.

1. $P(\text{down}) = 0.6$, $P(\text{cotton}) = 0$, $P(\text{polyester}) = 0.4$.
2. $P(\text{down}) = 0.2$, $P(\text{cotton}) = 0.3$, $P(\text{polyester}) = 0.4$.
3. $P(\text{down}) = 0.7$, $P(\text{cotton}) = -0.1$, $P(\text{polyester}) = 0.4$.
4. $P(\text{down}) = 0.2$, $P(\text{cotton}) = 0.5$, $P(\text{polyester}) = 0.4$.
5. $P(\text{down}) = 0.2$, $P(\text{cotton}) = 0.5$, $P(\text{polyester}) = 0.3$.

6-8. A recent survey asked people to identify their favorite sport. They were given the choices of baseball, basketball, football, tennis, skiing, hockey, swimming, and soccer. The percentages each sport was selected are: 18% baseball, 23% basketball, 16% football, 9% tennis, 4% skiing, 5% hockey, 12% swimming, and 13% soccer. If a person is randomly selected, what is the probability that the person's favorite sport is:

6. basketball?
7. a sport that involves a ball?
8. not baseball?

9-12. The highest degrees of the faculty of a private college include 231 Ph.D.'s, 17 ED.D.'s, 71 M.A.'s, 22 M.S.'s, 25 B.A.'s, 14 B.S.'s, and 13 with other qualifications. Suppose one faculty member is chosen randomly. What is the probability that this faculty member's highest degree will:

9. be a Ph.D.?
10. be a Master's?
11. be a B.S.?
12. not be an ED.D.?

Solutions to Practice Exercises 4-1

1. Valid.
2. Invalid, the probabilities sum to 0.9.
3. Invalid, there is a negative probability.
4. Invalid, the probabilities sum to 1.1.
5. Valid.
6. 23%
7. This would be the sum of the probabilities associated with baseball, basketball, football, tennis, and soccer: $18\% + 23\% + 16\% + 9\% + 13\% = 79\%$.
8. Since 18% prefer baseball and since the sum of all the probabilities is one, the probability of finding someone whose favorite sport is not baseball is 82%.
9. There are $231 + 17 + 71 + 22 + 25 + 14 + 13 = 393$ faculty members. The probability of a randomly selected faculty member having a Ph.D. is $\dfrac{231}{393} \approx 0.59$.
10. The probability of a randomly selected faculty member's highest degree being a Master's is $\dfrac{22 + 25}{393} = \dfrac{47}{393} \approx 0.129$.
11. The probability of a randomly selected faculty member's highest degree being a B.S. is $\dfrac{14}{393} \approx 0.04$.
12. The probability of a randomly selected faculty member's highest degree not being an ED.D. is $\dfrac{393 - 17}{393} = \dfrac{376}{393} \approx 0.96$.

4-2 Probabilities for Compound Events

Study Objective

You should be able to:

1. Calculate conditional probabilities.
2. Calculate the probability that both the events A and B occur.
3. Calculate the probability that either of the events A or B occurs.
4. Identify when two events are independent and when they are mutually exclusive.
5. Identify and calculate the probability of the complement of an event.

Section Overview

We often deal with events that have more than one characteristic. For example, we might be interested in finding a job that pays well and is in a good location. Or we might want to see a movie that is either a comedy or an adventure. Or we might want to move into a new apartment if it is near work, not on the ground level, and is not too close to a busy road. Each of these are examples of compound events, events that combine two or more events. Once we recognize that we are dealing with compound events, we have techniques and formulas that will help in calculating their probabilities.

A conditional probability, denoted $P(A|B)$, is the probability that the event A occurs if we find out that event B has occurred. For example, the probability that a single die roll results in a 2 is $\frac{1}{6}$. But if we roll a die where we cannot see it and someone says that the die roll resulted in an even number, the probability changes. Now the probability of a 2 is $\frac{1}{3}$, since the only possibilities are now 2 or 4 or 6. Often the calculation of a conditional probability is aided by the creation of a contingency table, a two dimensional table giving information about two characteristics simultaneously.

A compound event we often use is when both of two events occur. The probability of such a compound event is denoted $P(A \text{ and } B)$, where A and B are the two events involved. For example, we might be interested in knowing the probability that a student is a senior (A) and a biology major (B). Or we might like to know the probability that our favorite sports team wins two games (A = win first game, B = win second game). If we can somehow determine $P(A)$ and $P(B|A)$, then we can calculate the probability that both A and B occur by using $P(A \text{ and } B) = P(A) \times P(B|A)$. Sometimes knowing that one event occurs does not affect the probability that the second event occurs. When this is true, the two events are independent. When two events are independent, we can calculate the probability that both occur by using $P(A \text{ and } B) = P(A) \times P(B)$.

Another compound event of interest is when at least one of two events occurs. The probability of this compound event is denoted $P(A \text{ or } B)$. For example, we might be interested in knowing the probability that a student is a senior or a biology major. Or we might like to know the probability that our favorite sports team wins at least one of two games. Sometimes two events cannot occur during the same trial of an experiment. For example, suppose A = a person has two brown eyes and B = a person has two blue eyes. Then when randomly selecting a single person from a class, it would be impossible for both A and B to occur. Such events are mutually exclusive events. For mutually exclusive events, $P(A \text{ or } B) = P(A) + P(B)$. If A and B are not mutually exclusive, then $P(A \text{ or } B) = P(A) + P(B) - P(A \text{ and } B)$.

The complement of an event E, denoted $\overline{E}$, consists of all possible simple events not in the event E. Often we can simplify a probability calculation by determining the probability of $P(\overline{E})$ and using $P(E) = 1 - P(\overline{E})$.

Key Terms & Formulas

Compound Event Any event created by combining two or more other events.

Conditional Probability Denoted $P(A|B)$, the conditional probability of A given B is the probability that the event A occurs given the knowledge that the event B has occurred already.

$P(A \text{ and } B)$ The joint probability of A and B is the probability that both occur in the same trial of a probability experiment.

Hint: There is often difficulty in distinguishing a conditional probability from the probability of A and B. In both, two events are involved and the probabilities concern themselves with both occurring. The difference is that with conditional probabilities, we already know that one event has occurred. For example, suppose we roll a six sided die. Also suppose we define A = rolling an odd number and B = rolling a 4 or greater. One possible question you could be asked is "What is the probability that you roll an odd number that is 4 or more?" That is, what is $P(A \text{ and } B)$? (The answer is $\frac{1}{6}$). Compare this to "What is the probability of rolling an odd number if you know that the roll is a 4 or more?" That is, what is $P(A|B)$? (Now the answer is $\frac{1}{3}$).

Multiplication Rule for Computing $P(A \text{ and } B)$ $P(A \text{ and } B) = P(A) \times P(B|A)$.

Independent Events Two events A and B are independent if the occurrence or nonoccurrence of A does not change the probability that B will occur (and vice-versa). In formula form, if A and B are independent, $P(B) = P(B|A)$ and $P(A) = P(A|B)$.

Multiplication Rule for Computing $P(A \text{ and } B)$ for Independent Events $P(A \text{ and } B) = P(A) \times P(B)$.

Mutually Exclusive Events Two events, A and B, are mutually exclusive if they cannot occur in the same trial of an experiment, i.e., the occurrence of one event prevents the other from occurring.

Caution: There is often confusion about independent and mutually exclusive events. If two events are mutually exclusive, they cannot be independent. If two events are independent, they cannot be mutually exclusive. It is also possible that A and B are neither independent nor mutually exclusive. The one thing that is not possible is that A and B are simultaneously independent and mutually exclusive. (The one exception is the strange case where either $P(A)$ or $P(B) = 0$.)

$P(A \text{ or } B)$ for Mutually Exclusive Events $P(A \text{ or } B) = P(A) + P(B)$.

$P(A \text{ or } B)$ for Non-mutually Exclusive Events $P(A \text{ or } B) = P(A) + P(B) - P(A \text{ and } B)$.

Complement of an Event The complement of an event E, denoted $\overline{E}$, is the event that the event E does not occur.

Hint: There is nothing derogatory about being in the event "not E." For example, if the event E is that a person is a female, the event that a person is not a female (sometimes called a male). Or if the event E is rolling an odd number on a die, the event is rolling an even number.

Caution: There is a tendency to make "complement" synonymous with "opposite." For example, consider the probability experiment of flipping three coins and determining for each coin whether a head or tail faces upward. One possible event is rolling all heads. Most people would think that the "opposite" of this is to roll all tails. However, the complement of rolling all heads is to roll at least one tail.

Probability Rule for Complementary Events $P(E) = 1 - P(\overline{E})$.

Worked Examples

A 7-Card Deck

Suppose a person has three 3's and four 4's from a deck of playing cards. Two cards are to be selected randomly from this small deck.

1. What is the probability that the first card selected is a three?
2. What is the probability that the second card selected is a three if the first card selected is a three?
3. What is the probability that both cards selected are threes?

Solution

1. Because there is a total of seven cards and because three of them are threes, the probability of randomly selecting a three on the first draw is $\frac{3}{7}$.

2. Once it is known that the first card is a three, the probability that the second card is a three is calculated by taking the number of threes that remain over the number of cards that remain, $\frac{2}{6}$.

   Hint: The second part of the above illustration is an example of a very simple conditional probability. We could define A = the first card selected is a three and let B = second card selected is a three. Then we can write the second part of the illustration as $P(B|A)$.

3. The probability that both cards are threes can be represented by $P(A \text{ and } B)$. The multiplication rule says $P(A \text{ and } B) = P(A) \times P(B|A)$. Here we get $P(A \text{ and } B) = \dfrac{3}{7} \times \dfrac{2}{6} = \dfrac{6}{42} = \dfrac{1}{7}$.

Urban Forestry

A recent survey in California asked people to state whether they lived in an urban, suburban, or rural area. It also asked the participants to rate the importance of planting trees in urban areas. The results are contained in the following table.

	Important	Slightly Important	Not Important At All	Total
Urban	432	213	198	843
Suburban	277	124	67	468
Rural	97	117	265	479
Total	806	454	530	1790

A person is randomly selected from the respondents to the survey. Find the probability that the person:

1. is from an urban area.
2. believes the planting of trees in urban areas is important.
3. is not from a rural area.
4. is from an urban area or from a rural area.
5. is from an urban area if it is known that the person believes that the planting of trees in urban areas is important.
6. is from an urban area and believes that the planting of trees in urban areas is important.
7. is from an urban area or believes that the planting of trees in urban areas is important.

Solution

Hint: In the solutions that follow, each probability can be calculated by taking ratios of numbers from the above table.

$\Big($ We will represent the numerator of the ratios by ▢ and the denominator of the ratios by ▉.$\Big)$

1. To find the probability that the person is from an urban area, we take the number of people from urban areas and divide by the total number of observations. So if we let $A =$ the person is from an urban area, we have $P(A) = \dfrac{843}{1790} = 0.4709 \approx 0.47$.

	Important	Slightly Important	Not Important At All	Total
Urban	432	213	198	843
Suburban	277	124	67	468
Rural	97	117	265	479
Total	806	454	530	1790

2. If $B =$ a person believes that the planting of trees in urban areas is important, then $P(B) = \dfrac{806}{1790} = 0.4503 \approx 0.45$.

	Important	Slightly Important	Not Important At All	Total
Urban	432	213	198	843
Suburban	277	124	67	468
Rural	97	117	265	479
Total	806	454	530	1790

3. If $C =$ a person is from a rural area, then we are looking for $\overline{C}$.
    $P(\overline{C}) = \dfrac{843 + 468}{1790} = \dfrac{1311}{1790} = 0.7324 \approx 0.73$.

	Important	Slightly Important	Not Important At All	Total
Urban	432	213	198	843
Suburban	277	124	67	468
Rural	97	117	265	479
Total	806	454	530	1790

Hint: While the above table makes it easy to see how to calculate the correct answer, this is an example where we can also use a formula to calculate a probability. Since we are looking for $P(\overline{C})$, we can use

$$P(\overline{C}) = 1 - P(C) = 1 - \frac{479}{1790} = 0.7324.$$

4. We have previously let $A =$ the person is from an urban area and $C =$ the person is from a rural area. So this question is asking us to calculate $P(A \text{ or } C)$. But since a person obviously cannot live both in an urban and rural area, these two events are mutually exclusive. So we can use the formula that says for mutually exclusive events

$$P(A \text{ or } C) = P(A) + P(C) = \frac{843}{1790} + \frac{479}{1790} = \frac{1322}{1790} = 0.7385 \approx 0.74.$$

	Important	Slightly Important	Not Important At All	Total
Urban	432	213	198	843
Suburban	277	124	67	468
Rural	97	117	265	479
Total	806	454	530	1790

5. Using $A =$ the person is from an urban area and $B =$ the person believes that the planting of trees in urban areas is important, this question is asking us to calculate $P(A|B)$. This is $P(A|B) = \frac{432}{806} = 0.5360 \approx 0.53.$

	Important	Slightly Important	Not Important At All	Total
Urban	432	213	198	843
Suburban	277	124	67	468
Rural	97	117	265	479
Total	806	454	530	1790

6. Again we use $A =$ the person is from an urban area and $B =$ the person believes that the planting of trees in urban areas is important. Here we want $P(A \text{ and } B)$. There is only one cell in the table that involves people who simultaneously are from an urban area and believe that the planting of trees in urban areas is important. So we have $P(A \text{ and } B) = \frac{432}{1790} = 0.2413 \approx 0.24.$

	Important	Slightly Important	Not Important At All	Total
Urban	432	213	198	843
Suburban	277	124	67	468
Rural	97	117	265	479
Total	806	454	530	1790

7. Using the previous A and B, we are looking for $P(A \text{ or } B)$. We can calculate this by taking the ratio

$$P(A \text{ or } B) = \frac{432 + 213 + 198 + 277 + 97}{1790} = \frac{1217}{1790} = 0.6799 \approx 0.68.$$

	Important	Slightly Important	Not Important At All	Total
Urban	432	213	198	843
Suburban	277	124	67	468
Rural	97	117	265	479
Total	806	454	530	1790

Hint: Because in parts 1, 2, and 6 we had obtained $P(A)$, $P(B)$, and $P(A \text{ and } B)$, we could have obtained $P(A \text{ or } B)$ by using $P(A \text{ or } B) = P(A) + P(B) - P(A \text{ and } B) = \frac{843}{1790} + \frac{806}{1790} - \frac{432}{1790} = \frac{1217}{1790}.$

PC's

At a local library, the probability that someone is reading the current issue of *Sports Illustrated* is 0.5, the probability that someone is reading *Time* is 0.4, and these two events are independent of each other. What is the probability that

1. both of the magazines are being read?
2. neither of the two is being read?
3. at least one is being read?
4. *Sports Illustrated* is being read if we know *Time* is being read?

Solution

For all parts to this question, we will let $S = $ *Sports Illustrated* is being read and $T = $ *Time* is being read.

1. Both of the magazines being read can be written $P(S \text{ and } T)$. Because S and T are independent, we can use the multiplication rule for computing $P(S \text{ and } T)$ for independent events.
   $$P(S \text{ and } T) = P(S)P(T) = (0.5)(0.4) = 0.2.$$
2. Neither of the magazines being read can be written $P(\overline{S} \text{ and } \overline{T})$. When S and T are independent, so are their complements. Then we again can use the multiplication rule for computing $P(\overline{S} \text{ and } \overline{T})$ for independent events. Here we have
   $$P(\overline{S} \text{ and } \overline{T}) = P(\overline{S})\,P(\overline{T}) = (1 - 0.5)(1 - 0.4) = (0.5)(0.6) = 0.3.$$
3. At least one magazine being read can be written $P(S \text{ or } T)$. Having been given or previously calculated all the required parts, we can use the addition rule for calculating $P(S \text{ or } T)$ for events that are not mutually exclusive.
   $$P(S \text{ or } T) = P(S) + P(T) - P(S \text{ and } T) = 0.5 + 0.4 - 0.2 = 0.7.$$
4. This question wants us to obtain $P(S|T)$. Because S and T are independent, $P(S|T) = P(S) = 0.5$.

Coin Flipping

What is the probability that a person flipping a coin will flip two heads in a row?

Solution

The coin flips are independent. (Unless you believe that the coin has a memory and after being a head on the first flip it gets bored and decides to become a tail for the second flip.) We can define two events $A = $ the first flip is a head and $B = $ the second flip is a head. Then to get two flips in a row to be heads, we would need both A and B to occur. Because the events are independent, the probability of two heads in a row is $P(A \text{ and } B) = P(A)$

$$P(B) = \left(\frac{1}{2}\right)\left(\frac{1}{2}\right) = \left(\frac{1}{2}\right)^2 = \frac{1}{4}.$$

Coin Flipping + 1

What is the probability that a person flipping a coin will flip three heads in a row?

Solution

We can build on the previous problem. The third coin flip is independent of the first two. If there is a $\left(\frac{1}{2}\right)^2$ probability of the first two flips being heads and a $\frac{1}{2}$ probability that the third flip results in a head, then the probability of three heads in a row is $\left(\frac{1}{2}\right)^2\left(\frac{1}{2}\right) = \left(\frac{1}{2}\right)^3 = \frac{1}{8}$.

Hint: This pattern obviously continues. For example, the probability of 10 heads in a row is $\left(\frac{1}{2}\right)^{10}$. Similarly, with a fair die, the probability of rolling five 6's in a row is $\left(\frac{1}{6}\right)^5$.

Practice Exercises 4-2

1-14. The table below contains possible win amounts of a lottery with their associated probabilities.

Payoff	Probability
$2.00	0.05
$25.00	0.01
$100.00	0.005
$500.00	0.001
$5,000.00	0.0005
$10,000.00	0.0001

You buy a single lottery ticket. Consider the following events.
$A = \{$You win 100.00\}$
$B = \{$You win at least $5,000.00$\}$
$C = \{$You win at least $25.00 but no more than $5,000.00$\}$
$D = \{$You win$\}$
Find the probabilities of the following events.

1. $A = \{$You win 100.00\}$
2. $\overline{A} = \{$You do not win 100.00\}$
3. $D = \{$You win$\}$
4. A and $B = \{$You win 100.00\}$ and $\{$You win at least $5,000.00$\}$
5. A or $B = \{$You win 100.00\}$ or $\{$You win at least $5,000.00$\}$
6. A and $C = \{$You win 100.00\}$ and $\{$You win at least $25.00 but no more than $5,000.00$\}$
7. A or $C = \{$You win 100.00\}$ or $\{$You win at least $25.00 but no more than $5,000.00$\}$
8. B and $C = \{$You win at least $5,000.00$\}$ and $\{$You win at least $25.00 but no more than $5,000.00$\}$
9. B or $C = \{$You win at least $5,000.00$\}$ or $\{$You win at least $25.00 but no more than $5,000.00$\}$
10. $A|B = \{$You win 100.00\}|\{$You win at least $5,000.00$\}$
11. $A|D = \{$You win 100.00\}|\{$You win$\}$
12. $A|C = \{$You win 100.00\}|\{$You win at least $25.00 but no more than $5,000.00$\}$
13. $D|A = \{$You win$\}|\{$You win 100.00\}$
14. $B|C = \{$You win at least $5,000.00$\}|\{$You win at least $25.00 but no more than $5,000.00$\}$

15-17. New software products are highly successful 10% of the time, moderately successful 30% of the time, slightly successful 20% of the time, and failures 40% of the time.

15. What is the probability that a new software product is at least moderately successful?
16. What is the probability that a new software product is successful?
17. Given that the product is successful, what is the probability that it is only slightly successful?

18-20. New hardware stores show a profit in their first year 60% of the time. Suppose that four hardware stores open in locations where they do not compete with each other. (So that it is reasonable to believe that each store's ability to show a profit in the first year is independent of any other store's success.)

18. What is the probability that all four stores show a profit in their first year?
19. What is the probability that all four stores fail to show a profit in their first year?
20. What is the probability that at least one store shows a profit in their first year?

21-26. A poll in Texas asked people to identify their political party and their opinion on bilingual election ballots.

Opinion Party	Favor	Neutral	Oppose	Total
Republican	92	26	145	263
Democratic	132	54	61	247
Other	33	43	61	137
Total	257	123	267	647

A person is randomly selected from the respondents to the poll. Find the probability that the person:

21. is a Republican.

22. is not a Democrat.

23. is in favor of or neutral about bilingual ballots.

24. is a Democrat if it is known that the person is neutral.

25. is a Democrat and is neutral.

26. is a Democrat or is neutral.

Solutions to Practice Exercises 4-2

1. Reading directly from the table, we have $P(A) = 0.005$.

2. Using the probability rule for complementary events, we have $P(\overline{A}) = 1 - P(A) = 1 - 0.005 = 0.995$.

3. $P(D) = P(\text{You win}) = $ the sum of the probabilities in the table
$$= 0.05 + 0.01 + 0.005 + 0.001 + 0.0005 + 0.0001 = 0.0666.$$

4. There is no way in which you can win \$100.00 at the same time that you win at least \$5,000.00. Therefore, these events are mutually exclusive and $P(A \text{ and } B) = 0$.

5. As in 4., A and B are mutually exclusive, so
$$P(A \text{ or } B) = P(A) + P(B) = 0.005 + (0.0005 + 0.0001) = 0.0056.$$

6. Because A is part of C (a subset of C), the simple events that constitute A and C are identical to the simple events that constitute A. Therefore $P(A \text{ and } C) = P(A) = 0.005$.

7. Because A is part of C (a subset of C), the simple events that constitute A or C are identical to the simple events that constitute C. Therefore,
$$P(A \text{ or } C) = P(C) = 0.01 + 0.005 + 0.001 + 0.0005 = 0.0165$$

8. B and C consists of the simple event that a person wins exactly \$5,000.00. Examining the table, we find $P(B$ and $C) = 0.0005$.

9. B or C consists of the event that a person wins at least \$25.00. Examining the table, we find
$$P(B \text{ or } C) = 0.01 + 0.005 + 0.001 + 0.0005 + 0.0001 = 0.0166.$$

10. Because it is impossible to win exactly \$100.00 when you have won at least \$5,000.00 (i.e., A and B are mutually exclusive), $P(A|B) = 0$.

11. $P(A|D) = \dfrac{0.005}{0.0666} = 0.0751 \approx 0.08$.

12. $P(A|C) = \dfrac{0.005}{0.01650} = 0.0758 \approx 0.08$.

13. The event A is that you won \$100.00. So if you know that you won \$100.00, the probability that you won is 1.

14. Given that you win at least \$25.00 but no more than \$5,000.00, the only way you can win at least \$5,000.00 is to win exactly \$5,000.00. So $P(B|C) = \dfrac{0.0005}{0.0165} = 0.0303 \approx 0.03$.

15. The probability that a new software product is at least moderately successful is equal to the probability that it is moderately successful plus the probability that it is highly successful $= 30\% + 10\% = 40\%$.

16. We can calculate this probability two ways. One way is to sum the probabilities of the three levels of success, $10\% + 30\% + 20\% = 60\%$. Alternatively, we could recognize that the complement of being successful is to fail. Then $P(\text{successful}) = 1 - P(\text{failure}) = 1 - 0.40 = 0.60$ or 60%.

17. $P(\text{slightly successful}|\text{successful}) = \dfrac{20\%}{60\%} = \dfrac{1}{3} \approx 33\%$.

18. For the four stores to show a profit in the first year, the first store must show a profit in the first year and the second store must show a profit in the first year and the third store must show a profit in the first year and the fourth store must show a profit in the first year. Since they operate independently, the probability that they all show a profit in the first year is the probability that each shows a profit multiplied together,
$$(0.6)(0.6)(0.6)(0.6) = 0.6^4 = 0.1296 \approx 0.13.$$

19. If events are independent, so are their complements. So, using a logic similar to the previous exercise, the probability that all four stores fail to show a profit in their first year is $(0.4)^4 = 0.0256 \approx 0.03$.

20. The complement of at least one store showing a profit in their first year is that all four stores fail to show a profit in their first year. But we calculated the probability of this to be 0.0256 in the previous problem. Therefore, the probability of at least one profitable store is $1 - 0.0256 = 0.9744 \approx 0.97$.

21-26. We will represent the numerator of the ratios by ▢ and the denominator of the ratios by ▣.

21. The probability that the person is a Republican is $\frac{263}{647} = 0.4065 \approx 0.41$.

Opinion\Party	Favor	Neutral	Oppose	Total
Republican	92	26	145	263
Democratic	132	54	61	247
Other	33	43	61	137
Total	257	123	267	647

22. The probability that the person is not a Democrat can be calculated either of two ways. One is to sum the Republican and other totals, $\frac{263 + 137}{647} = \frac{400}{647} = 0.6182 \approx 0.62$. The other is to use the complementary rule, calculating the probability of the person being a Democrat and subtracting that result from 1, $1 - \frac{247}{647} = 1 - 0.3818 = 0.6182 \approx 0.62$.

Opinion\Party	Favor	Neutral	Oppose	Total
Republican	92	26	145	263
Democratic	132	54	61	247
Other	33	43	61	137
Total	257	123	267	647

23. Being in favor and neutral are mutually exclusive. Therefore to find the probability that a person is one or the other, we add their individual probabilities, $\frac{257}{647} + \frac{123}{647} = \frac{380}{647} = 0.5873 \approx 0.59$.

Opinion\Party	Favor	Neutral	Oppose	Total
Republican	92	26	145	263
Democratic	132	54	61	247
Other	33	43	61	137
Total	257	123	267	647

24. This conditional probability is calculated by taking the ratio of the number of neutral Democrats and dividing by the number of people who are neutral, $\frac{54}{123} = 0.4390 \approx 0.44$.

Opinion\Party	Favor	Neutral	Oppose	Total
Republican	92	26	145	263
Democratic	132	54	61	247
Other	33	43	61	137
Total	257	123	267	647

25. The probability that a person is both a Democrat and neutral is calculated by taking the single cell involving people that have both these characteristics and dividing by the number of observations, $\frac{54}{647} = 0.0835 \approx 0.08$.

Opinion\Party	Favor	Neutral	Oppose	Total
Republican	92	26	145	263
Democratic	132	54	61	247
Other	33	43	61	137
Total	257	123	267	647

26. The probability that a person is either a Democrat or neutral is calculated by taking the total of all cells involving people with either or both characteristics and dividing by the number of observations,
$$\frac{132 + 54 + 61 + 26 + 43}{647} = \frac{316}{647} = 0.4884 \approx 0.49.$$

Opinion Party	Favor	Neutral	Oppose	Total
Republican	92	26	145	263
Democratic	132	54	61	247
Other	33	43	61	137
Total	257	123	267	647

4-3 Random Variables, Probability Distributions, and Expected Values

Study Objective

You should be able to:
1. Define random variables.
2. Distinguish between discrete and continuous random variables.
3. Work with probability distributions for discrete random variables.
4. Calculate the expected value (mean), variance, and standard deviation of a discrete random variable.

Section Overview

We will be interested in analyzing the results of a probability experiment. Usually the result of interest will be some number or numbers. We will use random variables to generate these numbers. A random variable links with each outcome of a probability experiment a single number. For example, rolling two dice, a possible random variable will be the total of the two dice. Or another experiment would involve taking a statistics quiz and finding out what numerical grade we receive. As when we were talking about variables in general, there are discrete and continuous random variables. Discrete random variables have gaps or breaks in their possible values, continuous random variables do not.

A probability distribution gives a picture of the percent of time we should see each value of a discrete random variable. A probability distribution links a probability to each possible value of the random variable. We can obtain this linkage with our probability rules or by using a relative frequency approach. We can also create a probability graph of the probability distribution to provide easier access to the probability values.

Another way to describe a probability distribution is to calculate the expected value of a random variable. This is similar to the mean of a sample in that it tries to describe the center of a set of numbers. The expected value is what the mean would be if we performed the experiment an infinite number of times. This of course is silly. However, if we repeat an experiment often, we would "expect" the mean of the observed values of the random variable to be close to the expected value. Similarly, we can compute the variance and standard deviation of a discrete random variable.

Key Terms & Formulas

Random Variable A random variable assigns a single numerical value to each simple event of a probability experiment.

<u>Discrete Random Variable</u> A random variable whose possible values consist of a finite or countably infinite set of distinct values.

<u>Continuous Random Variable</u> A random variable whose possible values consist of a countless number of values along a line interval.

Hint: The two above definitions are similar to those describing discrete and continuous variables given in Chapter 3. The difference is that these are variables defined for a probability experiment. As a reminder, it is easy to judge whether a random variable is discrete or continuous. Determine reasonable lows and highs for the values of a variable. If every value between the low and high value is possible, then the random variable is continuous. If there are gaps in the possible values, the random variable is discrete.

<u>Probability Distribution</u> The probability distribution for a discrete random variable gives the probability for each value of the random variable.

Hint: A probability distribution for a discrete random variable has two common forms. Usually the probabilities are presented in a tabular form similar to a frequency distribution, or by giving a formula that can generate any probability. This chapter concentrates on the tabular form, but we will encounter the formula approach in future chapters.

<u>Relative Frequency Distribution</u> A probability distribution developed by finding the proportion of times each value of the random variable occurs.

<u>Probability Distribution Graph</u> A graph of a probability distribution, usually a histogram or frequency polygon.

<u>Expected Value</u> The mean of a probability distribution. The formula for an expected value of a discrete random variable is $E(x) = \Sigma x \cdot P(x)$.

<u>Standard Deviation</u> A measure of the spread of values of a probability distribution. For a discrete random variable, it can be calculated two ways: $\sigma = \sqrt{\Sigma[(x - \mu)^2 P(x)]} = \sqrt{\Sigma[x^2 \cdot P(x)] - \mu^2}$.

Worked Examples

Two Dice

Consider the experiment of rolling two dice. The sample space for the experiment is below. Define three different random variables for this experiment.

(1,1)	(1,2)	(1,3)	(1,4)	(1,5)	(1,6)
(2,1)	(2,2)	(2,3)	(2,4)	(2,5)	(2,6)
(3,1)	(3,2)	(3,3)	(3,4)	(3,5)	(3,6)
(4,1)	(4,2)	(4,3)	(4,4)	(4,5)	(4,6)
(5,1)	(5,2)	(5,3)	(5,4)	(5,5)	(5,6)
(6,1)	(6,2)	(6,3)	(6,4)	(6,5)	(6,6)

Solution

The most obvious random variable is the total of the two dice. A second random variable could be the absolute value of the difference of the two dice. A third would be the minimum of the two dice.

Discrete or Continuous

For the following random variables, decide whether each is discrete or continuous.

1. The yearly salary of an employee.
2. The amount of floor space occupied by a high speed printer.
3. The number of articles shipped to a customer that are returned due to defects.
4. The angle of a chair back that maximizes a person's comfort.

Solution

1. Because an employee could earn $33,000.00 or $33,000.01, but not $33,000.005 (i.e., there is a gap), the yearly salary of an employee is discrete.
2. The amount of floor space is an area and can take on any reasonable value within physical limits and is therefore continuous.
3. The number of articles shipped to a customer are counts and cannot take any fractional value. Therefore there are gaps and this random variable is discrete.
4. Angles can take on any value within the limits of the chair's architecture and is continuous.

Two Dice Rerolled

Reconsider the experiment of rolling two dice with the sample space below. Find the probability distribution for $X = $ the minimum of the two dice. Also, create a histogram for the probability distribution. Finally, calculate the expected value and standard deviation of X.

(1,1)	(1,2)	(1,3)	(1,4)	(1,5)	(1,6)
(2,1)	(2,2)	(2,3)	(2,4)	(2,5)	(2,6)
(3,1)	(3,2)	(3,3)	(3,4)	(3,5)	(3,6)
(4,1)	(4,2)	(4,3)	(4,4)	(4,5)	(4,6)
(5,1)	(5,2)	(5,3)	(5,4)	(5,5)	(5,6)
(6,1)	(6,2)	(6,3)	(6,4)	(6,5)	(6,6)

Solution

The possible minimums are 1 through 6. Let's start by figuring out the probability that the minimum is 1. Notice that the eleven simple events in the first row and the first column of the above sample space are the only simple events that have a minimum of 1. Also notice that each of the 36 dice throws are equally likely and have a probability of $\frac{1}{36}$. Since the simple events are mutually exclusive, we can use the addition rule for mutually exclusive events to calculate the probability of a minimum of 1. Summing $\frac{1}{36}$ eleven times, we get $\frac{11}{36}$. Next we can calculate the probability of a minimum of 2. We note that there are nine simple events that have a minimum of 2. These are the simple events in the second row and second column of the sample space, except those whose minimum is 1. Summing, we find the probability of a minimum of 2 is $\frac{9}{36}$. We can continue similarly to obtain the probabilities for 3, 4, 5, and 6. The result is the table below, with the associated histogram to its right.

Comment: Notice that the table does satisfy the two rules of probability. All the probabilities are between 0 and 1, and the sum of the probabilities does equal 1.

Minimum	Probability
1	$\frac{11}{36}$
2	$\frac{9}{36}$
3	$\frac{7}{36}$
4	$\frac{5}{36}$
5	$\frac{3}{36}$
6	$\frac{1}{36}$
Total	$\frac{36}{36} = 1$

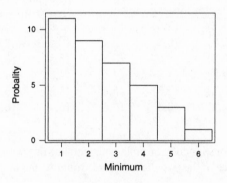

The expected value of X is obtained by using:

$$E(x) = 1\left(\frac{11}{36}\right) + 2\left(\frac{9}{36}\right) + 3\left(\frac{7}{36}\right) + 4\left(\frac{5}{36}\right) + 5\left(\frac{3}{36}\right) + 6\left(\frac{1}{36}\right)$$

$$= \frac{11}{36} + \frac{18}{36} + \frac{21}{36} + \frac{20}{36} + \frac{15}{36} + \frac{6}{36}$$

$$= \frac{91}{36} = 2.5278 \approx 2.53.$$

Finally, the standard deviation is computed by first calculating:

$$\Sigma x^2 \cdot P(x) = 1^2\left(\frac{11}{36}\right) + 2^2\left(\frac{9}{36}\right) + 3^2\left(\frac{7}{36}\right) + 4^2\left(\frac{5}{36}\right) + 5^2\left(\frac{3}{36}\right) + 6^2\left(\frac{1}{36}\right)$$

$$= \frac{11}{36} + \frac{36}{36} + \frac{63}{36} + \frac{80}{36} + \frac{75}{36} + \frac{36}{36} = \frac{301}{36} = 8.3611.$$

Then $\sigma = \sqrt{8.3611 - 2.5278^2} = \sqrt{1.9715} \approx 1.40.$

Practice Exercises 4-3

1. Consider the experiment of flipping a coin three times. The sample space for the experiment is below. Define three different random variables for this experiment.

$$\{HHH, HHT, HTH, THH, TTH, THT, HTT, TTT\}$$

2. A person randomly selects five cards from an ordinary deck of playing cards. Define three random variables appropriate to this probability experiment.

3-6. Identify the following as discrete or continuous random variables.

3. The age of your Statistics instructor.

4. The number of degrees your Statistics instructor has earned.

5. The number of names your instructor mispronounces on the first day of class.

6. The amount of time your instructor spends answering questions about homework problems.

7-10. A probability experiment involved dropping an egg from a building. Identify the following as discrete or continuous random variables.

7. The time it took for the egg to hit the ground.

8. The distance the egg traveled.

9. The number of pieces into which the egg shattered.

10. The speed the egg was traveling when it hit the ground.

11-14. Consider the experiment of flipping a coin three times from Exercise 1. The sample space for the experiment is reprinted below. Suppose the random variable of interest is the number of heads obtained.

$$\{HHH, HHT, HTH, THH, TTH, THT, HTT, TTT\}$$

11. Find the probability distribution for the number of heads.

12. Draw a histogram of the probability distribution.

13. Calculate the expected number of heads.

14. Compute the standard deviation.

15-18. Again reconsider the experiment of rolling two dice with the sample space below.

(1,1)	(1,2)	(1,3)	(1,4)	(1,5)	(1,6)
(2,1)	(2,2)	(2,3)	(2,4)	(2,5)	(2,6)
(3,1)	(3,2)	(3,3)	(3,4)	(3,5)	(3,6)
(4,1)	(4,2)	(4,3)	(4,4)	(4,5)	(4,6)
(5,1)	(5,2)	(5,3)	(5,4)	(5,5)	(5,6)
(6,1)	(6,2)	(6,3)	(6,4)	(6,5)	(6,6)

15. Find the probability distribution for $X = $ the absolute value of the difference of the two dice.

16. Create a histogram for the probability distribution.

17. Calculate the expected value of X.

18. Compute the standard deviation.

Solutions to Practice Exercises 4-3

1. Many random variables are possible. Included would be the number of heads, the number of tails, the flip that yielded the first head, the difference between the number of heads and the number of tails, etc.

2. Examples include the number of spades, the number of picture cards, the number of pairs obtained, the number of suits obtained, the total points of the cards where picture cards count as 10, etc.

3. While we usually truncate our age to our age on our most recent birthday, we age continuously and age is a continuous random variable.

4. Since there is no such thing as a fractional degree, this is a discrete random variable.

5. This results in integer values and is discrete.

6. Time is continuous, so the instructor could take any time between 0 minutes and the length of the entire period (and perhaps more).

7. The time it took for the egg to hit the ground has no gaps in its possible values and is continuous.

8. The distance the egg traveled could be any value within the limits of the height of the building and therefore is continuous.

9. The number of pieces into which the egg shattered can only be integer valued and must be discrete.

10. The speed the egg was traveling when it hit the ground has no gaps in its possible values and by that is continuous.

11. There are eight equally likely simple events in the sample space, so each has a probability of $\frac{1}{8}$. One has 3 heads, three have 2 heads, another three have 1 head, and one has 0 heads. Summing probabilities because the simple events are mutually exclusive, we obtain the following probability distribution.

Number of Heads	Probability
0	$\frac{1}{8}$
1	$\frac{3}{8}$
2	$\frac{3}{8}$
3	$\frac{1}{8}$
Total	$\frac{8}{8} = 1$

12.

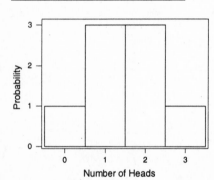

13. The expected value of X is obtained by using $E(x) = \Sigma x \cdot P(x)$.

$$E(x) = 0\left(\frac{1}{8}\right) + 1\left(\frac{3}{8}\right) + 2\left(\frac{3}{8}\right) + 3\left(\frac{1}{8}\right)$$
$$= 0 + \frac{3}{8} + \frac{6}{8} + \frac{3}{8}$$
$$= \frac{12}{8} = \frac{3}{2} = 1.5.$$

14. Using the definition, we compute

$$\sigma = \sqrt{\Sigma[(x-\mu)^2 P(x)]}$$
$$= \sqrt{\left(0 - \frac{3}{2}\right)^2 \frac{1}{8} + \left(1 - \frac{3}{2}\right)^2 \frac{3}{8} + \left(2 - \frac{3}{2}\right)^2 \frac{3}{8} + \left(3 - \frac{3}{2}\right)^2 \frac{1}{8}}$$
$$= \sqrt{\frac{9}{32} + \frac{3}{32} + \frac{3}{32} + \frac{9}{32}}$$
$$= \sqrt{\frac{24}{32}} = \sqrt{\frac{3}{4}} = \sqrt{0.75} \approx 0.87.$$

15. The possible values are 0 through 5. The six dice rolls of "doubles," which lie on what is called the main diagonal of the sample space given with the problem, give a difference of 0. The ten simple events on either side of this main diagonal give differences of 1 or -1, giving absolute values of 1. Continuing, we find eight

simple events whose absolute value of the difference is 2, six with 3, four with 4 and 2 with 5. Summing the individual probabilities, we obtain the probability distribution presented in the table that follows.

|Difference|	Probability
0	$\dfrac{6}{36} = \dfrac{3}{18}$
1	$\dfrac{10}{36} = \dfrac{5}{18}$
2	$\dfrac{8}{36} = \dfrac{4}{18}$
3	$\dfrac{6}{36} = \dfrac{3}{18}$
4	$\dfrac{4}{36} = \dfrac{2}{18}$
5	$\dfrac{2}{36} = \dfrac{1}{18}$
Total	$\dfrac{18}{18} = 1$

16.

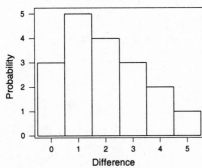

17. The expected value of X is obtained by using $E(x) = \Sigma x \cdot P(x)$.

$$E(x) = 0\left(\frac{3}{18}\right) + 1\left(\frac{5}{18}\right) + 2\left(\frac{4}{18}\right) + 3\left(\frac{3}{18}\right) + 4\left(\frac{2}{18}\right) + 5\left(\frac{1}{18}\right)$$

$$= 0 + \frac{5}{18} + \frac{8}{18} + \frac{9}{18} + \frac{8}{18} + \frac{5}{18}$$

$$= \frac{35}{18} = 1.9444 \approx 1.94.$$

18. Using the shortcut formula, we first calculate:

$$\Sigma\left[x^2 \cdot P(x)\right] = 0^2\left(\frac{3}{18}\right) + 1^2\left(\frac{5}{18}\right) + 2^2\left(\frac{4}{18}\right) + 3^2\left(\frac{3}{18}\right) + 4^2\left(\frac{2}{18}\right) + 5^2\left(\frac{1}{18}\right)$$

$$= 0 + \frac{5}{18} + \frac{16}{18} + \frac{27}{18} + \frac{32}{18} + \frac{25}{18}$$

$$= \frac{105}{18} = \frac{35}{6} = 5.8333.$$

Then, $\sigma = \sqrt{\Sigma[x^2 \cdot P(x)] - \mu^2} = \sqrt{5.8333 - 1.9444^2} = \sqrt{2.0525} \approx 1.43.$

Solutions to Odd-Numbered Exercises

1. Using the relative frequency formula this probability is $\dfrac{5,417}{7,581} = 0.7145$.

3. Since there is replacement of the first card before selection of the second card we use the formula
    $$P(\text{ace}_1 \text{ and ace}_2) = P(\text{ace}_1) \cdot P(\text{ace}_2) = \left(\frac{4}{52}\right)\left(\frac{4}{52}\right) = \frac{16}{2704} = 0.0059,$$ where ace_i means an ace on the ith selection.

5. Again, since there is replacement of the first card before selection of the second card we use the formula
    $$P(\text{king}_1 \text{ and ace}_2) = P(\text{king}_1) \cdot P(\text{ace}_2) = \left(\frac{4}{52}\right)\left(\frac{4}{52}\right) = \frac{16}{2704} = 0.0059.$$

7. Since the first card is not replaced before the second card is selected we use the formula
    $$P(\text{ace}_1 \text{ and ace}_2) = P(\text{ace}_1) \cdot P(\text{ace}_2 \mid \text{ace}_1) = \left(\frac{4}{52}\right)\left(\frac{3}{51}\right) = \frac{12}{2652} = 0.0045.$$

9. Since the first card is not replaced before the second card is selected we use the formula
    $$P(\text{king}_1 \text{ and ace}_2) = P(\text{king}_1) \cdot P(\text{ace}_2 \mid \text{king}_1) = \left(\frac{4}{52}\right)\left(\frac{4}{51}\right) = \frac{16}{2652} = 0.0060.$$

11. $P(\text{def}) = \dfrac{7}{26} = 0.2692$.

13. $P(\text{not def}_1 \text{ and not def}_2) = P(\text{not def}_1) \cdot P(\text{not def}_2 \mid \text{not def}_1)$
    $$= \left(\frac{19}{26}\right)\left(\frac{18}{25}\right) = \frac{342}{650} = 0.5262.$$

15. $P(1 \text{ or } 5) = P(1) + P(5) = 0.15 + 0.20 = 0.35$.

17. $P(\text{18-24}) = 0.165$.

19. $P(< 18 \text{ or } > 45) = P(< 18) + P(> 45) = 0.011 + 0.106 = 0.117$.

21. $P(\text{female} \| \text{expressed interest}) = \dfrac{46}{99} = 0.4646$.

23. $P(\text{pop. vote}) = \dfrac{18}{50} = 0.36$.

25. $P(\text{woman and pop. vote}) = P(\text{woman}) \cdot P(\text{pop. vote} \mid \text{woman}) = \left(\dfrac{9}{50}\right)\left(\dfrac{8}{9}\right) = \dfrac{72}{450} = 0.16$.

27. $P(\text{Center City}) = \dfrac{533}{1,706} = 0.3124$.

29. $P(\text{Center City and 3 bdr.}) = \dfrac{97}{1,706} = 0.0569$.

31. $P(\text{Center City} \mid \text{3 bdr.}) = \dfrac{97}{425} = 0.2282$.

33. The events are not mutually exclusive since being in Center City doesn't exclude three-bedroom apartments. That is, the intersection of Center City and 3 bdr. is not empty.

35. $P(\text{community college}) = \dfrac{3,458}{16,715} = 0.2069.$

37. $P(\text{4-yr. college} \mid \text{Philomath County}) = \dfrac{3,185}{4,514} = 0.7056.$

39. $P(\text{Coldchester or Morgan Counties}) = P(\text{Coldchester County}) + P(\text{Morgan County})$
$$= \dfrac{2,483}{16,715} + \dfrac{3,815}{16.715} = \dfrac{6,298}{16,715} = 0.3768.$$

41. $P(\text{not 4-yr. col.}) = 1 - P(\text{4-yr. col.}) = 1 - \dfrac{12,155}{16,715} = 1 - 0.7272 = 0.2728.$

43. $P(\text{Suburban}) = \dfrac{51}{268} = 0.1903.$

45. $P(\text{Rural} \mid \text{low poll.}) = \dfrac{33}{65} = 0.5077.$

47. $P(\text{Urban or high poll.}) = P(\text{Urban}) + P(\text{High poll.}) - P(\text{Urban and high poll.})$
$$= \dfrac{90}{268} + \dfrac{150}{268} - \dfrac{73}{268} = \dfrac{167}{268} = 0.6231.$$

49. $P(\text{commercial or mod. poll.}) = \dfrac{11}{268} = 0.0410.$

51. Discrete.   **53.** Continuous.

55. Continuous.   **57.** Continuous.

59. Discrete.   **61.** Continuous.

63. $P(x \geq 1) = 1 - P(x = 1) = 1 - 0.75 = 0.25.$

65. $E = \Sigma x \cdot P(x)0 \cdot 0.75 + 1 \cdot 0.10 + 2 \cdot 0.05 + 3 \cdot 0.10 = 0.50.$

67. $P(x \geq 1) = 1 - P(x = 1) = 1 - 0.3 = 0.7.$

69. $E = \Sigma x \cdot P(x) = 0 \cdot 0.3 + 1 \cdot 0.5 + 2 \cdot 0.1 + 3 \cdot 0.1 = 1.0.$

71. $P(\$1,000) = \dfrac{2}{9} = 0.2222.$

73. $E = 0 \cdot \left(\dfrac{2}{9}\right) + 100 \cdot \left(\dfrac{2}{9}\right) + 500 \cdot \left(\dfrac{2}{9}\right) + 1,000 \cdot \left(\dfrac{2}{9}\right) + 5,000 \cdot \left(\dfrac{1}{9}\right) = \$911.11.$

Chapter 5 Probability Distributions

Study Aids and Practice Exercises

5-1 Binomial Experiments

Study Objective
You should be able to:
Identify binomial experiments.

Section Overview
   We will often encounter a discrete random variable called a binomial random variable. It results from counting the number of successes in a binomial experiment. A binomial experiment has four characteristics:
1. The experiment involves "n" identical trials.
2. The trials are independent.
3. Each trial results in a success or failure.
4. The probability of success, p, and the probability of failure, q, are constant from trial to trial.

Because a binomial random variable, denoted r, is a count of the number of successes in n trials, its possible values are $0, 1, 2, \ldots, n$.

Key Terms & Formulas
Binomial Experiment An experiment that involves n identical trials each of which results in a success or failure. The trials are independent and each has the same probability of success.

Binomial Random Variable The number of successes in a binomial experiment. A binomial random variable is a type of discrete random variable.

Worked Example
Binomial?
Check to see if the following are binomial random variables.
   A. The number of heads in three flips of a coin.
   B. The number of sixes obtained in ten rolls of a die.
   C. The number of female births over a 24-hour period in a local hospital.
   D. The number of spades dealt to a player in a bridge game.
   E. In a room with 12 people, the number with brown, blue, hazel, and green eyes.
   F. In a Statistics class of 35 students, the number who pass the final exam.

Solution
We need to check the four conditions for a binomial experiment.

A. Yes. 1: There are $n = 3$ coin flips. 2: The coin flips are independent as the outcome on one flip does not influence the outcome of another. 3: Each trial results in either a success = "a head" or a failure = "a tail". 4: The probabilities of success and failure are constant, $p = \frac{1}{2}$ and $q = \frac{1}{2}$ for all three flips.

B. Yes. 1: There are $n = 10$ die rolls. 2: The die rolls are independent. 3: Each trial results in either a success = "a 6" or a failure = "a 1 or 2 or 3 or 4 or 5". 4: The probabilities of success and failure are constant, $p = \frac{1}{6}$ and $q = \frac{5}{6}$.

C. No. We do not know the number of births so we do not have a fixed number of identical trials in this experiment.

D. No. The trials are not independent. As the player receives cards, the suit received on the first card changes the probability of receiving a spade with the second card. If the first card is a spade, the probability of a spade on the second card goes down. If the first card is anything else, the probability of getting a spade on the second card increases. Whichever happens, the probability of a spade changes and the trials are not independent.

E. No. There are *four* rather than two categories in this experiment.

F. No. The probabilities of success and failure, p and q, are not constant. While students have an equal opportunity to pass (identical points for identical answers), the probabilities would vary from student to student, i.e., from trial to trial. A student who has read the book and done all the class assignments would have a greater probability of passing than someone on academic probation who has yet to purchase the wonderful text book (or the wonderful study guide).

Practice Exercises 5-1

1-3. Determine if the following satisfy the four conditions of a binomial experiment.
1. The number of cars that are speeding on a one mile stretch of a two-lane highway.
2. The number of cars that are speeding among the next 100 that pass a particular signpost on a two-lane highway.
3. The number of people in each of the next 100 cars that pass a particular signpost on a two-lane highway.

4-6. Determine if the following satisfy the four conditions of a binomial experiment.
4. The number of cars out of 60 that pass an inspection of their catalytic converters.
5. The number of cars that pass an inspection at the Simeon's inspection station next week.
6. The number of times 60 cars need to be inspected before they pass.

7-8. Determine if the following satisfy the four conditions of a binomial experiment.
7. The number of children among the next 20 born at Sierra Vista Hospital that are color blind.
8. The number of boys among the next 20 boys born at Sierra Vista Hospital that are color blind.

9-10. Determine if the following satisfy the four conditions of a binomial experiment.
9. The number of aces among four cards selected without replacement from a well-shuffled deck.
10. The number of aces among four cards selected from a deck. The selection involves selecting a card, learning if it is an ace, returning it to the deck, and reshuffling the deck before the next card is selected.

Solutions to Practice Exercises 5-1

1. No, there is not a fixed number of identical trials.
2. No, the trials are not independent. If one car is following another and the lead car is going slow, the second car could not be speeding.
3. No, there are more than two possible outcomes.
4. No, each car does not have the same probability of passing. For example, older cars with older equipment are less likely to pass than a later model auto.
5. No, there is not a fixed number of trials because we do not know the number of cars to be inspected next week.
6. No, a car could require any number of reinspections, so there are more than two possible outcomes.
7. No, boys are more likely to be color blind than girls, so the probabilities are not constant from trial to trial.
8. Yes. There are 20 trials, the color-blindness of one boy is independent of the color-blindness of another, there are only two outcomes for each boy, and the probability of color-blindness is constant.

9. No, the trials are not independent. Whether prior cards are aces will affect the probability that a subsequent card is an ace.

10. Yes. There are four trials; with the replacement in the deck and reshuffling, the trials are independent; by combining all nonaces into a single category, we have only two outcomes on each trial; and there is a constant probability of $\frac{4}{52} = \frac{1}{13}$ of an individual selection resulting in an ace.

5-2 Determining Binomial Probabilities

Study Objectives

You should be able to:
1. Calculate the number of ways r items can be selected from n.
2. Calculate binomial probabilities using a formula.
3. Use a binomial table to obtain probabilities.
4. Obtain the expected value, variance, and standard deviation of a binomial random variable.

Section Overview

We often encounter binomial random variables and want to be able to calculate binomial probabilities. We can do so either by using the binomial formula or employing binomial tables. The binomial formula involves combinations, the number of ways that r items can be chosen from n. This is denoted $_nC_r$ and is calculated with $\frac{n!}{r!(n-r)!}$. Then the binomial probability of r successes in n trials, where p and q are the probability of success and failure, is $_nC_r p^r q^{n-r}$.

It is convenient to use binomial tables to obtain binomial probabilities. They are available for selected values of n and p. Such tables usually only go as high as 0.5 for the value of p. We can get binomial probabilities for $p > 0.5$ by switching successes and failures before using the table. For example, suppose we want the probability of 7 successes in 10 trials where $p = 0.8$. To get 7 successes in 10 trials, we must get 3 failures, where the probability of failure is 0.2. By looking for the binomial probability of $x = 3$ when $p = 0.2$, we will get the correct answer.

Finally, there are simple formulas available to calculate the expected value, variance, and standard deviation for a binomial random variable.

Key Terms & Formulas

<u>n Factorial</u> Denoted $n!$, it is the product of all the integers from 1 to n, i.e., $n! = n \cdot (n-1) \cdot (n-2) \cdots 3 \cdot 2 \cdot 1$.

<u>Combination</u> Denoted $_nC_r$, it is the number of ways of selecting r items from a set of n where the order of selection is irrelevant: $_nC_r = \frac{n!}{r!(n-r)!}$.

<u>Binomial Formula</u> The formula that calculates binomial probabilities: $P(r) = {_nC_r} p^r q^{n-r}$.

<u>Binomial Expectation</u> The expected value of a binomial random variable: $E(x) = np$.

<u>Binomial Variance</u> The variance of a binomial random variable: $\sigma^2 = npq$.

Worked Examples

Senior Seminar

Suppose 12 students are enrolled in senior seminar. The instructor will select three students to make presentations during the first class meeting. In how many ways can the instructor select the three students?

Solution

$$_{12}C_3 = \frac{12!}{3!9!} = \frac{12 \cdot 11 \cdot 10 \cdot 9 \cdot 8 \cdot 7 \cdot 6 \cdot 5 \cdot 4 \cdot 3 \cdot 2 \cdot 1}{(3 \cdot 2 \cdot 1)(9 \cdot 8 \cdot 7 \cdot 6 \cdot 5 \cdot 4 \cdot 3 \cdot 2 \cdot 1)} = 220.$$

Trumm

The game of Trumm involves rolling a balanced four-sided die marked with the numbers 1, 2, 3, and 4. Suppose the die is rolled six times in succession. First calculate the probability that the six rolls will result in exactly two 1's by using the binomial formula. Compare this to the probability found in the binomial tables.

Solution

To find the probability that exactly two 1's are rolled, we calculate:

$$P(2) = {}_6C_2 p^2 q^{6-2} = \frac{6!}{2!4!}(0.25)^2(0.75)^4 = \frac{6 \cdot 5 \cdot 4 \cdot 3 \cdot 2 \cdot 1}{(2 \cdot 1)(4 \cdot 3 \cdot 2 \cdot 1)}(0.25)^2(0.75)^4 \approx 0.29663 \approx 0.30.$$

The equivalent probability is available from the binomial tables. We look in the $n = 6$ section, the $p = 0.25$ column, and the $r = 2$ row. We find 0.2966, correct to the fourth decimal.

Shocking

A psychotherapist believes that 30% of all manic-depressives have a positive reaction to shock therapy. The psychotherapist takes a random sample of seven manic-depressives and subjects them to shock therapy. Use the binomial formula to calculate the probability that exactly two have a positive reaction. Then use the binomial tables to calculate the probability that fewer than 2 have a positive reaction.

Solution

To find the probability that exactly two have a positive reaction, we use the binomial formula with $n = 7$, $p = 0.3$, $q = 0.7$, and $r = 2$:

$$P(2) = {}_7C_2 p^2 q^{7-2} = \frac{7!}{2!5!}(0.3)^2(0.7)^5 = \frac{7 \cdot 6 \cdot 5 \cdot 4 \cdot 3 \cdot 2 \cdot 1}{(2 \cdot 1)(5 \cdot 4 \cdot 3 \cdot 2 \cdot 1)}(0.3)^2(0.7)^5 \approx 0.31765 \approx 0.32.$$

To find the probability that fewer than two have a positive reaction, we add the probabilities that 0 or 1 have a positive reaction. We look in the binomial table in the column with $n = 7$ and $p = 0.3$. In the $r = 0$ row we find 0.0824, while in the $r = 1$ row we find 0.2471. So the probability that less than 2 have a positive reaction is $0.0824 + 0.2471 = 0.3295$.

New Moms

Eighty percent of new mothers leave the hospital within 48 hours of the delivery. If we take a random sample of 20 new mothers, find the probability that 15 of them leave the hospital within 48 hours of the delivery. Also find the probability that at least 17 of them leave the hospital within 48 hours of the delivery.

Solution

We want the probability of 15 successes in 20 trials with $p = 0.8$. The binomial tables only go as high as $p = 0.5$. But the probability of getting 15 successes in 20 trials with $p = 0.8$ is identical to the probability of getting 5 failures in 20 trials with $q = 0.2$. Looking in the binomial tables, we find that this probability is 0.1746.

The probability that at least 17 leave the hospital within 48 hours is identical to the probability that no more than 3 remain in the hospital past 48 hours. So we would add the probabilities the r is equal to 0, 1, 2, and 3 when $q = 0.2$. This is $0.0115 + 0.0576 + 0.1369 + 0.2054 = 0.4114$.

CMC

The warden at the California Men's Colony believes that 60% of the prisoner's sent to his institution are repeat offenders. If he takes a sample of 20 new arrivals, find the expected value, variance, and standard deviation in the number of repeat offenders.

Solution

This is a binomial experiment with $n = 20$ and $p = 0.6$. The expected value is $E = np = (20)(0.6) = 12$. The variance and standard deviation are $\sigma^2 = npq = 20(0.6)(0.4) = 4.8$ and

$$\sigma = \sqrt{npq} = \sqrt{(20)(0.6)(0.4)} = \sqrt{4.8} = 2.19.$$

Practice Exercises 5-2

1-5. The proportion of residents in Finley, Ohio, that favor sex education classes at the high school level is 40%. Suppose a sample of 9 residents is taken.

1. Use the binomial formula to calculate the probability that exactly four will favor sex education classes at the high school level.

2. Use the binomial formula to calculate the probability that none will favor sex education classes at the high school level.

3. Calculate the expected value, variance, and standard deviation of the number that favor sex education classes at the high school level.

4. Use the binomial tables to calculate the probability that exactly three will favor sex education classes at the high school level.

5. Use the binomial tables to calculate the probability that no more than three will favor sex education classes at the high school level.

6-11. A lawyer has a television ad that states she wins 90% of her personal injury cases. Suppose a sample of 20 of her personal injury cases is taken. Assuming the lawyer's claim is true,

6. use the binomial formula to calculate the probability that she wins exactly 18 cases.

7. use the binomial formula to calculate the probability that she wins all 20 of these cases.

8. calculate the expected value, variance, and standard deviation of the number of personal injury cases she wins.

9. use the binomial tables to calculate the probability that she wins exactly 17 of these cases.

10. use the binomial tables to calculate the probability that she wins at least 15 of these cases.

11. Suppose she wins only 14 of these cases. What might you infer?

Solutions to Practice Exercises 5-2

1. This is a binomial random variable with $n = 9$, $p = 0.4$, and $q = 0.6$. To calculate the probability that exactly 4 favor the high school sex education, we calculate:

$$P(4) = {}_9C_4 p^4 q^{9-4} = \frac{9!}{4!5!}(0.4)^4(0.6)^5 = \frac{9 \cdot 8 \cdot 7 \cdot 6 \cdot 5 \cdot 4 \cdot 3 \cdot 2 \cdot 1}{(4 \cdot 3 \cdot 2 \cdot 1)(5 \cdot 4 \cdot 3 \cdot 2 \cdot 1)}(0.4)^4(0.6)^5 \approx 0.25082 \approx 0.25.$$

2. To obtain the probability that none favor the high school sex education, we calculate:

$$P(0) = {}_9C_0 p^0 q^{9-0} = \frac{9!}{0!9!}(0.4)^0(0.6)^9 = \frac{9 \cdot 8 \cdot 7 \cdot 6 \cdot 5 \cdot 4 \cdot 3 \cdot 2 \cdot 1}{(1)(9 \cdot 8 \cdot 7 \cdot 6 \cdot 5 \cdot 4 \cdot 3 \cdot 2 \cdot 1)}(0.4)^0(0.6)^9 \approx 0.01008 \approx 0.01.$$

3. The expected value is $E(x) = np = (9)(0.4) = 3.6$. The variance and standard deviation are $\sigma^2 = npq = (9)(0.4)(0.6) = 2.16$ and $\sigma = \sqrt{npq} = \sqrt{(9)(0.4)(0.6)} = \sqrt{2.16} \approx 1.49$.

4. The probability is found in the binomial table in the $n = 9$ section, the $p = 0.40$ column, and the $r = 3$ row. We find 0.2787.

5. The probability is found in the binomial table by summing the $r = 0, 1, 2, 3$, rows in the $n = 9$ section, the $p = 0.40$ column. We find $0.0168 + 0.0896 + 0.2090 + 0.2787 = 0.5941$.

6. This is a binomial random variable with $n = 20$, $p = 0.9$, and $q = 0.1$. The probability that the lawyer wins exactly 18 of her personal injury cases is:

$$P(18) = {}_{20}C_{18} p^{18} q^{20-18} = \frac{20!}{18!2!}(0.9)^{18}(0.1)^2 = \frac{20 \cdot 19 \cdot \cdots \cdot 3 \cdot 2 \cdot 1}{(18 \cdot 19 \cdot \cdots \cdot 2 \cdot 1)(2 \cdot 1)}(0.9)^{18}(0.1)^2 \approx 0.28518 \approx 0.29.$$

7. The probability that the lawyer wins all 20 of her personal injury cases is:

$$P(20) = {}_{20}C_{20} p^{20} q^{20-20} = \frac{20!}{20!0!}(0.9)^{20}(0.1)^0 = \frac{20 \cdot 19 \cdot \cdots \cdot 3 \cdot 2 \cdot 1}{(20 \cdot 19 \cdot \cdots \cdot 3 \cdot 2 \cdot 1)(1)}(0.9)^{20}(0.1)^0 \approx 0.12158 \approx 0.12.$$

8. The expected value is $E(x) = np = (20)(0.9) = 18$ The variance and standard deviation are $\sigma^2 = npq = (20)(0.9)(0.1) = 1.80$ and $\sigma = \sqrt{npq} = \sqrt{(20)(0.9)(0.1)} = \sqrt{1.80} \approx 1.34$.

9. Obtaining 17 successes when $p = 0.90$ is equivalent to having 3 failures with $q = 0.10$. So we get the probability by looking in the binomial table in the $n = 20$ section, the $p = 0.10$ column, and the $r = 3$ row. We find 0.1901.

10. Obtaining 15 or more successes when $p = 0.90$ is equivalent to having 5 or fewer failures with $q = 0.10$. So we can get the probability by summing the probabilities for $r = 0, 1, 2, 3, 4,$ and 5 in the $n = 20$ section and the $p = 0.10$ column. We find $0.1216 + 0.2702 + 0.2852 + 0.1901 + 0.0898 + 0.0319 = 0.9888$.

11. The previous exercise indicates that, if the claim made by the lawyer is true, she is very likely to win at least 15 of 20 cases. Since she won less than that amount, we might infer that her claim is not true.

5-3 The Poisson Distribution

Study Objectives
You should be able to:
1. Recognize a Poisson experiment.
2. Calculate Poisson probabilities using the Poisson formula.
3. Obtain Poisson probabilities using the Poisson tables

Section Overview
While we often encounter the binomial distribution, slight variations in the conditions for a binomial experiment lead to different probability distributions. One common alternative to the binomial distribution is the Poisson distribution. It is appropriate in experiments similar to the binomial, except there is not a fixed number of identical trials. Rather, we are interested in the number of occurrences of some event (similar to a binomial success) over some continuum like time, length, or area. Examples include the number of people arriving in a bank in a 15-minute period (time) and the number of knots in a 10-foot length of 2×4 construction grade fir (area). When the Poisson distribution is appropriate, we can get the probability of x occurrences by calculating $P(x) = \dfrac{\mu^x e^{-\mu}}{x!}$ where μ is the mean number of successes per unit of time, length, area, etc. Poisson probabilities can also be obtained from Poisson tables for selected values of μ.

Key Terms & Formula
Poisson Experiment An experiment that counts the number of occurrences of some event over a continuum like time or area. The probability is the same for each identical unit of the continuum and the occurrences are independent.

Poisson Formula The formula that calculates Poisson probabilities: $P(x) = \dfrac{\mu^x e^{-\mu}}{x!}$.

Worked Examples
Satellite Transmissions
The number of hit signals in an Earth-to-satellite data transmission has a Poisson distribution with an average of three hits per minute. Use the Poisson formula to calculate the probability of two hits in one minute. Calculate the probability that at least one hit occurs in a one-minute period. Compute the probability of exactly one hit in a 30-second interval. Check each answer using a Poisson table.

Solution
To calculate the probability of two hits in one minute, we use the Poisson formula with $\mu = 3$ and $x = 2$:

$$P(2) = \frac{3^2 e^{-3}}{2!} = \frac{(9)(0.49787)}{2 \cdot 1} = 0.22404 \approx 0.22.$$ To check this answer using a Poisson table, we look in the $\mu = 3$ column and the $x = 2$ row. There we find 0.2240.

Caution: In doing calculations like this, it can be a mistake to round in the midst of the computations. We recommend that you use all the decimal places your calculator provides. When all the calculations are done, round to your heart's content.

To calculate the probability of at least one hit in a one-minute period would be difficult without complements. But the complement of at least one hit is to have no hits. We can calculate the probability that $x = 0$ and subtract this from 1 to get the correct answer. The probability of no hits is $P(0) = \dfrac{3^0 e^{-3}}{0!} = \dfrac{(1)(0.49787)}{1} = 0.049787 \approx 0.05.$

Then the probability of at least one hit is $1 - 0.05 = 0.95$. To check $P(0)$ using a Poisson table, we look in the $\mu = 3$ column and the $x = 0$ row. There we find 0.0498.

To calculate the probability of one hit in a 30-second interval, we need to take one extra but simple step. If there is an average of 3 hits in one minute (60 seconds), then the average number of hits in half that time (30 seconds) is 1.5 hits. With that, we can calculate the probability of one hit in 30 seconds using the Poisson formula with $\mu = 1.5$ and $x = 1$: $P(1) = \dfrac{1.5^1 e^{-1.5}}{1!} = \dfrac{(1.5)(0.223130)}{1} = 0.334694 \approx 0.33$. To check this answer using a Poisson table, we look in the $\mu = 1.5$ column and the $x = 1$ row. There we find 0.3347.

Practice Exercises 5-3

1-2. The number of bad sectors on a computer disk is a Poisson random variable with a mean of 0.25. Find the probability that

 1. there are two bad sectors on a disk.
 2. there is at least one bad sector on a disk.

3-8. The number of emergency 911 calls received by a police dispatcher is a Poisson random variable with a mean of 5 per hour. Use the Poisson

 3. formula to calculate the probability of receiving exactly 10 calls in one hour.
 4. table to obtain the probability of receiving exactly 10 calls in one hour.
 5. formula to calculate the probability of receiving 0 or 1 calls in one hour.
 6. table to obtain the probability of receiving between 5 and 8 (inclusive) calls in one hour.
 7. formula to calculate the probability of receiving 6 calls in two hours.
 8. table to obtain the probability of receiving 3 calls in a half hour.

Solutions to Practice Exercises 5-3

 1. To calculate the probability of two bad sectors on a disk, we use the Poisson formula with $\mu = 0.25$ and $x = 2$: $P(2) = \dfrac{0.25^2 e^{-0.25}}{2!} = \dfrac{(0.0625)(0.778801)}{2 \cdot 1} = 0.02434 \approx 0.024$.

 2. To find the probability of at least one bad sector, we will calculate the probability of no bad sectors and subtract this probability from 1. The probability of no bad sectors is calculated by the Poisson formula with $\mu = 0.25$ and $x = 0$: $P(0) = \dfrac{0.25^0 e^{-0.25}}{0!} = \dfrac{(1)(0.778801)}{1} = 0.778801 \approx 0.78$. Then the probability of at least one bad sector is $1 - 0.78 = 0.22$.

 3. To calculate the probability of 10 calls in one hour, we use the Poisson formula with $\mu = 5$ and $x = 10$: $P(10) = \dfrac{5^{10} e^{-5}}{10!} = \dfrac{(9,765,625)(0.006738)}{10 \cdot 9 \cdot 8 \cdot 7 \cdot 6 \cdot 5 \cdot 4 \cdot 3 \cdot 2 \cdot 1} = 0.01813 \approx 0.018$

 4. Looking in the Poisson table in the $\mu = 5$ column and the $x = 10$ row, we find 0.0181.

 5. To calculate the probability of 0 or 1 calls in one hour, we use the Poisson formula with $\mu = 5$ twice, once with $x = 0$ and once with $x = 1$: $P(0) = \dfrac{5^0 e^{-5}}{0!} = \dfrac{(1)(0.006738)}{1} = 0.006738 \approx 0.0067$, $P(1) = \dfrac{5^1 e^{-5}}{1!} = \dfrac{(5)(0.006738)}{1} = 0.03369 \approx 0.0337$. Then $P(x = 0 \text{ or } 1) = 0.0067 + 0.0337 = 0.0404$

 6. Looking in the Poisson table in the $\mu = 5$ column and the $x = 0$ and 1 rows, we find $0.0067 + 0.0337 = 0.0404$.

 7. Because the question involves two hours, $\mu = 2(5) = 10$. To calculate the probability of 6 calls in two hours, we use the Poisson formula with $\mu = 10$ and $x = 6$: $P(6) = \dfrac{10^6 e^{-10}}{6!} = \dfrac{(1,000,000)(0.0000454)}{6 \cdot 5 \cdot 4 \cdot 3 \cdot 2 \cdot 1} = 0.06306 \approx 0.063$.

 8. Because the question involves a half hour, we use $\mu = 0.5(5) = 2.5$. Looking in the Poisson table in the $\mu = 2.5$ column and the $x = 3$ row, we find 0.2138.

5-4 The Normal Distribution

Study Objectives
You should be able to:
1. Recognize the normal (or Gaussian) distribution.
2. Calculate z values.
3. Compute probabilities for normal random variables.

Section Overview
We handle probabilities for continuous and discrete probability distributions differently. With continuous distributions, we do not try to obtain the probability that $x = 10$. (It is 0.) Rather, we calculate the probability that a continuous variable will take on values within an interval. For example, we could calculate the probability that $5 < x < 10$.

The most commonly used continuous probability distribution is the normal (or Gaussian) distribution. It is a symmetrical distribution centered on the mean and whose shape resembles a bell (and is sometimes called a bell or mound shaped curve). There are an infinite number of normal distributions because there are an infinite number of possible values for the mean and standard deviation. However, any probability can be calculated by converting a general normal problem to one involving z scores or z values. A z value tells us how far a number is away from the mean, expressed in units of standard deviations. For example, suppose we have a normal random variable whose mean is 100 and standard deviation is 10. Then 120 has a z value of +2 because it is 2 standard deviations above the mean, while 90 has a z value of -1 because it is one standard deviation below the mean.

Available is a table of normal probabilities called the standard normal table. It gives us normal probabilities for various z values. Suppose we want to know the probability that a normal random variable, whose mean is 100 and standard deviation is 10, falls between 90 and 120. We find this by using the normal table to find the probability that a z value falls between -1 and $+2$. Conversely, suppose we want to know the value that is exceeded only 10% of the time. We can find that value in terms of z values and then convert back to the original units.

Key Terms & Formulas
Normal Probability Distribution A continuous probability distribution whose curve is symmetrical and resembles a bell.

Standard Normal Distribution A normal curve that has a mean of 0 and a standard deviation of 1. The random variable is denoted by z, where z represents the number of standard deviations a number is away from the mean.

Standard Normal Probability Table Contains normal probabilities for a standard normal z value.

Standard Score To use the standard normal probability table, we must convert a normal random variable x with mean μ and standard deviation σ to a z value (standard score). We do so using $z = \dfrac{x - \mu}{\sigma}$.

Hint: In some problems, we will want to convert back to x from z. To do so, we use $x = \mu + x\sigma$.

Worked Examples
Computing normal probabilities
Use the standard normal table to calculate the following probabilities.
1. $P(0 < z < 1.47)$
2. $P(-1.47 < z < 0)$
3. $P(z < 0.94)$
4. $P(z > -0.94)$

5. $P(z > 1.54)$
6. $P(z < -1.54)$
7. $P(-2.20 < z < 0.53)$
8. $P(-0.53 < z < 2.20)$
9. $P(0.41 < z < 1.88)$
10. $P(-1.88 < z < -0.41)$

Solution

1. This type of probability, the probability that z is between 0 and some positive number, is exactly what the standard normal table gives us. So to find $P(0 < z < 1.47)$, we look for 1.4 in the left column (row heading), 0.07 in the column heading. Where that row and column intersect, we find 0.4292.

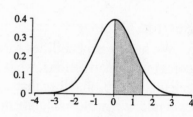

2. We want to calculate $P(-1.47 < z < 0)$. By symmetry, $P(-1.47 < z < 0) = P(0 < z < 1.47)$. In the previous part, we found that this probability is 0.4292.

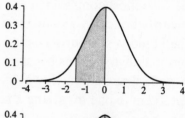

3. $P(z < 0.94)$ is found by taking the total of the two areas in the diagram. The right side in the diagram is $P(0 < z < 0.94)$. We find that by looking for 0.9 in the left column (row heading), 0.04 in the column heading. Where that row and column intersect, we find 0.3264. To this we want to add the probability represented by the left side of the diagram, $P(z < 0)$. But this is an easy probability to find. The The standard normal is a symmetric curve centered at 0. The total area or probability is 1. Combining these two facts, we get $P(z < 0) = 0.5$. Therefore, $P(z < 0.94) = P(z < 0) + P(0 < z < 0.94) = 0.5 + 0.3264 = 0.8264$.

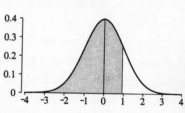

4. Again using the symmetry of the normal curve, we have $P(z > -0.94) = P(z < 0.94) = 0.8264$.

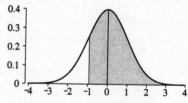

5. To find $P(z > 1.54)$, we recognize that $0.5 = P(z > 0) = P(0 < z < 1.54) + P(z > 1.54)$. Therefore, $P(z > 1.54) = 0.5 - P(0 < z < 1.54)$. $P(0 < z < 1.54)$ is found by by locating the row heading of 1.5 and the column heading of 0.04. Where these intersect, we find a probability of 0.4382. Then $P(z > 1.54) = 0.5 - 0.4382 = 0.0618$.

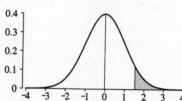

6. By symmetry, we have $P(z < -1.54) = P(z > 1.54) = 0.0618$.

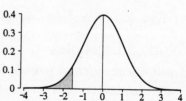

Caution: Notice in these last four calculations that there is not one thing you can look at to know whether you should add 0.5 to the table probability or subtract the table probability from 0.5. Which to do depends both on the sign of the z value and the direction of the inequality. However, if you do not want to memorize these four combinations, you can always sketch a normal curve and shade areas to decide what you should do.

7. To find this probability, we can consider $P(-2.20 < z < 0.53)$ to be a combination of $P(-2.20 < z < 0) + P(0 < z < 0.53)$
$= P(0 < z < 2.20) + P(0 < z < 0.53)$. Each of these can be found form the normal tables. $P(0 < z < 2.20) = 0.4861$ and $P(0 < z < 0.53) = 0.2019$. Therefore,
$P(-2.20 < z < 0.53) = 0.4861 + 0.2019 = 0.6880$.

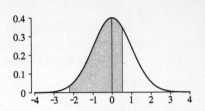

8. By symmetry,
$P(-0.53 < z < 2.20) = P(-2.20 < z < 0.53) = 0.6880$.

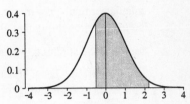

9. To find $P(0.41 < z < 1.88)$, we would first find $P(0 < z < 1.88)$. When we do so, we have more probability than we want. By how much is it too much? By the area between 0 and 0.41, i.e. by $P(0 < z < 0.41)$. Thus
$P(0.41 < z < 1.88) = P(0 < z < 1.88) - P(0 < z < 0.41)$. Finding both from the normal table, we get $0.4699 - 0.1591 = 0.3108$.

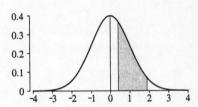

10. By symmetry, $P(-1.88 < z < -0.41) = 0.3108$.

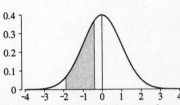

Hint: Parts 7 -10 are easier to remember than parts 3-6. In parts 7-10, if the signs on the z values match, i.e., you have two positive signs or two negative signs, you subtract the two table probabilities. If the signs are different, i.e., you want the probability that a z value is between a negative and positive value, you add the two table probabilities.

Baseball Mitts Strike 1
The time required to sew the rawhide stitching of a baseball mitt is a normal random variable with a mean of 150 seconds and a standard deviation of 10 seconds. What is the z value for a worker who required 165 minutes to complete the assembly? What about 127 seconds?

Solution
To convert from any normal random variable, designated x, to a standard normal, designated z, we use $z = \dfrac{x - \mu}{\sigma}$.

For 165: $z = \dfrac{165 - 150}{10} = \dfrac{15}{10} = 1.5$.

For 127: $z = \dfrac{127 - 150}{10} = \dfrac{-23}{10} = -2.3$.

Baseball Mitts Strike 2
Still working with the time required to stitch a baseball mitt, which is a normal random variable with $\mu = 150$ seconds and $\sigma = 10$ seconds, find the probability that a worker
1. takes more than 165 minutes to complete the stitching.
2. takes more than 127 minutes to complete the stitching.
3. will finish the mitt between 127 and 140 minutes.
4. will finish the mitt between 140 and 165 minutes.

Solution

1. To find the probability that a worker takes more than 165 minutes to complete the stitching, we need to find $P(x > 165)$ where $\mu = 150$ and $\sigma = 10$. First we convert the question to one that involves z scores: $P(x > 165) = P(z > \dfrac{165 - 150}{10}) = P(z > 1.5)$. Then, either based on the diagram or on similarity to a previous example, we would calculate a final answer with: $P(z > 1.5) = 0.5 - 0.4332 = 0.0668.$

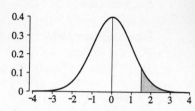

2. To get the probability that a worker takes more than 127 minutes to complete the stitching, we use: $P(x > 127) = P(z > \dfrac{127 - 150}{10}) = P(z > -2.30)$. Then, either based on the diagram or on similarity to a previous example, we get: $P(z > -2.30) = 0.5 + 0.4893 = 0.9893.$

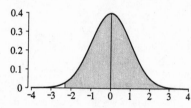

3. The probability that a worker will finish the mitt between 127 and 140 minutes is written: $P(127 < x < 140) =$ $P(\dfrac{127 - 150}{10} < z < \dfrac{140 - 150}{10}) = P(-2.30 < z < -1.00)$. This is: $P(-2.30 < z < -1.00) = 0.4893 - 0.3413 = 0.1480.$

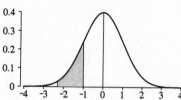

4. The probability that a worker will finish the mitt between 140 and 165 minutes is: $P(140 < x < 165) =$ $P(-1.00 < z < 1.50) = 0.3413 + 0.4332 = 0.7745.$

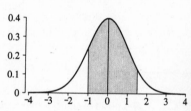

Finding z scores with given probabilities

Find the z value such that a standard normal random variable has a probability of
1. 0.1700 of being between 0 and this positive z score.
2. 0.1700 of being between 0 and this negative z score.
3. 0.85 of being below this z value.
4. 0.15 of being above this z value.
5. 0.20 of being below this z value.
6. 0.99 of being above this z value.

Solution

1. The probability of 0.1700 is shown by the shaded area. From this we see that we would simply look for a probability of 0.1700 in the main body of the standard normal. That value is found in the 0.4 row and the 0.04 column, meaning the z value is 0.44.

    Caution: There is a tendency to look for 0.17 in the row and column headings. Make sure when working with probabilities, that you look in the main body of the table.

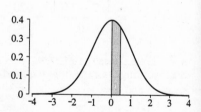

2. The probability of 0.1700 is similar to the previous graph but is to the left of 0. By symmetry, the appropriate z value is -0.44.

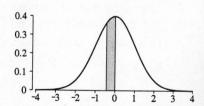

3. If a z value is such that a standard normal random variable has a probability of 0.85 of being below it, then this z value must be above 0 (because a standard normal random variable has a probability of 0.50 of being below 0). To reach a total of 0.85, the probability between 0 and the z value to the right must be 0.3500. The closest we can come to this is 0.3508. This is in the 1.0 row and the 0.04 column, so the z value is 1.04.

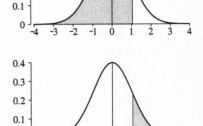

4. If there is a probability of 0.15 that a standard normal random variable is above a z value, then there must be a probability of 0.85 of being below it. But we just found that this z value is 1.04.

5. If there is a probability of 0.20 that a standard normal random variable is below a z value, it must be negative. The total probability below 0 is 0.5. If the probability of being below this z value is 0.2, then the probability of being between this z value and 0 must be 0.3 (to make the total below 0 equal to 0.5). Finally, by symmetry, we can find its positive counterpart, i.e., the z value that has a 0.3 probability between 0 and itself. So we look for 0.3000 in the body of the normal table. The closest we find is 0.2995, which is given by a z value of 0.84. So the final answer is -0.84.

6. If there is a probability of 0.99 of a standard normal being above a z value, it must be below 0 because there is a 0.5 probability of being above 0. In fact, there must be an additional 0.49 probability between 0 and this negative z value. We look for its positive counterpart by looking for 0.4900 in the body of the table. We find 0.4901, which is given by a z value of 2.33. So the answer is -2.33.

Baseball Mitts Strike 3

One last time working with baseball mitt stitching (normal with $\mu = 150$ seconds and $\sigma = 10$ seconds), find the following times.

1. The time below which there is only a 0.10 probability (i.e., a fairly fast time, since small numbers imply high speed).
2. The time that only has a 0.02 probability of being exceeded.

Solution

1. First we must find the z value that only has 0.10 probability below it. It would be a negative z value that had a 0.4 probability between it and 0. Looking for 0.4000 in the body of the normal table, the closest we can come is 0.3997. This is found in the 1.2 row and the 0.08 column. The z value we want is -1.28. To convert this back to times, we use
    $x = \mu + z\sigma = 150 + (-1.28)(10) = 150 - 12.8 = 137.2$ seconds.

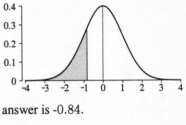

2. If there is a z value that has only a 0.02 probability of being exceeded, it is a positive z value. Additionally, since the total area above 0 is 0.5, there must be a probability of 0.48 between 0 and this z value. Looking for 0.4800 in the normal table, the nearest probability we find is 0.4798. This is in the 2.0 row and the 0.05 column, therefore the z value is 2.05. To convert this back to times, we again use:
    $x = \mu + z\sigma = 150 + (2.05)(10) = 150 + 20.5 = 170.5$ seconds.

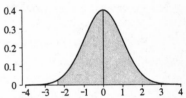

Practice Exercises 5-4

1-4. For a standard normal random variable, find the following.
  1. $P(z < 1.97)$.
  2. $P(z < -0.55)$.
  3. $P(-2.08 < z < 1.11)$.
  4. $P(0.72 < z < 2.41)$.

5-8. For a standard normal random variable, find the following.
  5. $P(z > 2.02)$.
  6. $P(z > -0.98)$.
  7. $P(-1.96 < z < 1.96)$.
  8. $P(-2.47 < z < -0.65)$.

9-11. Find the z value such that a standard normal random variable has a probability of
  9. 0.46 of being between 0 and this positive z score.
  10. 0.40 of being below this z value.
  11. 0.68 of being above this z value.

12-14. Find the z value such that a standard normal random variable has a probability of
  12. 0.22 of being between 0 and this negative z score.
  13. 0.97 of being below this z value.
  14. 0.11 of being above this z value.

15-19. The waiting time for a teller on the last day of the month at a local bank is a normal random variable with a mean of 11 minutes and a standard deviation of 2 minutes.
  15. Find the probability that a person waits more than 14 minutes for a teller.
  16. Find the probability that a person waits between 6 and 10 minutes for a teller.
  17. Find the probability that a person waits more than 7 minutes for a teller.
  18. The 30 percent who spend the least time waiting for a teller will wait for less than what time?

19-21. The weight of Australian shepherds has a normal distribution with a mean of 63 pounds and a standard deviation of 8 pounds.
  19. What percent of Australian shepherds weigh between 65 and 85 pounds?
  20. What is the probability that an Australian shepherd weighs less than 55 lbs.?
  21. What weight divides the heaviest 20% from the lighter 80%?

22-24. The time to failure of components of an electrical system has a normal distribution with a mean of 400 days and a standard deviation of 100 days.
  22. What is the probability that a component will last between 333 and 554 days?
  23. What is the probability that a component will last less than 627 days?
  24. By how many days will 15% fail?

Solutions to Practice Exercises 5-4

1. $P(z < 1.97) = 0.5 + 0.4756 = 0.9756$

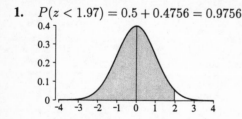

2. $P(z < -0.55) = 0.5 - 0.2088 = 0.2912$

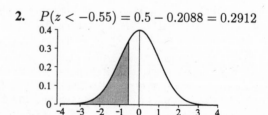

3. $P(-2.08 < z < 1.11) = 0.4812 + 0.3665$
$$= 0.8477$$

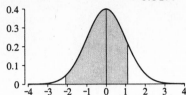

4. $P(0.72 < z < 2.41) = 0.4920 - 0.2642$
$$= 0.2278$$

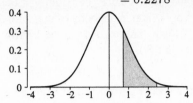

5. $P(z > 2.02) = 0.5 - 0.4783 = 0.0217$

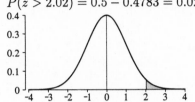

6. $P(z > -0.98) = 0.5 + 0.3365 = 0.8365$

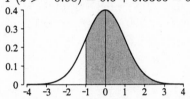

7. $P(-1.96 < z < 1.96) = 0.4750 + 0.4750$
$$= 0.9500$$

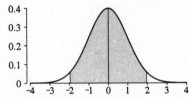

8. $P(-2.47 < z < -0.65) = 0.4932 - 0.2422$
$$= 0.2510$$

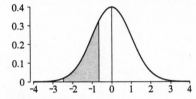

9. To find the z value that has a probability of 0.46 between 0 and itself, we look for 0.4600 in the body of the table. The closet value we can find is 0.4599. This is in the 1.7 row and the 0.05 column, so the z value is 1.75.

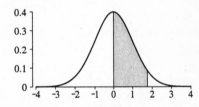

10. If there is a probability of 0.40 of being below a z value, the z value is negative and has a 0.10 probability between it and 0. We look for 0.1000 in the body of the table. The closest probability we can find is 0.0987. This is in the 0.2 row and the 0.05 column. The z value is below zero, so the final answer is -0.25.

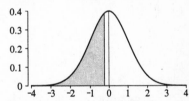

11. If there is a probability of 0.68 of being above a z value, it must be negative and such that there is a 0.18 probability of being between it and 0. We look for 0.1800 in the body of the normal table. The closest we can find is 0.1808, which is in the 0.4 row and the 0.07 column. Therefore, the z value is -0.47.

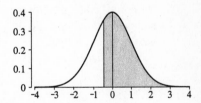

12. By symmetry, the z value is the negative of the z value with the identical probability. We look for a probability of 0.2200 in the body of the normal table. The nearest value is 0.2190. This is in the 0.5 row and the 0.08 column, so the z value is -0.58.

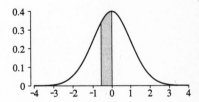

13. If here is a probability of 0.97 of being below this z value, then it is positive and there is a 0.47 probability of being between it and 0. Looking for 0.4700 in the body of the normal table, the nearest we can come is 0.4699. This is found in the 1.8 row and the 0.08 column, so the z value is 1.88.

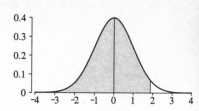

14. If there is a probability of 0.11 of being above this z value, then it is positive and the probability of being between it and 0 is 0.39. Looking for 0.3900 in the body of the normal table, the closest value we can find is 0.3907. This is in the 1.2 row and the 0.03 column, so the z value is 1.23.

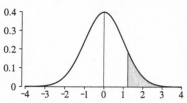

15. $P(x > 14) = P\left(z > \dfrac{14 - 11}{2}\right) = P(z > 1.50)$

$\qquad = 0.5 - 0.4336 = 0.0668.$

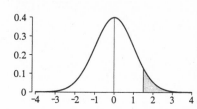

16. $P(6 < x < 10) = P\left(\dfrac{6 - 11}{2} < z < \dfrac{10 - 11}{2}\right)$

$\qquad = P(-2.50 < z < -0.50)$

$\qquad = 0.4938 - 0.1915 = 0.3023.$

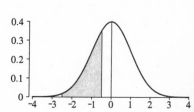

17. $P(x > 7) = P\left(z > \dfrac{7 - 11}{2}\right)$

$\qquad = P(z > -2.00) = 0.5 + 0.4772 = 0.9772.$

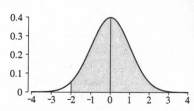

18. This is a negative z value that has a 0.3 probability below it and therefore a 0.2 probability between itself and 0. Looking for 0.2000 in the normal table, the closest we come is 0.1985, found in the 0.5 row and the 0.02 column. So the z value is -0.52. Converting this back to waiting times, we get: $x = \mu + z\sigma = 11 + (-0.52)(2) = 11 - 1.04 = 9.96$ minutes.

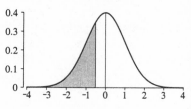

19. $P(65 < x < 85) = P\left(\dfrac{65 - 63}{8} < z < \dfrac{85 - 63}{8}\right)$

$\qquad = P(0.25 < z < 2.75) = 0.4970 - 0.0987$

$\qquad = 0.3983.$

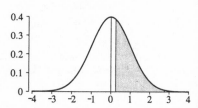

20. $P(x < 55) = P\left(z < \dfrac{55 - 63}{8}\right) = P(z < -1.00)$

$= 0.5 - 0.3413 = 0.1587$

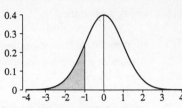

21. This is a positive z value. There is a 0.20 probability of exceeding this value, so there must be a 0.30 probability of being between it and 0. We look for 0.3000 in the normal table, finding 0.2995 to be the nearest value. It is found in the 0.8 row and 0.04 column, so the z value is 0.84. Next we convert to weights, getting: $x = \mu + z\sigma = 63 + (0.84)(8) = 63 + 6.72 = 69.72$ pounds.

22. $P(333 < x < 554) = P\left(\dfrac{333 - 400}{100} < z < \dfrac{554 - 400}{100}\right)$

$= P(-0.67 < z < 1.54)$

$= 0.4382 + 0.2486 = 0.6868$

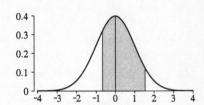

23. $P(x < 627) = P\left(z < \dfrac{627 - 400}{100}\right)$

$= P(z < 2.27) = 0.5 + 0.4884 = 0.9884.$

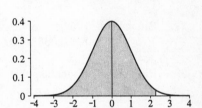

24. This is a negative z value that has a 0.15 probability below it and therefore a 0.35 probability between itself and 0. Looking for 0.3500 in the normal table, the closest we come is 0.3508, found in the 1.0 row and the 0.04 column. So the z value is -1.04. Converting this back to failure times, we get:

$x = \mu + z\sigma = 400 + (-1.04)(100) = 400 - 104 = 296$ days.

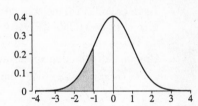

Solutions to Odd-Numbered Exercises

1. Using Appendix 1 in back of the main text with $n = 13$, $r = 5$, and $p = 0.35$ we obtain $P(r = 5) = 0.2154$.

3. Using Appendix 1 with $n = 6$ and $p = 0.40$ we obtain $P(r = 0) = 0.0467$.

5. Using Appendix 10 with $\mu = 3.9$ we obtain $P(x = 4) = 0.1951$.

7. Again using Appendix 10 with $\mu = 3.9$ we obtain
$P(x = 3, 4) = P(x = 3) + P(x = 4) = 0.2001 + 0.1951 = 0.3952$.

9. Using Appendix 10 with $\mu = 6.7$ we obtain $P(x \leq 1) = 1 - P(x = 0) = 1 - 0.0012 = 0.9988$.

11. Using Appendix 2, $P(0 \leq z \leq 1.53) = 0.4370$.

13. Using Appendix 2, $P(z \leq 1.96) = P(z \leq 0) + P(0 \leq z \leq 1.96) = 0.5 + 0.4750 = 0.9750$.

15. Using Appendix 2, $P(z \geq 2.58) = 0.5 - P(0 \leq z \leq 2.58) = 0.5 - 0.4951 = 0.0049$.

17. Binomial distribution applies with $n = 15, r = 10$, and $p = 0.80$. Since $p > 0.50$, to use the table in Appendix 1 we consider the equivalent probability with $n = 15, q = 1 - p = 0.20$ and $r = 5$; $P(r = 5) = 0.1032$.

19. As in Exercise 17, we consider the equivalent probability with $n = 15, q = 0.20$ and $r > 9$. So $P(r > 9) = P(r \geq 10) = 0.0001$.

21. Binomial distribution applies with $n = 20, r = 20$ and $p = 0.60$. Since p > 0.50, to use the table in Appendix 1 we consider the equivalent probability with $n = 20, q = 0.40$ and $r = 0$. So $P(r = 0) = 0.0000$. Note: $0.60^{20} = 0.0000366$.

23. Using Appendix 10 with $\mu = 5$ we obtain $P(x = 4) = 0.1755$.

25. Binomial distribution applies with $n = 7, r > 4$, and $p = 0.80$. Since $p > 0.50$, to use the table in Appendix 1 we consider the equivalent probability with $n = 7, q = 1 - p = 0.20$ and $r < 3$; $P(r < 3) = 0.2097 + 0.3670 + 0.2753 = 0.8520$.

27. Binomial distribution applies with $n = 17, 5 \leq r \leq 8$ and $p = 0.15$. So $P(5 \leq r \leq 8) = 0.0668 + 0.0236 + 0.0065 + 0.0014 = 0.0983$.

29. Binomial distribution applies with $n = 10, r > 7$ and $p = 0.60$. The equivalent probability occurs with $n = 10, q = 0.40$ and $r < 3$. So $P(r < 3) = 0.0060 + 0.0403 + 0.1209 = 0.1672$.

31. As in Exercise 29, the equivalent probability is with $n = 10$, q $= 0.40$ and $r < 1$. So $P(r < 1) = 0.0060 + 0.0403 = 0.0463$.

33. Binomial distribution applies with $n = 28, p = 0.43$, and $q = 0.57$. Using the formula for standard deviation $\sigma = \sqrt{npq} = \sqrt{28 \cdot 0.43 \cdot 0.57} = \sqrt{6.8628} = 2.62$.

35. Binomial distribution applies with $n = 9, r < 9$ and $p = 0.40$ so $P(r < 9) = 1 - 0.0003 = 0.9997$. Note that 'success' is defined as "not living with both natural parents", so "at least 1 is living with both natural parents" means at least one 'failure', or $r < 9$.

37. This is a Poisson probability distribution with $\mu = 2.4$ and $x \geq 1$. Using Appendix 10, $P(x \geq 1) = 1 - P(x = 0) = 1 - 0.0907 = 0.9093$.

39. Binomial distribution applies with $n = 18, r < 6$, and $p = 0.55$; the equivalent probability is when $n = 18, q = 0.45$, and $r > 12$. Thus $P(r > 12) = 0.0134 + 0.0039 + 0.0009 + 0.0001 = 0.0183$.

41. Using $\mu = 26$ and $\sigma = 6$, $z = \frac{30-26}{6} = 0.67$. Thus $P(30 < x) = P(0.67 < z) = 0.5 - P(0 < z < 0.67) = 0.5 - 0.2486 = 0.2514$.

43. Using $\mu = 26$ and $\sigma = 6$ in the formula $z = \dfrac{x - \mu}{\sigma}$ to find the corresponding z-values for $x_1 = 15$ and $x_2 = 30$. Substituting x_1, we get $z_1 = \dfrac{15 - 26}{6} = -1.83$ and substituting x_2, we get $z_2 = \dfrac{30 - 26}{6} = 0.67$. Thus $P(15 < x < 30) = P(-1.83 < z < 0.67) =$ $P(-1.83 < z < 0) + P(0 < z < 0.67) = 0.4664 + 0.2486 = 0.7150$.

45. Binomial distribution applies with $n = 10$, $r \geq 5$, and $p = 0.65$; the equivalent probability is when $n = 10$, $q = 0.35$, and $r \leq 5$. Thus $P(r \leq 5) = 0.1536 + 0.2377 + 0.2522 + 0.1757 + 0.0725 + 0.0135 = 0.9052$. (If you calculate $P(r \leq 5)$ as $1 - P(r \geq 5)$ you will get 0.9051.)

47. This is a Poisson probability distribution with $\mu = 4.60$ and $x = 4$. Using Appendix 10, $P(x = 4) = 0.1875$.

49. Binomial distribution applies with $n = 20$, $r < 10$, and $p = 0.70$; the equivalent probability is when $n = 20$, $q = 0.30$, and $r > 10$. Thus $P(r > 10) = 0.0120 + 0.0039 + 0.0010 + 0.0002 = 0.0171$.

51. Using $\mu = 12$ and $\sigma = 2$ in the formula $z = \dfrac{x - \mu}{\sigma}$ to get $z = \dfrac{7 - 12}{2} = -2.50$. Thus
$P(7 < x) = P(-2.50 < z) = P(-2.50 < z < 0) + 0.5 = 0.4938 + 0.5 = 0.9938$.

53. Using $\mu = 12$ and $\sigma = 2$ in the formula $z = \dfrac{x - \mu}{\sigma}$ to find the corresponding z-values for $x_1 = 7$ and $x_2 = 14$. Substituting x_1, we get $z_1 = \dfrac{7 - 12}{2} = -2.50$ and substituting x_2, we get $z_2 = \dfrac{14 - 12}{2} = 1.00$. Thus $P(7 < x < 14) = P(-2.50 < z < 1.00) = P(-2.50 < z < 0) + P(0 < z < 1.00)$
$$= 0.4938 + 0.3413 = 0.8351.$$

55. Using $\mu = 675$ and $\sigma = 75$ in the formula $z = \dfrac{x - \mu}{\sigma}$ to get $z = \dfrac{700 - 675}{75} = 0.33$. Thus
$P(700 < x) = P(0.33 < z) = 0.5 - P(0 < z < 0.33) = 0.5 - 0.1293 = 0.3707$.

57. Binomial distribution applies with $n = 10$, $5 \leq r < 8$, and $p = 0.30$. Thus
$P(5 \leq r < 8) = 0.1029 + 0.0368 + 0.0090 + 0.0014 = 0.1501$.

59. Using $\mu = 30.6$ and $\sigma = 9.1$ in the formula $z = \dfrac{x - \mu}{\sigma}$ to get $z = \dfrac{45 - 30.6}{9.1} = 1.58$. Thus
$P(x < 45) = P(z < 1.58) = 0.5 + P(0 < z < 1.58) = 0.5 + 0.4429 = 0.9429$.

61. Using the values $\mu = 23.3$ and $\sigma = 7.7$ to find $z_1 = \dfrac{10 - 23.3}{7.7} = -1.73$ when $x_1 = 10$ and
$z_2 = \dfrac{30 - 23.3}{7.7} = 0.87$ when $x_2 = 30$. Thus $P(10 < x < 30) = P(-1.73 < z < 0.87) =$
$P(-1.73 < z < 0) + P(0 < z < 0.87) = 0.4582 + 0.3078 = 0.7660$.

63. Using the values $\mu = 118.3$ and $\sigma = 43.1$ to find $z_1 = \dfrac{180 - 118.3}{43.1} = 1.43$ when $x_1 = 180$ minutes (3 hours) and $z_2 = \dfrac{240 - 118.3}{43.1} = 2.82$ when $x_2 = 240$ minutes (4 hours). Thus
$P(10 < x < 30) = P(1.43 < z < 2.82) = P(0 < z < 2.82) - P(0 < z < 1.43)$
$$= 0.4976 - 0.4236 = 0.0740.$$

65. Binomial distribution applies with $n = 250$, $p = 0.39$ and $q = 0.61$. Using the formula for standard deviation $\sigma = \sqrt{npq} = \sqrt{250 \cdot 0.39 \cdot 0.61} = \sqrt{59.475} = 7.71$

67. Binomial distribution applies with $n = 530$, $p = 0.26$ and $q = 0.74$. Using the formula for standard deviation $\sigma = \sqrt{npq} = \sqrt{530 \cdot 0.26 \cdot 0.74} = \sqrt{101.972} = 10.098$.

69. Here we want to find z_0 so that $P(z_0 < z) = 0.10$. Since $P(z_0 < z) = 0.5 - P(0 < z < z_0) = 0.10$, solving yields $P(0 < z < z_0) = 0.40$. Looking in Appendix 2 for 0.40 we find $P(0 < z < 1.28) = 0.3997$ and $P(0 < z < 1.29) = 0.4014$. Since 0.3997 is closer to 0.40 than 0.4014 we chose $z_0 = 1.28$.

71. Here we want to find z_0 so that $P(z_0 < z) = 0.20$. Since $P(z_0 < z) = 0.5 - P(0 < z < z_0) = 0.20$, solving yields $P(0 < z < z_0) = 0.30$. Looking in Appendix 2 for 0.30 we find $P(0 < z < 0.84) = 0.2995$ and $P(0 < z < 0.85) = 0.3023$. Since 0.2995 is closer to 0.30 than 0.3023 we chose $z_0 = 0.84$.

73. First find z_0 so that $P(z_0 < z) = 0.05$. Since $P(z_0 < z) = 0.5 - P(0 < z < z_0) = 0.05$, implies $P(0 < z < z_0) = 0.45$. Since 0.45 is halfway between 1.64 and 1.65 we chose $z_0 = 1.645$. Substituting $\mu = 100$ and $\sigma = 15$ into formula $z = \dfrac{x - \mu}{\sigma}$ and solving for x yields: $1.645 = \dfrac{x - 100}{15}$ $\Rightarrow$ $x = (1.645)(15) + 100 = 124.675$.

Chapter 6 Sampling Concepts

Study Aids and Practice Exercises

6-1 Sampling: The Need and The Advantages

Study Objectives
You should be able to:
1. Understand the need to take samples from populations.
2. Recognize the advantages of taking a sample over performing a census.

Section Overview
Sampling is the process of getting a look at part of a population. Whether it is the life span of smokers or the birth weight of babies, populations are something that concern us. If it was reasonable, we would make a complete examination of a population, called a census. However, usually a population is too large for a census to be sensible. So taking a sample becomes the feasible alternative.

Sampling has several advantages over a census. It is more cost effective to take a sample if it gives us all the information we need for reasonable decisions. It is faster to take a sample than to do a census of a population. What is more important, we can achieve results that, within a narrow margin of error, are as accurate as the results we could achieve through a census. Finally, sometimes analyzing a member of a population is destructive to it. For example, suppose we wanted to know the average lifetime of light bulbs. To do a census would leave us completely in the dark.

Key Terms & Formulas
Finite Population A population that consists of a fixed number of members, e.g., GPA's of students currently enrolled at UConn, incomes of all current employees at Motorola, lengths of stays for patients at the Betty Ford clinic.

Infinite Population A population that has an unlimited number of members, e.g., the total of two dice thrown endlessly, the number of heads obtained when three coins are flipped repeatedly, the birth weights of all children, including those not yet born.

Census A complete examination of a population.

Sample A part of a population.

Worked Examples
Alcoholics
Suppose a medical researcher is interested in the percent of liver damage among alcoholics. He plans to examine 40 alcoholics and assess their degree of liver damage. What are the population and sample?

<u>Solution</u>
The population is the liver damage for all alcoholics. The sample would consist of the degree of liver damage among the 40 sampled alcoholics.

Practice Exercises 6-1

1. A manufacturer of steel girders wants to measure the amount that the girders warp under stress. Twenty girders are stressed and the amount of warpage is measured. What are the population and sample?
2. A highway patrol captain wants to know the amount of alcohol in drivers arrested for drunk driving. The captain examines 200 arrest reports and discovers the amount of alcohol from the blood tests for the drivers. What are the population and sample?

Solutions to Practice Exercises 6-1

1. The population is the amount of warpage under stress shown by steel girders. The amount of warpage in the twenty girders is the sample.
2. The population is the amount of alcohol in the bloodstream's of drunk drivers. The sample is the alcohol amount for the 200 drivers whose records were examined.

6-2 Sampling Distribution of Means--A Pattern of Behavior

Study Objectives

You should be able to:
1.   Understand a sampling distribution.
2.   Specify the sampling distribution of means
3.   Calculate probabilities about $\bar{x}$.

Section Overview

We want to know a population mean. When, as often happens, census information is unavailable, we will use the sample mean to estimate the population mean. It is unlikely that the sample mean will be exactly equal to the population mean. We hope, however, that it will be close. The question we need to address is what we mean by "close." To answer that question, we need to examine how a sample mean will change from sample to sample, i.e., we want to examine the sampling distribution of means.

We need to distinguish between three types of distributions. There is a distribution for the population. If we did a census and produced a histogram of the results, that would be a picture of the population distribution. If we take a sample and drew a histogram for that sample, that is a picture of that sample's distribution. Because we can take many different samples, there would be many sample distributions. Neither of these is what we call the sampling distribution of means. Think of taking a sample and calculating $\bar{x}$. Then taking a second sample and calculating a second $\bar{x}$. Think of taking every possible sample and calculating every possible sample mean. If we were to create a histogram of all those possible $\bar{x}$'s, that would be a picture of the sampling distribution of the means.

The sampling distribution of the means has certain characteristics. First, the mean of the sampling distribution of means is equal to the mean of the population. In symbolic form, this is written $\mu_{\bar{x}} = \mu$. Second, the standard deviation of the sampling distribution of means, called the standard error of the mean and denoted $\sigma_{\bar{x}}$, is $\dfrac{\sigma}{\sqrt{n}}$ for infinite populations and $\dfrac{\sigma}{\sqrt{n}}\sqrt{\dfrac{N-n}{N-1}}$ for finite populations. Third, if the population is normally distributed, then the sampling distribution of means is normally distributed. Last, even if the population is not normal, when the sample size is large (usually $n > 30$ is considered large), the sampling distribution of means is approximately normal.

Key Terms & Formulas

Sampling Distribution of Means The distribution of the sample means from all possible samples of size n taken from a population.

Mean of the Sampling Distribution of Means Denoted $\mu_{\bar{x}}$, it is the average of the $\bar{x}$'s from all possible samples taken upon a population. It is equal to the mean of the population sampled.

Standard Error of the Sampling Distribution of Means Denoted $\sigma_{\bar{x}}$, it is the standard deviation of the $\bar{x}$'s from all possible samples taken upon a population. It is equal to $\dfrac{\sigma}{\sqrt{n}}$ for infinite populations and $\dfrac{\sigma}{\sqrt{n}}\sqrt{\dfrac{N-n}{N-1}}$ for finite populations.

Finite Population Correction Factor It is $\sqrt{\dfrac{N-n}{N-1}}$, the amount by which the standard errors for infinite and finite populations differ.

Central Limit Theorem While technically the Central Limit Theorem is only the last of the four statements below, all four statements are needed to completely describe the sampling distribution of the means.

1. The mean of the sampling distribution of means is equal to the mean of the population, i.e., $\mu_{\bar{x}} = \mu$.
2. The standard deviation of the sampling distribution of means, called the standard error of the mean and denoted $\sigma_{\bar{x}}$, is $\dfrac{\sigma}{\sqrt{n}}$ for infinite populations and $\dfrac{\sigma}{\sqrt{n}}\sqrt{\dfrac{N-n}{N-1}}$ for finite populations.
3. If the population is normally distributed, then the sampling distribution of means is normally distributed no matter the sample size n.
4. If the sample size is large ($n > 30$), the sampling distribution of means is approximately normal (even when the population is not normal).

Worked Examples

Easy as 1, 2, 3

Suppose we have a small population that consists of the numbers 1, 2, and 3. Perhaps we have a spinner on which any of these three values are equally likely. Then the probability distribution of the population is given in the table that follows:

x	$P(x)$
1	$\dfrac{1}{3}$
2	$\dfrac{1}{3}$
3	$\dfrac{1}{3}$

For this population:
1. Calculate the mean and standard deviation.
2. List all possible samples of size 2.
3. Obtain the means of each of the samples.
4. For these means, calculate the mean and standard deviation, i.e., the mean and standard error of the sampling distribution of the means.
5. Verify that the formulas for the mean and standard error of the sampling distribution of means.

Solution

1. $\mu = \Sigma x P(x) = 1\left(\dfrac{1}{3}\right) + 2\left(\dfrac{1}{3}\right) + 3\left(\dfrac{1}{3}\right) = \dfrac{1}{3} + \dfrac{2}{3} + \dfrac{3}{3} = 2$

$$\sigma = \sqrt{\Sigma(x-\mu)^2 P(x)} = \sqrt{(1-2)^2\left(\dfrac{1}{3}\right) + (2-2)^2\left(\dfrac{1}{3}\right) + (3-1)^2\left(\dfrac{1}{3}\right)}$$

$$= \sqrt{\dfrac{1}{3} + 0 + \dfrac{1}{3}} = \sqrt{\dfrac{2}{3}} = 0.8165.$$

2-3. The samples and their means their means are in the following table:

Sample		$\bar{x}$
1	1	1
1	2	1.5
1	3	2
2	1	1.5
2	2	2
2	3	2.5
3	1	2
3	2	2.5
3	3	3

4. We can obtain the mean of the sample means by taking the average of the nine values in the previous table:
$\mu_{\bar{x}} = \dfrac{1 + 1.5 + 2 + 1.5 + 2 + 2.5 + 2 + 2.5 + 3}{9} = \dfrac{18}{9} = 2.0.$ Or by working with the sampling distribution of the sample means given in the following table:

Sample Means $\bar{x}$	$P(\bar{x})$
1	$\dfrac{1}{9}$
1.5	$\dfrac{2}{9}$
2	$\dfrac{3}{9}$
2.5	$\dfrac{2}{9}$
3	$\dfrac{1}{9}$

Then, $\mu_{\bar{x}} = \Sigma \bar{x} P(\bar{x}) = 1\left(\dfrac{1}{9}\right) + 1.5\left(\dfrac{2}{9}\right) + 2\left(\dfrac{3}{9}\right) + 2.5\left(\dfrac{2}{9}\right) + 3\left(\dfrac{1}{9}\right) = \dfrac{1}{9} + \dfrac{3}{9} + \dfrac{6}{9} + \dfrac{5}{9} + \dfrac{3}{9} = 2.$

The standard error is:

$$\sigma_{\bar{x}} = \sqrt{\Sigma(\bar{x}-\mu)^2 P(\bar{x})}$$

$$= \sqrt{(1-2)^2\left(\dfrac{1}{9}\right) + (1.5-2)^2\left(\dfrac{2}{9}\right) + (2-2)^2\left(\dfrac{3}{9}\right) + (2.5-2)^2\left(\dfrac{2}{9}\right) + (3-2)^2\left(\dfrac{1}{9}\right)}$$

$$= \sqrt{\dfrac{1}{9} + \dfrac{1}{18} + 0 + \dfrac{1}{18} + \dfrac{1}{9}} = \sqrt{\dfrac{1}{3}} = 0.5774.$$

5. Using the formulas, we get $\mu_{\bar{x}} = \mu = 2$ and $\sigma_{\bar{x}} = \dfrac{\sigma}{\sqrt{n}} = \dfrac{0.8165}{\sqrt{2}} = 0.5774.$

Cabbies

The distance driven by cabbies during an eight-hour evening shift has a mean of 220 miles with a standard deviation of 60 miles. Suppose a sample of 36 cabbies is taken and their mileages recorded. What is the mean and standard error of the sampling distribution of $\bar{x}$? What is the probability that the sample mean will be less than 200 miles? More than 225 miles? Between 210 and 230 miles?

Solution
First we calculate the mean and standard error of the sampling distribution of $\bar{x}$. $\mu_{\bar{x}} = \mu = 220$ and

$$\sigma_{\bar{x}} = \frac{\sigma}{\sqrt{n}} = \frac{60}{\sqrt{36}} = \frac{60}{6} = 10.$$

Because the sample size is large, $\bar{x}$ is approximately normal. So we can use the normal table to calculate probabilities about the sample mean.

$$P(\bar{x} < 200) = P\left(\frac{\bar{x} - \mu}{\sigma_{\bar{x}}} < \frac{200 - 220}{10}\right) = P(z < -2.00) = 0.5 - 0.4772 = 0.0228.$$

$$P(\bar{x} > 225) = P\left(\frac{\bar{x} - \mu}{\sigma_{\bar{x}}} > \frac{225 - 220}{10}\right) = P(z > 0.50) = 0.5 - 0.1915 = 0.3085.$$

$$P(210 < \bar{x} < 230) = P\left(\frac{210 - 220}{10} < \frac{\bar{x} - \mu}{\sigma_{\bar{x}}} < \frac{230 - 220}{10}\right) = P(-1.00 < z < 1.00)$$
$$= 0.3413 + 0.3413 = 0.6826.$$

Practice Exercises 6-2

1-6. Residents are bothered by the volume of noise coming from an airport. The county manager plans to take a sample of 100 measurements on the noise level. Suppose that the mean volume is 30 dB with a standard deviation of 20 dB.

1. Find the mean of the sampling distribution of means.
2. Find the standard error of the sampling distribution of means.
3. Is there any reason to believe that the sampling distribution of means will be normally distributed?
4. What is the probability that the mean will be between 24 and 32 dB?
5. What is the probability that the mean will be less than 25 dB?
6. What is the probability that the mean will be more than 38 dB?

7-12. A water company claims that the average water pressure they provide is 80 gpm with a standard deviation of 36 gpm. A consumer group plans to take a random sample of 81 homes and determine the water pressure at each. Answer the following questions assuming the claim of the water company is true.

7. Find the mean of the sampling distribution of means.
8. Calculate the standard error of the sampling distribution of means.
9. Will the sampling distribution of means be normally distributed?
10. What is the probability that the mean will be between 70 and 75 gpm?
11. What is the probability that the mean will be less than 90 gpm?
12. What is the probability that the mean will be more than 76 gpm?

Solutions to Practice Exercises 6-2

1. $\mu_{\bar{x}} = \mu = 30$ dB.

2. $\sigma_{\bar{x}} = \dfrac{\sigma}{\sqrt{n}} = \dfrac{20}{\sqrt{100}} = \dfrac{20}{10} = 2.$

3. Because the sample size is $100 > 30$, the central limit theorem says that $\bar{x}$ is approximately normal.

4. $P(24 < \bar{x} < 32) = P\left(\dfrac{24 - 30}{2} < \dfrac{\bar{x} - \mu}{\sigma_{\bar{x}}} < \dfrac{32 - 30}{2}\right) = P(-3.00 < z < 1.00)$
   $$= 0.4987 + 0.3413 = 0.8400.$$

5. $P(\bar{x} < 25) = P\left(\dfrac{\bar{x} - \mu}{\sigma_{\bar{x}}} < \dfrac{25 - 30}{2}\right) = P(z < -2.50) = 0.5 - 0.4938 = 0.0062.$

6. $P(\bar{x} > 38) = P\left(\dfrac{\bar{x} - \mu}{\sigma_{\bar{x}}} > \dfrac{38 - 30}{2}\right) = P(z > 4.00) = 0.5 - 0.4999683 = 0.0000317.$

7. $\mu_{\bar{x}} = \mu = 80$ gpm.

8. $\sigma_{\bar{x}} = \dfrac{\sigma}{\sqrt{n}} = \dfrac{36}{\sqrt{81}} = \dfrac{36}{9} = 4.$

9. Because the sample size is $81 > 30$, the central limit theorem says that $\bar{x}$ is approximately normal.

10. $P(70 < \bar{x} < 75) = P\left(\dfrac{70 - 80}{4} < \dfrac{\bar{x} - \mu}{\sigma_{\bar{x}}} < \dfrac{75 - 80}{4}\right) = P(-2.50 < z < -1.25)$
    $$= 0.4938 - 0.3944 = 0.0994.$$

11. $P(\bar{x} < 90) = P\left(\dfrac{\bar{x} - \mu}{\sigma_{\bar{x}}} < \dfrac{90 - 80}{4}\right) = P(z < 2.50) = 0.5 + 0.4938 = 0.9938.$

12. $P(\bar{x} > 76) = P\left(\dfrac{\bar{x} - \mu}{\sigma_{\bar{x}}} > \dfrac{76 - 80}{4}\right) = P(z > -1.00) = 0.5 + 0.3414 = 0.8413.$

6-3 Sampling Distribution of Percentages

Study Objectives
You should be able to:
1. Specify the sampling distribution of percentages.
2. Calculate probabilities about p.

Section Overview

When dealing with attribute or categorical data, we will be interested in making inferences about a population percentage, π. To do so, we work with the sample percentage, p. We define p as $\dfrac{x}{n} \cdot 100$, where x is the number of items that possess the characteristic of interest and n is the sample size. Because we will be basing our inferences on p, we need to learn about the sampling distribution of percentages. The sampling distribution of percentages is the sampling distribution of p from all possible random samples of size n taken from a population. The mean of the sampling distribution of percentages is π, i.e., $\mu_p = \pi$. The standard error of the sampling distribution of means, denoted σ_p, is $\sqrt{\dfrac{\pi(100 - \pi)}{n}}$. With percentages, it is impossible that the population is normal. But the Central Limit Theorem also applies to percentages. If the sample size n is sufficiently large, then the sampling distribution of p is approximately normal. However, what constitutes a large sample is no longer $n > 30$. For this type of data, attribute data, a large sample is where both np and $n(100 - p)$ are at least 500. In the large sample case, we can use the normal tables to solve probability problems about p.

Key Terms & Formulas

Sampling Distribution of Percentages   The distribution of the sample percentages from all possible samples of size n taken from a population.

Mean of the Sampling Distribution of Percentages   Denoted μ_p, it is the average of the values of p calculated from all possible samples taken from a population. It is equal to the population percentage, i.e., $\mu_p = \pi$.

Standard Error of the Sampling Distribution of Percentages   Denoted σ_p, it is the standard deviation of the percentages from all possible samples taken from a population. It is equal to $\sqrt{\dfrac{\pi(100 - \pi)}{n}}$.

Central Limit Theorem for Percentages   If the sample size n is such that both np and $n(100 - p)$ are at least 500, then the sampling distribution of p is approximately normal.

Worked Examples

Large Samples for Percentages

Which of the following would be a large enough sample to ensure that the sample percentage would be approximately normal?

1. $n = 10$, $p = 50\%$
2. $n = 50$, $p = 10\%$
3. $n = 50$, $p = 5\%$
4. $n = 50$, $p = 95\%$

Solution

1. If $n = 10$ and $p = 50\%$, then $np = 10(50) = 500$ and $n(100 - p) = 10(100 - 50)500$. Since both are 500, this is a large sample.
2. For $n = 50$ and $p = 10\%$, $np = 50(10) = 500$ and $n(100 - p) = 50(100 - 10) = 4500$. Since both are at least 500, this is a large sample.
3. With $n = 50$ and $p = 5\%$, we have $np = 50(5) = 250$ and $n(100 - p) = 50(100 - 95) = 4750$. Since $250 < 500$, this is not a large sample.
4. Now with $n = 50$, $p = 95\%$, we have $np = 50(95) = 4750$ and $n(100 - p) = 50(100 - 95) = 250$. Since $250 < 500$, this is not a large sample.

   Caution: There is a tendency to check only np to see if the sample size is large when dealing with percentages. As 4 above suggests, it is important to check both np and $n(100 - p)$.

Preteen Smoking

A doctor is concerned with the percentage of teenagers who smoke cigarettes. She plans to take a sample of 100 teenagers and find out if they are smokers. Suppose the true percentage of teenage smokers is 10%. Calculate the mean and standard error of the sampling distribution of p. What is the probability that the sample percentage will be within 3% of the population percentage? What is the probability that the sample percentage will be over 4%? Less than 15%?

Solution

First we calculate the mean and standard error of p.

$$\mu_p = \pi = 10\% \text{ and } \sigma_p = \sqrt{\frac{\pi(100 - \pi)}{n}} = \sqrt{\frac{10(100 - 10)}{100}} = \sqrt{\frac{900}{100}} = 3\%.$$

Because $np = 1000$ and $n(100 - p) = 9000$ are both greater than 500, the sample size is large and p is approximately normal. Therefore, we can use the normal table to calculate probabilities about p. To find the probability that p is within 3% of π, we calculate the probability that p is between 7 and 13%.

$$P(7 < p < 13) = p\left(\frac{7 - 10}{3} < \frac{p - \mu_p}{\sigma_p} < \frac{13 - 10}{3}\right) = P(-1.00 < z < 1.00)$$
$$= 0.3413 + 0.3413 = 0.6826.$$

The remaining two questions are straightforward.

$$P(p > 4) = P\left(\frac{p - \mu_p}{\sigma_p} > \frac{4 - 10}{3}\right) = P(z > -2.00) = 0.5 + 0.4772 = 0.9772.$$

$$P(p < 15) = P\left(\frac{p - \mu_p}{\sigma_p} < \frac{15 - 10}{3}\right) = P(z < 1.67) = 0.5 + 0.4525 = 0.9525.$$

Practice Exercises 6-3

1-5. Which of the following samples would be a large enough to ensure that the sample percentage would be approximately normal?

1. $n = 10$, $p = 25\%$
2. $n = 50$, $p = 20\%$
3. $n = 200$, $p = 2\%$
4. $n = 50$, $p = 80\%$

5. $n = 100, p = 98\%$

6-9. A pollster is gathering people's opinions on whether a shopping center should be built on open range land outside Salt Lake City. Unbeknownst to her, 50% favor the construction of the shopping center. Suppose she samples 400 people.

6. What are the mean and standard error of the sampling distribution of p?

7. What is the probability that her sample percentage will be within 5% of the population percentage?

8. What is the probability that the sample percentage will be over 44%?

9. What is the probability that the sample percentage will be below 48%?

10-13. In New York City, an assistant subdirector aide-de-camp to a vice-muckity-muck in the mayor's office is going to survey 100 residents and ask them if they have used public transportation in the last month. Eighty percent of all residents of New York City use some form of public transportation at least once a month.

10. What are the mean and standard error of the sampling distribution of p?

11. What is the probability that the sample percentage will be within 5% of the population percentage?

12. What is the probability that the sample percentage will be over 90%?

13. What is the probability that the sample percentage will be below 88%?

Solutions to Practice Exercises 6-3

1. $np = 10(25) = 250$ and $n(100 - p) = 10(100 - 25) = 750$. Since $250 < 500$, this is not a large sample.

2. $np = 50(20) = 1000$ and $n(100 - p) = 50(100 - 20) = 4000$. Since both are at least 500, this is a large sample.

3. $np = 200(2) = 400$ and $n(100 - p) = 200(100 - 2) = 19600$. Since $400 < 500$, this is not a large sample.

4. $np = 50(80) = 4000$ and $n(100 - p) = 50(100 - 80) = 1000$. Since both are at least 500, this is a large sample.

5. $np = 100(98) = 9800$ and $n(100 - p) = 100(100 - 98) = 200$. Since $200 < 500$, this is not a large sample.

6. $\mu_p = \pi = 50\%$ and $\sigma_p = \sqrt{\dfrac{\pi(100 - \pi)}{n}} = \sqrt{\dfrac{50(100 - 50)}{400}} = \sqrt{\dfrac{2500}{400}} = 2.5.$

7. To find the probability that p is within 5% of π, we calculate the probability that p is between 45 and 55%.

$$P(45 < p < 55) = P\left(\frac{45 - 50}{2.5} < \frac{p - \mu_p}{\sigma_p} < \frac{45 - 50}{2.5}\right) = P(-2.00 < z < 2.00)$$
$$= 0.4772 + 0.4772 = 0.9544.$$

8. $P(p > 44) = P\left(\dfrac{p - \mu_p}{\sigma_p} > \dfrac{44 - 50}{2.5}\right) = P(z > -2.40) = 0.5 + 0.4893 = 0.9893.$

9. $P(p < 48) = P\left(\dfrac{p - \mu_p}{\sigma_p} < \dfrac{48 - 50}{2.5}\right) = P(z < -0.80) = 0.5 - 0.2881 = 0.2119.$

10. $\mu_p = \pi = 80\%$ and $\sigma_p = \sqrt{\dfrac{\pi(100 - \pi)}{n}} = \sqrt{\dfrac{80(100 - 80)}{100}} = \sqrt{\dfrac{1600}{100}} = 4.$

11. To find the probability that p is within 5% of π, we calculate the probability that p is between 75 and 85%.

$$P(75 < p < 85) = P\left(\frac{75 - 80}{4} < \frac{p - \mu_p}{\sigma_p} < \frac{85 - 80}{4}\right) = P(-1.25 < z < 1.25)$$
$$= 0.3944 + 0.3944 = 0.7888.$$

12. $P(p > 90) = P\left(\dfrac{p - \mu_p}{\sigma_p} > \dfrac{90 - 80}{4}\right) = P(z > 2.50) = 0.5 - 0.4938 = 0.0062.$

13. $P(p < 88) = P\left(\dfrac{p - \mu_p}{\sigma_p} < \dfrac{88 - 80}{4}\right) = P(z < 2.00) = 0.5 + 0.4772 = 0.9772.$

Solutions to Odd-Numbered Exercises

1. This is a single item, so we calculate the z-value as we did in Chapter 5: $z = \dfrac{4.7 - 4.5}{0.7} = 0.29$. Thus $P(x > 4.7) = 0.5000 - P(0 < z < 0.29) = 0.5000 - 0.1141 = 0.3859$.

3. The population is infinite in size because these are items from an assembly line so we calculate the standard error of the mean using Formula 6.4, $\sigma_{\bar{x}} = \dfrac{\sigma}{\sqrt{n}} = \dfrac{0.7}{\sqrt{100}} = 0.07$. Using $\sigma_{\bar{x}}$ in the formula $z = \dfrac{\bar{x} - \mu}{\sigma_{\bar{x}}}$ from page 199 of the text, we find $z = \dfrac{4.7 - 4.5}{0.07} = 2.86$. Thus $P(4.7 < \bar{x}) = 1 - P(0 < z < 2.86)$ $= 0.5000 - 0.4979 = 0.0021$.

5. Again, this is a single item, so we calculate the z-value as in Chapter 5: $z = \dfrac{9 - 10.43}{3.72} = -0.38$. So $P(x < 9) = 0.5000 - P(-0.38 < z < 0) = 0.5000 - P(0 < z < 0.38) = 0.5000 - 0.1480 = 0.3520$.

7. First find $\sigma_{\bar{x}}$ and use it to find z as in Exercise 3. $\sigma_{\bar{x}} = \dfrac{3.72}{\sqrt{32}} = 0.658$, so $z = \dfrac{9 - 10.43}{0.658} = -2.17$. Thus $P(9 < \bar{x}) = P(z < -2.17) = 0.5000 - P(-2.17 < z < 0) = 0.5000 - 0.4850 = 0.0150$.

9. $\mu_{\bar{x}} = \mu = 26$ mg. and using Formula 6.4, $\sigma_{\bar{x}} = \dfrac{6}{\sqrt{100}} = 0.6$ mg.

11. Since we center this region about $z = 0$, we divide the probability by 2 and find the corresponding z-value. Here $\dfrac{0.954}{2} = 0.477$ and 0.477 corresponds to a z-value of 2. Using this z-value with the standard deviation of the sampling distribution from Exercise 9, the sample mean has a 95.4 percent chance of falling between $26 \pm 2(0.6)$ mg., that is between 24.8 mg. and 27.2 mg.

13. Using the technique as in Exercise 11, $\dfrac{0.683}{2} = 0.3415$ and 0.3415 corresponds to a z-value of 1. Using this value with the standard deviation of the sampling distribution of 0.52 days from Exercise 12, the sample mean has a 68.3 percent chance of falling between 12 ± 0.52 days, or between 11.48 days and 12.52 days.

15. From Exercise 11, 95.4 percent corresponds to a z-value of 2. The standard deviation of the sampling distribution is $\sigma_{\bar{x}} = \dfrac{10}{\sqrt{10}} = 3.162$. Thus the sample mean of a sample of size 10 has a 95.4 percent chance of falling between $132 \pm 2(3.162)$ degrees, or between 125.68 degrees and 138.32 degrees.

17. The standard deviation of the sampling distribution for this size of a sample is $\sigma_{\bar{x}} = \dfrac{10}{\sqrt{1000}} = 0.3162$. The sample mean of a sample of size 1000 has a 95.4 percent chance of falling between $132 \pm 2(0.3162)$ degrees, or between 131.37 and 132.63 degrees.

19. Using the same technique as in Exercise 11, 99.7 percent corresponds to a z-value of 3 and $\sigma_{\bar{x}} = \dfrac{5}{\sqrt{25}} = 1$. Thus the sample mean of a sample of size 25 has a 99.7 percent chance of having a BMI between $24 \pm 3(1)$, that is a BMI between 21 and 27.

21. Using a z-value of 3 from Exercise 19 and $\sigma_{\bar{x}} = \dfrac{5}{\sqrt{81}} = 0.556$, the sample mean of a sample of size 81 has a 99.7 percent chance of having a BMI between $24 \pm 3(0.556)$, or a BMI between 22.33 and 25.67.

23. Here we are asked to find $P(|\bar{x} - \mu_{\bar{x}}| < 0.05)$. Determine $\sigma_{\bar{x}}$ and the corresponding z-value; $\sigma_{\bar{x}} = \dfrac{0.16}{\sqrt{50}} = 0.0226$ and $z = \pm\dfrac{0.05}{0.0226} = \pm 2.21$. Thus $P(|\bar{x} - \mu_{\bar{x}}| < 0.05) = P(|z| < 2.21) = 2 \cdot P(0 < z < 2.21) = 2 \cdot 0.4864 = 0.9728$.

25. Here we are asked to find $P(\bar{x} < 90)$. Calculating $\sigma_{\bar{x}}$ and then z, we obtain: $\sigma_{\bar{x}} = \dfrac{18}{\sqrt{12}} = 5.196$ and
$z = \dfrac{5}{5.196} = 0.96$. So,
$P(\bar{x} < 90) = P(z < 0.96) = 0.5000 + P(0 < z < 0.96) = 0.5000 + 0.3315 = 0.8315$.

27. Determine σ_p and use to compute the corresponding z. $\sigma_p = \sqrt{\dfrac{0.67 \cdot 0.37}{174}} = 0.0356 = 3.56$ percent and
$z = \pm\dfrac{p - \mu_{\bar{x}}}{\sigma_p} = \pm\dfrac{0.02}{0.0356} = \pm 0.56$. Thus
$P(|\bar{x} - \mu_{\bar{x}}| < 0.02) = 2 \cdot P(0 < z < 0.56) = 2 \cdot 0.2123 = 0.4246$.

29. Determine $\sigma_{\bar{x}}$ and use it to calculate z_1 and z_2 and then to find the desired probability. $\sigma_{\bar{x}} = \dfrac{17}{\sqrt{37}} = 2.795$.
So $z_1 = \dfrac{60 - 62}{2.795} = -0.72$ and $z_2 = \dfrac{70 - 62}{2.795} = 2.86$.
$P(60 < \bar{x} < 70) = P(-0.72 < z < 2.86) = P(-0.72 < z < 0) + P(0 < z < 2.86)$
$= 0.2642 + 0.4979 = 0.7621$.

31. Determine σ_p and use to compute the corresponding z. $\sigma_p = \sqrt{\dfrac{0.46 \cdot 0.54}{25}} = 0.0997 = 9.97$ percent.
$z = \dfrac{0.50 - 0.46}{0.0997} = 0.40$. Thus
$P(\bar{x} > 50) = P(z > 0.40) = 0.5000 - P(0 < z < 0.40) = 0.5000 - 0.1554 = 0.3446$.

33. Determine $\sigma_{\bar{x}}$ and use it to calculate z_1 and z_2 and then to find the desired probability.
$\sigma_{\bar{x}} = \dfrac{\$98}{\sqrt{45}} = \$14.61$. So $z_1 = \dfrac{375 - 406}{14.61} = -2.12$ and $z_2 = \dfrac{450 - 406}{14.61} = 3.01$.
$P(375 < \bar{x} < 450) = P(-2.12 < z < 3.01) = P(-2.12 < z < 0) + P(0 < z < 3.01)$
$= 0.4830 + 0.4987 = 0.9817$.

35. Determine $\sigma_{\bar{x}}$ and use it to calculate z_1 and z_2 and then to find the desired probability.
$\sigma_{\bar{x}} = \dfrac{13.7}{\sqrt{42}} = 2.11$. So $z_1 = \dfrac{25 - 30.6}{2.11} = -2.65$ and $z_2 = \dfrac{35 - 30.6}{2.11} = 2.08$.
$P(25 < \bar{x} < 35) = P(-2.65 < z < 2.08) = P(-2.65 < z < 0) + P(0 < z < 2.08)$
$= 0.4960 + 0.4812 = 0.9772$.

37. In Exercise 11 we found that 95.4 percent corresponds to $z = \pm 2$. Now $\sigma_{\bar{x}} = \dfrac{\$5,000}{\sqrt{32}} = \$883.88$. Thus 95.4 percent of the average of 32 creative directors salary lie between $\$85,000 \pm 2(\$883.88) = \$83,232.23$ to $\$86,767.77$.

Chapter 7 Estimating Parameters

Study Aids and Practice Exercises

7-1 Estimate, Estimation, Estimator, et Cetera

Study Objectives
You should be able to:
1. Understand the difference between an estimate and an estimator.
2. Define the purpose of estimation.
3. Distinguish between point and interval estimation.

Section Overview
One of the main techniques in statistics is estimation. In estimation, we are trying to provide a reasonable guess for the value of a population parameter. For example, we might want to know the mean response time to a 911 call or the proportion of cases of malnutrition among third world nations. Being unable or unwilling to do a census of the population, we use a sample to produce an estimate of the population parameter. The sample statistic used to supply the estimate is the estimator. For example, we might decide to sample the response times to twenty-five 911 calls. From this data we could calculate the sample mean and use it as an estimate of the population mean. In this example, the estimator is the sample mean, the number it generates from the sample, say 12 minutes, is the estimate.

We want our estimators to be unbiased. An unbiased estimator is one that, on the average, is correct. More formally, an unbiased estimator is a statistic with a sampling distribution that has a mean equal to the population parameter being estimated.

A point estimate is a single number used to estimate a parameter. One problem with a point estimate is that it is usually wrong. While it might be close the population parameter, it is unlikely that a point estimate will be exactly correct, i.e., equal to the population parameter. So as an alternative to point estimates, we have interval estimates. Interval estimates do not try to estimate intervals. Like point estimates, they are estimates of population parameters. The difference is interval estimates give a spread of values or an interval that is likely to contain the population parameter. This likeliness comes from using methods that have known probabilities of yielding correct intervals.

Key Terms & Formulas
Estimate An estimate is the numerical value of a statistic (such as a sample mean, sample percentage, or sample variance) used as our best guess for the value of a parameter (such as a population mean, population percentage or population variance).

Estimator An estimator is a statistic (such as a sample mean, sample percentage, or sample variance) used to provide an estimate for the value of a parameter (such as a population mean, population percentage or population variance).

Hint: These two definitions sound just about the same. The difference is best explained by an example. Suppose we want to estimate μ, the mean of a population. Out best estimator is the statistic $\bar{x} = \dfrac{\Sigma x}{n}$. Once we take a random sample, we can use this formula to calculate a number, say 10, which is our estimate of μ. So the estimator is the statistic $\bar{x}$ and the estimate is the value of the sample mean, 10.

Unbiased Estimator An estimator whose sampling distribution has a mean equal to the population parameter being estimated.

Hint: We have previously looked at the sampling distributions of $\bar{x}$ and p. We discovered that $\mu_{\bar{x}} = \mu$ and $\mu_p = \pi$, so that $\bar{x}$ is an unbiased estimator of μ and p is an unbiased estimator of π.

Estimation The process of using an estimator to provide an estimate of a population parameter.

Point Estimate A single number used to estimate a population parameter.

Point Estimation The process of estimating a population parameter with a single number.

Hint: Point estimation sounds reasonable--what could make more sense than to use one number, a point estimate, to provide a guess for another individual number, a population parameter? The problem is that the probability that the guess is perfectly correct, i.e., exactly equal to the value of the parameter, is exceedingly small (in most practical cases, 0). We, of course, hope that our point estimate is "close," i.e., nearly equal to the population parameter. The meaning of the word close is troublesome. One person who says he is close to being done with a project means he has one more sentence to write. A second person saying he is close means that he has a few days of work left. So we want to know what it means if it is said that an estimate is close to the population parameter. Point estimates does not provide this information but interval estimates, our next topic, do.

Interval Estimate A spread of values or interval used to estimate a population parameter.

Interval Estimation The process of estimating a population parameter with a spread of values or interval.

Hint: An interval estimate's starting point (no pun intended) is a point estimate. Once we obtain the point estimate, we add and subtract an amount based on the sampling error of the statistic. This creates an interval intended to enclose the population parameter.

Worked Examples

IRS Help

A few years ago, the IRS received some bad publicity about the quality of information they were providing to phone inquiries. IRS employees were giving incorrect answers to questions coming from people trying to complete their tax returns. Suppose we want to see if the situation has changed. We call the IRS with 100 questions for which we have the correct answers and see how the agents respond. We find that 38% of the answers were incorrect. For this scenario, what is the parameter, sample, estimator, and estimate?

Solution

To identify the parameter, we must first know the population. Here the population is the answers given by IRS to various questions (not just the 100). The parameter is the percent of incorrect answers that would be given by IRS employees. The sample is the 100 IRS agents queried. The estimator is the sample proportion and the estimate is the value calculated for the estimator, 38%.

Practice Exercises 7-1

1-4. The manager of a department store wants to know the mean time customers wait for service. She observes the times 22 customers wait for service. She calculates the mean for these 22, and it is 4 minutes, 27 seconds.
   1. What is the parameter?
   2. What is the sample?
   3. What is the estimator?
   4. What is the estimate?

5-8. A dietitian is interested in the percentage of people who could explain the dietary recommendations of the food pyramid. Interviewing 120 people selected at random from the phone book, he finds that 18 could explain the food pyramid.
   5. What is the parameter?
   6. What is the sample?
   7. What is the estimator?
   8. What is the estimate?

Solutions to Practice Exercises 7-1

1. The parameter is the mean time customers have to wait for service.
2. The 22 customers.
3. The sample mean.
4. The value of the sample mean, four minutes and twenty-seven seconds.
5. The percentage of people who could explain the food pyramid dietary recommendations.
6. The 120 people selected at random from the phone book.
7. The sample percentage.
8. The value of the sample percentage, $\frac{18}{120} = 15\%$.

7-2 Interval Estimation of the Population Mean: Some Basic Concepts

Study Objectives

You should be able to:
1. Understand what a confidence interval is.
2. Recognize the difference between probability and confidence.

Section Overview

In this section we discuss estimation of a population mean. Many ideas we talk about in this section will carry over into other estimation situations, including the estimation of a population percentage and a population variance.

We would like to estimate the value of μ, the mean of a population. Because we want an indication of how precise our estimate is, we choose to use interval estimates. The interval estimate is based on the sample mean, $\overline{x}$. Because of this, the sampling distribution of means is involved in the creation of the interval estimate. In those cases where the sample mean is normal or approximately normal, the interval estimate of μ is: $\overline{x} - z_{\alpha/2}\sigma_{\overline{x}} < \mu < \overline{x} + z_{\alpha/2}\sigma_{\overline{x}}$. The value of z depends on the confidence coefficient $(1 - \alpha)$ or level of confidence we chose. Typically 90, 95, or 99% confidence is chosen. While 99% confidence sounds better than either of the other two, it leads to wider, i.e., less precise, interval estimates. Once we have selected our confidence level and calculated our interval estimate, it is called a confidence interval. The amount we add and subtract, in the normal case $z_{\alpha/2}\sigma_{\overline{x}}$, is denoted E and called the error bound or the maximum error of estimate.

Key Terms & Formulas

Interval Estimate of μ If the sampling distribution of $\overline{x}$ is normal, an interval estimate of μ is given by:
$\overline{x} - z_{\alpha/2}\sigma_{\overline{x}} < \mu < \overline{x} + z_{\alpha/2}\sigma_{\overline{x}}$, where $z_{\alpha/2}$ is a standard normal value determined by the confidence coefficient.

Level of Confidence (Confidence Coefficient) The probability that the interval estimator produces an interval estimate that includes the population parameter being estimated.

Hint: There is a probability for the interval estimator being correct. The estimator is the formula that yields correct answers at a certain rate equal to the confidence coefficient. However, there is not a probability for an interval estimate being correct. Any one interval estimate is either right or wrong, it is not correct a certain percent of the time. For example, suppose your statistics instructor says that she gives out grades randomly (bad news). Also suppose that she says she randomly selects 95% of the class to receive an A (good news) and the remaining 5% will get an F (bad news, but probably won't happen to you). Before the grades are assigned, you can be 95% confident that you will receive an A. This is equivalent to the confidence we have that an interval estimator will yield a correct answer. Now the instructor gives out her grades to the 40 people in her class. Jay Devore and Roxy Peck receive F's, the rest of the class receives A's. These grades are like 40 different interval estimates. Thirty-eight of them are A's (similar to an interval estimate containing the parameter being estimated), and two of them are F's (similar to an interval estimate missing the parameter). Your grade is an A. It is not an A 95% of the time, it is an A. You were confident that you would get an A because the success rate was high, and you were right. However, you should recognize that Jay and Roxy probably thought the same thing and ended up being wrong. But most of the time (38 out of 40), the results were satisfactory.

Confidence Intervals Intervals estimates based on specified confidence levels.

Confidence Limits The upper and lower limits of a confidence interval.

Maximum Error of Estimate (or Error Bound) Designated E, the maximum error of estimate is the amount added and subtracted to a point estimate to create the confidence interval. When estimating μ in cases where the sampling distribution of the sample mean is normal or approximately normal, $E = z_{\alpha/2}\sigma_{\bar{x}}$.

Hint: Since a confidence interval is formed by adding and subtracting the amount E, the width of the confidence interval is twice the value of E.

Worked Examples

Plants In Space

Scientists sent forty seeds on a recent space shuttle flight. During the flight, a crew member recorded their germination times. The scientists found that the mean germination time was 7.2 days. They were also able to be 90% confident that the mean germination time for all such seeds is somewhere between 6.8 and 7.6 days. What are the point estimate, the confidence interval, the confidence level and the maximum error of estimate?

Solution

The point estimate is the calculated value of the statistic, 7.2 days. The confidence interval is from 6.8 and 7.6 days. The confidence level is 90%. The maximum error of estimate is half the width of the confidence interval, so

$$E = \frac{7.6 - 6.8}{2} = 0.4 \text{ days.}$$

Practice Exercises 7-2

1-4. The manager of a department store, who wanted to know the mean time customers waited for service, sampled 22 customers and calculated a mean wait time of 4 minutes, 27 seconds. Using this information, she is 95% confident that the population mean is between 4 minutes, 2 seconds and 4 minutes, 52 seconds. What is the

1. point estimate?
2. confidence interval?
3. confidence level?
4. maximum error of estimate?

5-8. The dietitian interested in the percentage of people who could explain the food pyramid interviewed 120 people and found that 15% could explain it. Using that information, the dietitian could be 99% sure that the population percentage is between 7% and 23%. What is the

5. point estimate?
6. confidence interval?
7. confidence level?
8. maximum error of estimate?

Solutions to Practice Exercises 7-2

1. The point estimate is the value of the sample mean, 4 minutes and 27 seconds.
2. The confidence interval is from 4 minutes 2 seconds to 4 minutes 52 seconds.
3. 95%.
4. The error of estimate is half the distance from 4 minutes 2 seconds to 4 minutes 52 seconds, i.e., it is half of 50 seconds, 25 seconds.
5. The point estimate is the value of the sample percentage, 15% seconds.
6. The confidence interval is from 7% to 23%.
7. 99%.
8. The error of estimate is half the distance from 7% to 23%, $\frac{23 - 7}{2} = 8\%$.

7-3 Estimating the Population Mean

Study Objectives
You should be able to:
    Calculate a confidence interval for μ when $\bar{x}$ is (approximately) normal.

Section Overview
    There are several methods for calculating a confidence interval for a population mean. The method that is appropriate depends on the sample size, the distribution of the population, and whether the population standard deviation is known. If either the sample size is over 30 or the population is normal with a known standard deviation, the confidence intervals are based on the normal distribution. If we have a small sample from a normal population with σ unknown, then the Student t distribution is the basis of the intervals. There are $n-1$ degrees of freedom associated with the latter confidence interval.

Key Terms & Formulas
Standard Error of the Mean   For an infinite population $\sigma_{\bar{x}} = \dfrac{\sigma}{\sqrt{n}}$.

Confidence Interval for μ, σ known and $n > 30$ $\bar{x} - z_{\alpha/2}\sigma_{\bar{x}} < \mu < \bar{x} + z_{\alpha/2}\sigma_{\bar{x}}$.
    Hint: The ends of the confidence interval are called the lower confidence limit (LCL) and the upper confidence limit (UCL).
Estimated Standard Error   This is the estimate of the standard error of the mean that we use in inferences when σ is unknown: $\hat{\sigma}_{\bar{x}} = \dfrac{s}{\sqrt{n}}$.

Confidence Interval for μ; σ unknown and $n > 30$ $\bar{x} - z_{\alpha/2}\hat{\sigma}_{\bar{x}} < \mu < \bar{x} + z_{\alpha/2}\hat{\sigma}_{\bar{x}}$.

Student's t Distribution   The sampling distribution of the means when a small sample is taken from a normal population when σ is not known. Similar in form to a standard normal, i.e., to $z = \dfrac{\bar{x} - \mu}{\sigma/\sqrt{n}}$, its form is

$$t = \frac{\bar{x} - \mu}{s/\sqrt{n}}.$$

Confidence Interval for μ; σ unknown, population normal, and $n < 30$ $\bar{x} - t_{\alpha/2}\hat{\sigma}_{\bar{x}} < \mu < \bar{x} + t_{\alpha/2}\hat{\sigma}_{\bar{x}}$.

Degrees of Freedom   A characteristic of a variety of statistical distributions including the t distribution.
    Hint: The degrees of freedom must be known to use the t table. For different uses of the t table, the degrees of freedom will change. For the confidence interval for the mean in this section, the degrees of freedom are $n-1$.

Worked Examples

Flying Eyes
An Air Force trainer is interested in the mean test score achieved by flight candidates on a test of visual perception. This test has a known standard deviation of 75. A random sample of 55 flight candidates yielded a mean of 304. Construct a 90% confidence interval for the mean test score.

Solution
Because we have a large sample and σ is known, we can calculate a confidence interval using the formula

$\bar{x} - z_{\alpha/2}\sigma_{\bar{x}} < \mu < \bar{x} + z_{\alpha/2}\sigma_{\bar{x}}$. We begin by calculating the standard error: $\sigma_{\bar{x}} = \dfrac{\sigma}{\sqrt{n}} = \dfrac{75}{\sqrt{55}} = 10.11$.

With a 90% confidence interval, the z value equals 1.645. (This can be found by looking for $\dfrac{0.90}{2} = 0.45$ in the probability portion of the standard normal table.) Thus the confidence interval is:

$$\bar{x} - z_{\alpha/2}\sigma_{\bar{x}} < \mu < \bar{x} + z_{\alpha/2}\sigma_{\bar{x}}$$
$$304 - (1.645)(10.11) < \mu < 304 + (1.645)(10.11)$$
$$304 - 16.6 < \mu < 304 + 16.6$$
$$287.4 < \mu < 320.6$$

Des Moines Debts

The manager of a mortgage company in Des Moines would like to estimate the mean balance on homeowners' first deeds of trust. She randomly selects 49 homeowners and learns that the average balance is $120,000$ with a standard deviation of $35,000$. Construct a 99% confidence interval for μ.

Solution

Because we have a large sample but σ is unknown, we will calculate a confidence interval using the formula $\bar{x} - z_{\alpha/2}\hat{\sigma}_{\bar{x}} < \mu < \bar{x} + z_{\alpha/2}\hat{\sigma}_{\bar{x}}$. We first estimate the standard error:

$$\hat{\sigma}_{\bar{x}} = \frac{s}{\sqrt{n}} = \frac{35,000}{\sqrt{49}} = 5,000$$

For 99% confidence, the z value equals 2.58. (This is found by looking for $\frac{0.99}{2} = 0.495$ in the probability portion of the normal table.) Then the confidence interval is:

$$\bar{x} - z_{\alpha/2}\hat{\sigma}_{\bar{x}} < \mu < \bar{x} + z_{\alpha/2}\hat{\sigma}_{\bar{x}}$$
$$120,000 - (2.58)(5,000) < \mu < 120,000 + (2.58)(5,000)$$
$$120,000 - 12,900 < \mu < 120,000 + 12,900$$
$$\$107,100 < \mu < \$132,900$$

Cereal

A nutritionist is interested in the amount of carbohydrates in breakfast cereals. He samples 8 different packages of cereal and finds that the amounts of carbohydrates in a 1-oz. serving are (in grams):

$$24 \quad 26 \quad 23 \quad 20 \quad 24 \quad 25 \quad 24 \quad 25$$

Calculate a 95% confidence interval for μ, assuming that the amount of carbohydrates in a serving of cereal is normal.

Solution

Because the sample size is small and σ is unknown, we want to be able to use the t distribution to compute the confidence interval. If the amount of carbohydrates in a serving of cereal is normal, this is a valid approach. We will calculate the confidence interval using the formula $\bar{x} - t_{\alpha/2}\hat{\sigma}_{\bar{x}} < \mu < \bar{x} + t_{\alpha/2}\hat{\sigma}_{\bar{x}}$.

Hint: If the assumption of normality is not appropriate, our alternative is to use a nonparametric confidence interval. We will consider some nonparametric techniques in the last chapter.

The first step we will undertake is the calculation of s. As a reminder, there are two methods for calculating s. To provide a review, we will calculate s using both methods. The first formula we have for s is the original definition: $s = \sqrt{\dfrac{\Sigma(x - \bar{x})^2}{n - 1}}$. We do the calculations in the following table.

x	$x - \bar{x}$	$(x - \bar{x})^2$
24	0.125	0.0156
26	2.125	4.5156
23	−0.875	0.7656
20	−3.875	15.0156
24	0.125	0.0156
25	1.125	1.2656
24	0.125	0.0156
25	1.125	1.2656
$\Sigma x = 191$		$\Sigma(x - \bar{x})^2 = 22.875$

Using the first column in the table, we compute the sample mean: $\bar{x} = \dfrac{191}{8} = 23.875$. With this value, we can compute the remainder of the table. Then we can use the total of the last column to calculate the sample standard deviation: $s = \sqrt{\dfrac{\Sigma(x - \bar{x})^2}{n - 1}} = \sqrt{\dfrac{22.875}{8 - 1}} = \sqrt{3.26786} = 1.80772 \approx 1.808$. Alternatively, we can use

$s = \sqrt{\dfrac{n(\Sigma x^2) - (\Sigma x)^2}{n(n - 1)}}$ to calculate the sample standard deviation. We again set up the calculations in a table.

x	x^2
24	576
26	676
23	529
20	400
24	576
25	625
24	576
25	625
$\Sigma x = 191$	$\Sigma x^2 = 4583$

With the two totals in this table and knowing $n = 8$, we calculate the sample standard deviation

$$s = \sqrt{\frac{n(\Sigma x^2) - (\Sigma x)^2}{n(n-1)}} = \sqrt{\frac{8(4583) - (191)^2}{8(8-1)}} = \sqrt{\frac{183}{56}} = \sqrt{3.26786} \approx 1.808. \text{ Whatever the method used to}$$

calculate the standard deviation, we use it to estimate the standard error: $\hat{\sigma}_{\bar{x}} = \dfrac{s}{\sqrt{n}} = \dfrac{1.808}{\sqrt{8}} = 0.639$. To find the

appropriate t table value, we need the degrees of freedom. They are $n - 1 = 8 - 1 = 7$. In addition, to use the t table, we must have $\alpha/2$. Because $\alpha = 0.05$, $\alpha/2 = 0.025$. So we look in the t table at the df $= 7$ row and the 0.025 column, finding 2.365. Then the confidence interval is:

$$\bar{x} - t_{\alpha/2}\hat{\sigma}_{\bar{x}} < \mu < \bar{x} + t_{\alpha/2}\hat{\sigma}_{\bar{x}}$$
$$23.875 - (2.365)(0.639) < \mu < 23.875 + (2.365)(0.639)$$
$$23.875 - 1.512 < \mu < 23.875 + 1.512$$
$$22.363 < \mu < 25.387$$

Hint: Using a statistical package would simplify this process. Typically, using a package entails entering the data and providing the commands that ask for the required analysis. For example, MINITAB has a spreadsheet-like data editor into which we would type the eight carbohydrate numbers into "c1." Immediately after that, we could type the following command and receive the subsequent output:

```
MTB > tint c1
Variable      N      Mean     StDev   SE Mean        95.0 % CI
C1            8     23.875    1.808    0.639    (  22.363,   25.387)
```

Practice Exercises 7-3

1. Florida tax officials want to estimate the mean amount bet at the Ocala Jai Alai fronton. Based on tax records, they figure out that the standard deviation has consistently been around $18,000, and they are willing to assume that it is still the same. They sample 40 days and learn that the mean amount bet on those days was $194,034.55. Calculate a 99% confidence interval for μ.

2. A quality control engineer is interested in the mean length of sheet insulation being produced by an automated process. The machine involved is designed to cut the insulation into 50-foot lengths, but there is some variation in the actual lengths. In fact, the standard deviation in the cutting length is 0.08 feet. A sample of 60 sheets yielded a mean length of 50.111 feet. Calculate an 80% confidence interval for μ.

3. A high school counselor is coordinating evening college preparatory classes. She wants to estimate the mean GPA of all students in such classes. A sample of 45 students has a mean GPA of 3.21 and a standard deviation of 0.39. Compute a 95% confidence interval for μ.

4. A container is designed to be able to handle a maximum weight of 250 lb. The head of a shipping department wants to estimate the mean weight of materials being placed in this type container. He samples 60 containers and finds that the average weight of the material in these containers is 272.3 lb. with a standard deviation of 57.3 lb. Obtain a 90% confidence interval for the mean weight of materials being placed into this type container.

5. To become an actuary, it is necessary to pass a series of 10 exams, including the most important one, an exam in probability and statistics. An insurance company wants to estimate the mean score on this exam for actuarial students who have enrolled in a special study program. They take a sample of 11 actuarial students in this program and determine their scores:

$$2 \; 5 \; 8 \; 8 \; 7 \; 6 \; 5 \; 7 \; 9 \; 5 \; 6.$$

Compute a 95% confidence interval for μ.

6.   It takes a variable amount of time for a Statistics professor to bicycle from Los Osos to San Luis Obispo. Four instances when he managed to remember to time himself on the trip, the times were 55, 42, 47, and 54 minutes. Calculate a 90% confidence interval for the mean amount of time it takes him to make the trip.

Solutions to Practice Exercises 7-3

1.   Because we have a large sample and σ is known, we will calculate the confidence interval using the formula $\overline{x} - z_{\alpha/2}\sigma_{\overline{x}} < \mu < \overline{x} + z_{\alpha/2}\sigma_{\overline{x}}$. First we calculate the standard error:

$$\sigma_{\overline{x}} = \frac{\sigma}{\sqrt{n}} = \frac{18,000}{\sqrt{40}} \approx 2,846.05.$$ For a 99% confidence interval, we look in the standard normal table for a

z value with a table probability of $\dfrac{0.99}{2} = 0.495$. This z value is 2.58. Then the confidence interval is:

$$\overline{x} - z_{\alpha/2}\sigma_{\overline{x}} < \mu < \overline{x} + z_{\alpha/2}\sigma_{\overline{x}}$$
$$194{,}034.55 - (2.58)(18{,}000) < \mu < 194{,}034.55 + (2.58)(18{,}000)$$
$$194{,}034.55 - 7{,}342.81 < \mu < 194{,}034.55 + 7{,}342.81$$
$$\$186{,}691.74 < \mu < \$201{,}377.36.$$

2.   Because we have a large sample and σ is known, we will use the formula $\overline{x} - z_{\alpha/2}\sigma_{\overline{x}} < \mu < \overline{x} + z_{\alpha/2}\sigma_{\overline{x}}$. The

standard error: $\sigma_{\overline{x}} = \dfrac{\sigma}{\sqrt{n}} = \dfrac{0.08}{\sqrt{60}} \approx 0.010328.$ For an 80% confidence interval, we look in the standard

normal table for a z value with a table probability of $\dfrac{0.80}{2} = 0.40$. This z value is 1.28. Then the confidence interval is:

$$\overline{x} - z_{\alpha/2}\sigma_{\overline{x}} < \mu < \overline{x} + z_{\alpha/2}\sigma_{\overline{x}}$$
$$50.111 - (1.28)(.010328) < \mu < 50.111 + (1.28)(.010328)$$
$$50.111 - 0.013 < \mu < 50.111 + 0.013$$
$$50.098 < \mu < 50.124.$$

3.   Because we have a large sample but σ is unknown, we calculate a confidence interval using the formula

$\overline{x} - z_{\alpha/2}\widehat{\sigma}_{\overline{x}} < \mu < \overline{x} + z_{\alpha/2}\widehat{\sigma}_{\overline{x}}$. We begin by estimating the standard error. $\widehat{\sigma}_{\overline{x}} = \dfrac{s}{\sqrt{n}} = \dfrac{0.39}{\sqrt{45}} \approx 0.0503.$

For a 95% confidence interval, we look in the standard normal table for a z value with a probability of

$\dfrac{0.95}{2} = 0.475$. This z value is 1.96. The confidence interval is:

$$\overline{x} - z_{\alpha/2}\sigma_{\overline{x}} < \mu < \overline{x} + z_{\alpha/2}\sigma_{\overline{x}}$$
$$3.21 - (1.96)(0.0503) < \mu < 3.21 + (1.96)(0.0503)$$
$$3.21 - 0.099 < \mu < 3.21 + 0.099$$
$$3.111 < \mu < 3.309.$$

4.   We have a large sample but σ is unknown, so we will use $\overline{x} - z_{\alpha/2}\widehat{\sigma}_{\overline{x}} < \mu < \overline{x} + z_{\alpha/2}\widehat{\sigma}_{\overline{x}}$ to obtain the

confidence interval. First we calculate: $\widehat{\sigma}_{\overline{x}} = \dfrac{s}{\sqrt{n}} = \dfrac{57.3}{\sqrt{60}} \approx 7.397.$ For a 90% confidence interval, we

look in the standard normal table for a z value with a probability of $\dfrac{0.90}{2} = 0.45$. This z value is 1.645. The

confidence interval is:

$$\overline{x} - z_{\alpha/2}\widehat{\sigma}_{\overline{x}} < \mu < \overline{x} + z_{\alpha/2}\widehat{\sigma}_{\overline{x}}$$
$$272.3 - (1.645)(7.397) < \mu < 272.3 + (1.645)(7.397)$$
$$272.3 - 12.2 < \mu < 272.3 + 12.2$$
$$260.1 < \mu < 284.5$$

5.   Because the sample size is small and σ is unknown, we will use the t distribution to compute the confidence interval. For this to be valid, we must believe that the scores on the actuarial exams are normal. Then we calculate the confidence interval using the formula $\overline{x} - t_{\alpha/2}\widehat{\sigma}_{\overline{x}} < \mu < \overline{x} + t_{\alpha/2}\widehat{\sigma}_{\overline{x}}$. First we calculate s. Both

formulas are demonstrated, starting with the definition: $s = \sqrt{\dfrac{\Sigma(x - \overline{x})^2}{n - 1}}.$

x	$x - \bar{x}$	$(x - \bar{x})^2$
2	−4.18182	17.4876
5	−1.18182	1.3967
8	1.81818	3.3058
8	1.81818	3.3058
7	0.81818	0.6694
6	−0.18182	0.0331
5	−1.18182	1.3967
7	0.81818	0.6694
9	2.81818	7.9421
5	−1.18182	1.3967
6	−0.18182	0.0331
$\Sigma x = 68$		$\Sigma(x - \bar{x})^2 = 37.6364$

From the first column in the table, we can compute the sample mean: $\bar{x} = \dfrac{68}{11} = 6.18182$. Then using this to obtain the remainder of the table, we take the total of the third column to calculate the sample standard deviation: $s = \sqrt{\dfrac{37.6364}{11 - 1}} = \sqrt{3.76364} = 1.94001 \approx 1.940$. The computational alternative for the sample

standard deviation is: $s = \sqrt{\dfrac{n(\Sigma x^2) - (\Sigma x)^2}{n(n - 1)}}$. We can again do the calculations in a table.

x	x^2
2	4
5	25
8	64
8	64
7	49
6	36
5	25
7	49
9	81
5	25
6	36
$\Sigma x = 68$	$\Sigma x^2 = 458$

With the two totals from this table and knowing $n = 11$, we calculate:

$$s = \sqrt{\dfrac{n(\Sigma x^2) - (\Sigma x)^2}{n(n - 1)}} = \sqrt{\dfrac{11(458) - (68)^2}{11(11 - 1)}} = \sqrt{\dfrac{414}{110}} = \sqrt{3.7636} \approx 1.940.$$ Now we can

estimate the standard error: $\hat{\sigma}_{\bar{x}} = \dfrac{s}{\sqrt{n}} = \dfrac{1.940}{\sqrt{11}} = 0.585$. The degrees of freedom are

$n - 1 = 11 - 1 = 10$. We look in the t table at the df $= 10$ row and the $\dfrac{1 - 0.95}{2} = 0.025$ column, finding 2.228. The confidence interval is:

$$\bar{x} - t_{\alpha/2}\hat{\sigma}_{\bar{x}} < \mu < \bar{x} + t_{\alpha/2}\hat{\sigma}_{\bar{x}}$$
$$6.182 - (2.228)(0.585) < \mu < 6.182 + (2.228)(0.585)$$
$$6.182 - 1.303 < \mu < 6.182 + 1.303$$
$$4.878 < \mu < 7.485.$$

6. Because the sample size is small and σ is unknown, we will use the t distribution to compute the confidence interval. For this to be valid, we must believe that the riding times are normal. Then we calculate the confidence interval using the formula $\bar{x} - t_{\alpha/2}\hat{\sigma}_{\bar{x}} < \mu < \bar{x} + t_{\alpha/2}\hat{\sigma}_{\bar{x}}$.

First we calculate s. We will show how to use both formulas, starting with the definition: $s = \sqrt{\dfrac{\Sigma(x - \overline{x})^2}{n - 1}}$.

x	$x - \overline{x}$	$(x - \overline{x})^2$
55	5.5	30.25
42	7.5	56.25
47	2.5	6.25
54	4.5	20.25
$\Sigma x = 198$		$\Sigma(x - \overline{x})^2 = 113.00$

From the first column in the table, we compute the sample mean: $\overline{x} = \dfrac{198}{4} = 49.5$. Then using this to obtain the remainder of the table, we take the total of the third column to calculate the sample standard deviation:

$s = \sqrt{\dfrac{\Sigma(x - \overline{x})^2}{n - 1}} = \sqrt{\dfrac{113.00}{4 - 1}} = \sqrt{37.6667} \approx 6.137$. The computational

alternative for the sample standard deviation is $s = \sqrt{\dfrac{n(\Sigma x^2) - (\Sigma x)^2}{n(n - 1)}}$. We can again set up the calculations in a table.

x	x^2
55	3025
42	1764
47	2209
54	2916
$\Sigma x = 198$	$\Sigma x^2 = 9,914$

With the two totals from this table and knowing $n = 4$, we calculate:

$s = \sqrt{\dfrac{n(\Sigma x^2) - (\Sigma x)^2}{n(n - 1)}} = \sqrt{\dfrac{4(9914) - (198)^2}{4(4 - 1)}} = \sqrt{\dfrac{452}{12}} = \sqrt{37.6667} \approx 6.137$. Now we can estimate the

standard error. $\hat{\sigma}_{\overline{x}} = \dfrac{s}{\sqrt{n}} = \dfrac{6.137}{\sqrt{4}} = 3.069$. Here the degrees of freedom are $n - 1 = 4 - 1 = 3$. We look

in the t table at the df $= 3$ row and the $\dfrac{1 - 0.90}{2} = 0.05$ column, finding 2.353. The confidence interval is:

$$\overline{x} - t_{\alpha/2}\hat{\sigma}_{\overline{x}} < \mu < \overline{x} + t_{\alpha/2}\hat{\sigma}_{\overline{x}}$$
$$49.50 - (2.353)(3.069) < \mu < 49.50 + (2.353)(3.069)$$
$$49.50 - 7.22 < \mu < 49.50 + 7.22$$
$$42.28 < \mu < 56.72.$$

7-4 Estimating the Population Percentage

Study Objectives
You should be able to:
1. Recognize when it is appropriate to obtain an interval estimate for a population percentage.
2. Calculate a confidence interval for π.

Section Overview
Rather than means, we may be interested in the percentage of times an event occurs. For example, we might want to know the percentage of smokers who have lung disease or the percentage of voters that plan to vote in favor of a school bond. We have used π to designate this population percentage and p to represent its sample counterpart, the sample percentage. In the previous chapter we examined the sampling distribution of p. There we discovered that $\mu_p = \pi$. This means that p is an unbiased estimator of π. We also found that the standard error of p is

$\sigma_p = \sqrt{\dfrac{\pi(100 - \pi)}{n}}$ and that p is approximately normal in the large sample case. (A large sample is np and $n(100 - p)$ both being at least 500.) From these facts we can develop a large sample confidence interval for π that is $p - z_{\alpha/2}\widehat{\sigma}_p < \pi < p + z_{\alpha/2}\widehat{\sigma}_p$. The z table values we use are identical to those we use to calculate a confidence interval for μ. The estimate of the standard error, $\widehat{\sigma}_p$, substitutes p for the value of π.

Key Terms & Formulas

Population Percentage π $\pi =$ the percentage of occurrence of some event in a population.

Sample Percentage p $p =$ the percentage of occurrence of some event in a sample.

Confidence Interval for π If np and $n(100 - p)$ are both at least 500, then a confidence interval for π is obtained by calculating: $p - z_{\alpha/2}\widehat{\sigma}_p < \pi < p + z_{\alpha/2}\widehat{\sigma}_p$.

> Hint: In the previous section, we could calculate a confidence interval if either the sample size was large ($n > 30$) or the sample was taken from a normal population. Here, confidence intervals are valid only if we have a large sample. It is impossible for attribute data, i.e., the occurrence or nonoccurrence of some event, to be normal--the data is not even numerical.

Estimate of the Standard Error of Percentages (or Estimate of the Standard Error of the Sampling Distribution of Percentages) $\widehat{\sigma}_p = \sqrt{\dfrac{p(100 - p)}{n}}$.

Worked Examples

Quality Control

A quality control engineer wants to estimate the percentage of defective computer memory boards coming off an assembly line. In a sample of 200 boards, 28 are defective. Use a 90% confidence interval to estimate the percentage of defective computer memory boards.

Solution

We have $p = \dfrac{28}{200} \times (100 \text{ percent}) = 14\%$, $n = 200$, and the confidence coefficient is 90 percent. We need to estimate the standard error of p:

$$\widehat{\sigma}_p = \sqrt{\frac{p(100 - p)}{n}} = \sqrt{\frac{14(100 - 14)}{200}} = \sqrt{6.02} = 2.45357 \approx 2.45.$$

For a confidence coefficient of 90%, the z value of 1.645 is found in the standard normal table by searching for a probability of $\dfrac{0.90}{2} = 0.45$. So the confidence interval for the percentage of defective computer memory boards coming off an assembly line is:

$$p - z_{\alpha/2}\widehat{\sigma}_p < \mu < p + z_{\alpha/2}\widehat{\sigma}_p$$
$$14 - (1.645)(2.45) < \mu < 14 + (1.645)(2.45)$$
$$14 - 4.0 < \mu < 14 + 4.0$$
$$10.0 < \mu < 18.0.$$

Condos

The owners of an apartment complex in Manhattan want to estimate the percentage of their renters that would buy a unit if the complex converted to condominiums. Of the 54 apartments surveyed, 19 said that they would buy a unit if the complex switched to condominiums. Calculate a 95% confidence interval for π.

Solution

We have $p = \dfrac{19}{54} \times (100 \text{ percent}) = 35.2\%$, $n = 54$, and the confidence coefficient is 95 percent. We estimate the standard error of p:

$$\widehat{\sigma}_p = \sqrt{\frac{p(100 - p)}{n}} = \sqrt{\frac{35.2(100 - 35.2)}{278}} = 2.8644 \approx 2.86.$$

For a confidence coefficient of 95%, the z value of 1.96 is found in the standard normal table by searching for a probability of $\dfrac{0.95}{2} = 0.475$. So the confidence interval for the percentage of renters willing to buy a condominium unit is:

$$p - z_{\alpha/2}\hat{\sigma}_p < \mu < p + z_{\alpha/2}\hat{\sigma}_p$$
$$35.2 - (1.96)(2.86) < \mu < 35.2 - (1.96)(2.86)$$
$$35.2 - 5.6 < \mu < 35.2 + 5.6$$
$$29.6 < \mu < 40.8.$$

Practice Exercises 7-4

1. A prison official at California Men's Colony wants to estimate the percentage of cases of recidivism among inmates. Examining the records of 200 convicts, the official determines that there are 68 cases of recidivism. Estimate the percentage of cases of recidivism with a 99% confidence interval.

2. The head of the biology department at a state college wants to estimate the percentage of entering biology majors who will choose the field of environmental biology as their area of concentration. She samples 60 majors and finds that 18 have chosen environmental biology as their area of concentration. Calculate a 90% confidence interval for π.

3. A sociologist surveyed 150 college graduates and found that 102 believed in some type of afterlife. Compute a 95% confidence interval for π.

4. A philatelist is interested in the proportion of people who know the meaning of the word "philatelist." She takes a sample of 300 people and finds that 99 know that a philatelist is a stamp collector. Use this information to form an 80% confidence interval for π.

Solutions to Practice Exercises 7-4

1. We have $p = \dfrac{68}{200} \times (100 \text{ percent}) = 34\%$, $n = 200$, and the confidence coefficient is 99 percent. We estimate the standard error of p:

$$\hat{\sigma}_p = \sqrt{\frac{p(100 - p)}{n}} = \sqrt{\frac{34(100 - 34)}{200}} = 3.3496 \approx 3.35.$$

For a confidence coefficient of 99%, the z value of 2.58 is found in the standard normal table by searching for a probability of $\dfrac{0.95}{2} = 0.495$. So the confidence interval for the percentage of recidivism at CMC is:

$$p - z_{\alpha/2}\hat{\sigma}_p < \mu < p + z_{\alpha/2}\hat{\sigma}_p$$
$$34 - (2.58)(3.35) < \mu < 34 + (2.58)(3.35)$$
$$34 - 8.6 < \mu < 34 + 8.6$$
$$25.4 < \mu < 42.6.$$

2. We have $p = \dfrac{18}{60} \times (100 \text{ percent}) = 30\%$, $n = 60$, and the confidence coefficient is 90 percent. The standard error of p:

$$\hat{\sigma}_p = \sqrt{\frac{p(100 - p)}{n}} = \sqrt{\frac{30(100 - 30)}{60}} = 5.9161 \approx 5.92.$$

For a confidence coefficient of 90%, the z value of 1.645 is found in the standard normal table by searching for a probability of $\dfrac{0.90}{2} = 0.45$. So the confidence interval for the percentage of entering biology majors that will chose environmental biology as their area of concentration is:

$$p - z_{\alpha/2}\hat{\sigma}_p < \mu < p + z_{\alpha/2}\hat{\sigma}_p$$
$$30 - (1.645)(5.92) < \mu < 30 + (1.645)(5.92)$$
$$30 - 9.7 < \mu < 30 + 9.7$$
$$20.3 < \mu < 39.7.$$

3. We have $p = \dfrac{102}{150} \times (100 \text{ percent}) = 68\%$, $n = 150$, and the confidence coefficient is 95 percent. The estimate of the standard error of p:

$$\hat{\sigma}_p = \sqrt{\frac{p(100-p)}{n}} = \sqrt{\frac{68(100-68)}{150}} = 3.80876 \approx 3.81.$$

For a confidence coefficient of 95%, the z value of 1.96 is found in the standard normal table by searching for a probability of $\frac{0.95}{2} = 0.475$. So the confidence interval for the percentage of college graduates that believe in an after life is:

$$p - z_{\alpha/2}\hat{\sigma}_p < \mu < p + z_{\alpha/2}\hat{\sigma}_p$$
$$68 - (1.96)(3.81) < \mu < 68 + (1.96)(3.81)$$
$$68 - 7.5 < \mu < 68 + 7.5$$
$$60.5 < \mu < 75.5.$$

4. We have $p = \frac{99}{300} \times (100 \text{ percent}) = 33\%$, $n = 300$, and the confidence coefficient is 80 percent. We first estimate the standard error of p:

$$\hat{\sigma}_p = \sqrt{\frac{p(100-p)}{n}} = \sqrt{\frac{33(100-33)}{300}} = 2.71477 \approx 2.71$$

For a confidence coefficient of 80%, the z value of 1.28 is found in the standard normal table by searching for a probability of $\frac{0.80}{2} = 0.40$. So the confidence interval for the percentage of people who know the meaning of philatelist is:

$$p - z_{\alpha/2}\hat{\sigma}_p < \mu < p + z_{\alpha/2}\hat{\sigma}_p$$
$$33 - (1.28)(2.71) < \mu < 33 + (1.28)(2.71)$$
$$33 - 3.5 < \mu < 33 + 3.5$$
$$29.5 < \mu < 36.5.$$

7-5 Estimating the Population Variance

Study Objectives

You should be able to:
1. Recognize the need for a confidence interval for a population variance.
2. Calculate a confidence interval for a population variance.

Section Overview

In several situations, we want to know how consistent some phenomenon is. Does a machine fill packages with nearly the same amount of material? Do the ingredients in a multiple vitamin have a high degree of consistency? Are the threads of a screw consistently within acceptable tolerances? When faced with this type of situation, we want to be able to place limits on the value of σ^2, the population variance. This is possible if the population of interest, such as the machine fills or thread diameters, is normally distributed. If so, it is possible to obtain a confidence interval by calculating $\frac{(n-1)s^2}{\chi^2_{\alpha/2}} < \sigma^2 < \frac{(n-1)s^2}{\chi^2_{1-\alpha/2}}$. This involves the use of a new distribution, the χ^2 distribution. It is similar to the t distribution in that the χ^2 distribution involves $n-1$ degrees of freedom. However, the distribution consists only of positive values and is not a symmetric distribution. Because of this, we do not obtain the table values for the lower confidence limit by putting a minus sign in front of the table value for the upper confidence limit. Rather, we need to examine the χ^2 table for unique lower and upper table values.

Once we obtain a confidence interval for a population variance, it is possible to obtain a confidence interval for the population standard deviation by taking the square root of the lower and upper confidence limits for the variance.

Key Terms & Formulas

Chi-Square Distribution The χ^2 distribution is the sampling distribution of $\dfrac{(\text{df})s^2}{\sigma^2} = \dfrac{(n-1)s^2}{\sigma^2}$ when the population is normally distributed. The distribution is asymmetric and only positive values are possible. The distribution depends on the number of degrees of freedom, which is $n-1$.

Confidence Interval for a Population Variance If a random sample is taken from a normal population, a confidence interval for a population variance can be calculated using: $\dfrac{(n-1)s^2}{\chi^2_{\alpha/2}} < \sigma^2 < \dfrac{(n-1)s^2}{\chi^2_{1-\alpha/2}}$

Worked Examples

χ^2 Table

What χ^2 table values would we use to obtain a 95% confidence interval for a population variance based on a random sample of size 10 from a normal population?

Solution

To obtain a 95% confidence interval, we use $\alpha = 0.05$. The table value for the lower confidence limit is found by looking in the df $= n - 1 = 10 - 1 = 9$ row and the $\alpha/2 = 0.05/2 = 0.025$ column. There we find $\chi^2_{\alpha/2} = 19.02$. The table value for the upper confidence limit is located in the df $= 9$ row and the $1 - \alpha/2 = 1 - 0.05/2 = 0.975$ column. This is $\chi^2_{1-\alpha/2} = 2.70$.

Plastic

An engineer is concerned with the variability in the amount of plastic used in the manufacture of machine casings. She decides to estimate the variance in the weight of the plastic used in the casings (in ounces). A sample of 12 casings yielded a sample variance of 0.008 oz. Compute a 99% confidence interval for the population variance.

Solution

First we will obtain the table values. For a 99% confidence interval, we use $\alpha = 0.01$. The table value for the lower confidence limit is in the df $= n - 1 = 12 - 1 = 11$ row and the $\alpha/2 = 0.01/2 = 0.005$ column. This table value is $\chi^2_{\alpha/2} = 26.8$. For the upper confidence limit we use df $= 11$ row and the $1 - \alpha/2 = 1 - 0.005/2 = 0.995$ column. This is $\chi^2_{1-\alpha/2} = 2.60$. Then the confidence interval is:

$$\frac{(n-1)s^2}{\chi^2_{\alpha/2}} < \sigma^2 < \frac{(n-1)s^2}{\chi^2_{1-\alpha/2}}$$

$$\frac{(12-1)(0.08)}{26.8} < \sigma^2 < \frac{(12-1)(0.08)}{2.60}$$

$$0.0033 < \sigma^2 < 0.0338$$

Practice Exercises 7-5

1. The manager of the international sales division of a firm wants to estimate the variance in the number of orders received per day from outside the North American continent. Based on his experience, he believes that the number of orders received would be approximately normal. Sampling 17 days randomly from the past year and determining the number of orders received on each, he calculated a sample variance of 5.7. Obtain a 90% confidence interval for σ^2.

2. A forester is concerned with the consistency of the growth rate of Ponderosa pines. Taking a sample of 25 such trees, she finds that the yearly growth had a variance of 19.2 inches. Assuming the growth rate is normally distributed, obtain a 95% confidence interval for σ^2.

Solutions to Practice Exercises 7-5

1. For a 90% confidence interval, we use $\alpha = 0.10$. The table value for the lower confidence limit is in the $df = n - 1 = 17 - 1 = 16$ row and the $\alpha/2 = 0.10/2 = 0.05$ column. This table value is $\chi^2_{\alpha/2} = 26.3$. For the upper confidence limit we use $df = 16$ row and the $1 - \alpha/2 = 1 - 0.10/2 = 0.95$ column. This is $\chi^2_{1-\alpha/2} = 7.96$. Then the confidence interval is:

$$\frac{(n-1)s^2}{\chi^2_{\alpha/2}} < \sigma^2 < \frac{(n-1)s^2}{\chi^2_{1-\alpha/2}}$$

$$\frac{(17-1)(5.7)}{26.3} < \sigma^2 < \frac{(17-1)(5.7)}{7.96}$$

$$3.47 < \sigma^2 < 11.46.$$

2. For a 95% confidence interval, we use $\alpha = 0.05$. The table value for the lower confidence limit is in the $df = n - 1 = 25 - 1 = 24$ row and the $\alpha/2 = 0.05/2 = 0.025$ column. This table value is $\chi^2_{\alpha/2} = 39.4$. For the upper confidence limit we use $df = 24$ row and the $1 - \alpha/2 = 1 - 0.05/2 = 0.975$ column. This is $\chi^2_{1-\alpha/2} = 12.40$. Then the confidence interval is:

$$\frac{(n-1)s^2}{\chi^2_{\alpha/2}} < \sigma^2 < \frac{(n-1)s^2}{\chi^2_{1-\alpha/2}}$$

$$\frac{(25-1)(19.2)}{39.4} < \sigma^2 < \frac{(25-1)(19.2)}{12.40}$$

$$11.70 < \sigma^2 < 37.16.$$

7-6 Determining Sample Size to Estimate μ or π

Study Objectives

You should be able to:

1. Understand the desired maximum error of estimate.
2. Calculate sample size for estimating μ or π.

Section Overview

An early question faced by researchers is "How large a sample should we take?" In estimating a population mean or percentage, we can determine the sample size necessary to estimate the parameter with a given precision at a chosen level of confidence. For the mean of a population, we use: $n = \left(\dfrac{(z_{\alpha/2})\sigma}{E} \right)^2$, where $E = $ the desired maximum error of estimate. For a population percentage, the appropriate sample size is calculated using: $n = \dfrac{z^2_{\alpha/2}}{E^2}\pi(100 - \pi)$. Both formulas require us to select the level of confidence and the maximum error of estimate before we collect any data. The more stringent we set our requirements, i.e., the higher the confidence and/or the smaller the maximum error of estimate we want, the larger the necessary sample size will be.

Key Terms & Formulas

Desired Maximum Error of Estimate $E = $ the amount a researcher wants to add and subtract in forming a confidence interval.

Sample Size for Estimation of μ $n = \left(\dfrac{(z_{\alpha/2})\sigma}{E} \right)^2$.

Sample Size for Estimation of π $n = \dfrac{z_{\alpha/2}^2}{E^2}\pi(100 - \pi)$.

Worked Examples
Computer Manuals

A computer software company wants to estimate the mean time necessary for workers to learn a new operating system using a self-paced training manual. Previous studies have shown that the standard deviation in this time is 80 minutes. If the company wants to estimate the mean time necessary to learn the operating system to within 15 minutes with 95% confidence, how large a sample should they take?

Solution

The z value for a 95% confidence interval is 1.96. To calculate the appropriate sample size, we use:

$$n = \left(\frac{(z_{\alpha/2})\sigma}{E}\right)^2 = \left(\frac{1.96 \cdot 80}{15}\right)^2 = 109.3 \approx 110.$$

Solar Heating

The managers of an engineering design firm are interested in estimating the percentage of homes in Wasco that have sufficient support for a heavy solar heating system. How large a sample should they take if they wish to be 99% confident that their estimate is within 10% of the true value of π? How large should the sample be if they believe that the percentage is roughly 20%?

Solution

For a 99% confidence interval, the z value is 2.58. To compute the sample size, we calculate:

$$n = \frac{z_{\alpha/2}^2\pi(100 - \pi)}{E^2} = \frac{(2.58^2)\,(50)(100 - 50)}{10^2} = 166.41 \approx 167.$$

With the additional information that the population percentage is roughly 20%, we would substitute 20% for 50% in the previous calculations. Then the sample size becomes:

$$n = \frac{z_{\alpha/2}^2\pi(100 - \pi)}{E^2} = \frac{(2.58^2)\,(20)(100 - 20)}{10^2} = 106.50 \approx 107.$$

Practice Exercises 7-6

1. Theoretically, a circuit design has a known resistance. But due to variability in the quality of materials, the actual resistance may vary. It is known that the resistance in a particular circuit design has a standard deviation of 12 ohms. If the mean resistance of this design is to be estimated to within 0.5 ohms with 90% confidence, what sample size would be required?

2. A physician is interested in estimating the mean pulse rate of patients after they have been subjected to a stress test. She is interested in estimating this amount to within 1, using a 99% confidence interval. She believes that the standard deviation in the pulse rates is about 10. Determine the sample size necessary to satisfy the physician's requirements.

3-4. An industrial psychologist would like to estimate the percentage of employees in a large company who believe that they will continue to work with that company until retirement. He wants to estimate this percentage to within 4% with 95% confidence.

3. How large a sample should he take?

4. How large a sample should he take if he believes this percentage is roughly 75%?

5-6. A doctor is interested in estimating the percentage of surgeries performed which are unnecessary. She wishes to estimate this proportion to within 2% with a 90% confidence interval.

5. How large a sample should she obtain?

6. How large a sample should she take if she believes this percentage is roughly 10%?

Solutions to Practice Exercises 7-6

1. For a 90% confidence interval, the z value is 1.645. Then
$$n = \left(\frac{(z_{\alpha/2})\sigma}{E}\right)^2 = \left(\frac{1.645 \cdot 12}{0.5}\right)^2 = 1558.7 \approx 1559.$$

2. For a 99% confidence interval, the z value is 2.58. Then
$$n = \left(\frac{(z_{\alpha/2})\sigma}{E}\right)^2 = \left(\frac{2.58 \cdot 10}{1}\right)^2 = 665.64 \approx 666.$$

3. For a 95% confidence interval, the z value is 1.96. Then
$$n = \frac{z_{\alpha/2}^2\,\pi(100 - \pi)}{E^2} = \frac{(1.96^2)\,(50)(100 - 50)}{4^2} = 600.25 \approx 601.$$

4. With the information that the population percentage is roughly 75% we would substitute 75% for 50% in the previous calculations. Then the sample size becomes:
$$n = \frac{z_{\alpha/2}^2\,\pi(100 - \pi)}{E^2} = \frac{(1.96^2)\,(75)(100 - 75)}{4^2} = 450.19 \approx 451.$$

5. For a 90% confidence interval, the z value is 1.645. Then
$$n = \frac{z_{\alpha/2}^2\,\pi(100 - \pi)}{E^2} = \frac{(1.645^2)\,(50)(100 - 50)}{2^2} = 1691.27 \approx 1692.$$

6. With the information that the population percentage is roughly 10% we would substitute 10% for 50% in the previous calculations. Then the sample size becomes:
$$n = \frac{z_{\alpha/2}^2\,\pi(100 - \pi)}{E^2} = \frac{(1.645^2)\,(10)(100 - 10)}{2^2} = 608.86 \approx 609.$$

Solutions to Odd-Numbered Exercises

1. Using the techniques of Chapter 6, 90 percent corresponds to a z-value of 1.645.

3. Using the techniques of Chapter 6, 95 percent corresponds to a z-value of 1.96.

5. With a 95 percent confidence level $\alpha = 0.05$ (5 percent). With a sample size of 21 there are 20 degrees of freedom. Therefore using Appendix 4 under $t_{.025}$ we find t value of 2.086.

7. With a 90 percent confidence level $\alpha = 0.10$ (10 percent). With a sample size of 14 there are 13 degrees of freedom. Therefore using Appendix 4 under $t_{.05}$ we find t value of 1.771.

9. We have the following data: $n = 47$, $\bar{x} = 83.2$, $s = 5.5$, population distribution shape unknown, σ unknown. Because $n > 30$ we can assume that the sample distribution of means can be approximated by a normal distribution. So use the z distribution. Since we want a 95 percent confidence level we use $z = \pm 1.96$ (found in Exercise 3).

11. We have the following data: $n = 17$, $\bar{x} = 3.2$, $s = 1.4$, population distribution shape normal, σ unknown. Since $n < 30$ and the population shape is normal, we use the t distribution. Since we want a 95 percent confidence level with df $= 16$, we use $t_{.025} = 2.120$.

13. We have the following data: $n = 15$, $\bar{x} = 23.1$, $s = 7.5$, population distribution shape is unknown, σ unknown. Because $n < 30$ and the population distribution shape is unknown, this is neither a z distribution nor a t distribution.

15. Here $\alpha = .10$ so $\chi^2_{\alpha/2} = \chi^2_{.05}$ and $\chi^2_{1-\alpha/2} = \chi^2_{.95}$. Using Appendix 6 with $n - 1 = 21$ degrees of freedom we have $\chi^2_{.95} = 11.59$ and $\chi^2_{.05} = 32.7$.

17. We have the following data: $n = 36$, $\overline{x} = 76.1$, $s = 14.2$. Because $n > 30$ we use the z distribution with $z = 1.96$. Thus $\widehat{\sigma}_{\overline{x}} = \dfrac{s}{\sqrt{n}} = \dfrac{14.2}{\sqrt{36}} = 2.367$. Hence the interval is $76.1 \pm 1.96(2.367)$ or 71.46 to 80.739.

19. We have the following data: $n = 64$, $\overline{x} = 76.1$, $s = 14.2$. Because $n > 30$ we use the z distribution with $z = 1.96$. Thus $\widehat{\sigma}_{\overline{x}} = \dfrac{s}{\sqrt{n}} = \dfrac{14.2}{\sqrt{64}} = 1.775$. Hence the interval is $76.1 \pm 1.96(1.775)$ or 72.621 to 79.58.

21. We have the following data: $n = 100$, $\overline{x} = 364.1$, $s = 61.7$. Because $n > 30$ we use the z distribution with $z = 1.645$. Thus $\widehat{\sigma}_{\overline{x}} = \dfrac{s}{\sqrt{n}} = \dfrac{61.7}{\sqrt{100}} = 6.17$. Hence the interval is $364.1 \pm 1.645(6.17)$ or 353.95 to 374.25.

23. We have the following data: $n = 100$, $\overline{x} = 364.1$, $s = 61.7$. Because $n > 30$ we use the z distribution with $z = 2.575$. Thus $\widehat{\sigma}_{\overline{x}} = \dfrac{s}{\sqrt{n}} = \dfrac{61.7}{\sqrt{100}} = 6.17$. Hence the interval is $364.1 \pm 2.575(6.17)$ or 348.21 to 379.99.

25. Find the appropriate χ^2 values and substitute into the formula $\dfrac{(n-1)s^2}{\chi^2_{\alpha/2}} < \sigma^2 < \dfrac{(n-1)s^2}{\chi^2_{1-\alpha/2}}$. The data given: $n = 17$ and $s^2 = 31.8$. Since $\alpha = .05$ using, Appendix 6 with 16 df we have $\chi^2_{\alpha/2} = \chi^2_{0.025} = 28.8$ and $\chi^2_{1-\alpha/2} = \chi^2_{0.975} = 6.91$. Substituting: $\dfrac{(n-1)s^2}{\chi^2_{\alpha/2}} < \sigma^2 < \dfrac{(n-1)s^2}{\chi^2_{1-\alpha/2}} \Rightarrow \dfrac{16 \cdot 31.8}{28.8} < \sigma^2 < \dfrac{16 \cdot 31.8}{6.91}$

$\Rightarrow \quad 17.666 < \sigma^2 < 73.632$.

27. Point estimate $= \dfrac{\text{observed}}{\text{sample size}} = \dfrac{74}{116} \times (100 \text{ percent}) = 63.79 \text{ percent}$.

29. Here we wish to find $p \pm z\,\widehat{\sigma}_p$. Since $p = 63.79$ percent from Exercise 27 and
$\widehat{\sigma}_p = \sqrt{\dfrac{63.79 \cdot 36.21}{116}} = 4.462$ percent. For a 90 percent confidence interval $z = 1.645$. Substituting to find the confidence interval $p \pm z\,\widehat{\sigma}_p = 63.79 \pm 1.645(4.462)$ or 56.45 to 71.13 percent.

31. $p = 26.81$ from Exercise 30, so $\widehat{\sigma}_p = \sqrt{\dfrac{26.81 \cdot 73.19}{1,190}} = 1.284$ percent.

33. $\overline{x} = \dfrac{\Sigma \text{ scores}}{n} = \dfrac{939}{21} = 44.71$.

35. Since we are assuming that the scores on this test are normally distributed and $n < 30$, we use the t distribution, so with 20 df, $t_{0.05} = 1.725$. Using $\widehat{\sigma}_{\overline{x}} = 3.42$ from Exercise 34, the interval is $\overline{x} \pm t_{\alpha/2} \cdot \widehat{\sigma}_{\overline{x}} = 44.71 \pm 1.725(3.42)$ or test scores between 38.81 and 50.61.

37. $\overline{x} = \dfrac{\Sigma \text{ ml/min per kg}}{n} = \dfrac{158.9}{12} = 13.24$ ml/min per kg.

39. Since we are assuming that the population values are normally distributed and $n < 30$, we use the t distribution with 11 df; so $t_{0.005} = 3.106$. Using $\widehat{\sigma}_{\overline{x}} = 1.35$ from Exercise 38, the interval is $\overline{x} \pm t_{\alpha/2} \cdot \widehat{\sigma}_{\overline{x}} = 13.24 \pm 3.106(1.35)$ or 9.05 to 17.43 ml/min per kg.

41. $\bar{x} = \dfrac{\Sigma \text{ mean HMO cost}}{n} = \dfrac{35,960}{12} = \$2,996.67$

43. Since we are assuming that the population values are normally distributed and $n < 30$, we use the t distribution with 11 df; so $t_{.025} = 2.201$. Using $\hat{\sigma}_{\bar{x}} = \$88.122$ from Exercise 42, the interval is $\bar{x} \pm t_{\alpha/2} \cdot \hat{\sigma}_{\bar{x}} = \$2,996.67 \pm 2.201(88.122)$ or \$2802.71 to \$3190.63.

45. We have the following data: $n = 32$, $\bar{x} = 12.4$, $s = 3$, population distribution shape unknown, $\sigma = ?$. Because $n > 30$ we use the z distribution with $z = \pm 1.96$. $\sigma_{\bar{x}} = \dfrac{s}{\sqrt{n}} = \dfrac{3}{\sqrt{32}} = 0.53$ years. Hence the interval is $\bar{x} \pm z\sigma_{\bar{x}} = 12.4 \pm 1.96(0.53)$ or 11.36 to 13.44 years.

47. We have the following data: $n = 431$ and 99 expressed an interest in international business. Since the number in the study is larger than 30 we use a normal distribution and find the required values.
$\bar{x} = \dfrac{99}{431} = 22.97$ percent, $\hat{\sigma}_p = \sqrt{\dfrac{22.97 \cdot 77.03}{431}} = 2.03$, and $z = 1.645$. Thus $\bar{x} \pm z \cdot \hat{\sigma}_p = 22.97 \pm 1.645(2.03)$ or an interval between 19.64 to 26.30 percent.

49. We have the following data: $n = 141$, $\bar{x} = 106.9$, $s = 68.2$. Because $n > 30$ we use the z distribution with $z = \pm 1.645$. $\sigma_{\bar{x}} = \dfrac{s}{\sqrt{n}} = \dfrac{68.2}{\sqrt{141}} = 5.74$. Hence the interval is $\bar{x} \pm z \cdot \sigma_{\bar{x}} = 106.9 \pm 1.645(5.74)$ or 97.45 to 116.35 hours.

51. Here we are asked to find bounds for σ. We are given the following data: $n = 10$ and $s = 31.2$ minutes. So $s^2 = 973.44$. Since $\alpha = .05$ and with 9 df we have $\chi^2_{\alpha/2} = \chi^2_{0.025} = 19.02$ and $\chi^2_{1-\alpha/2} = \chi^2_{0.975} = 2.70$.
Substituting: $\dfrac{(n-1)s^2}{\chi^2_{\alpha/2}} < \sigma^2 < \dfrac{(n-1)s^2}{\chi^2_{1-\alpha/2}} \Rightarrow \dfrac{9 \cdot 973.44}{19.02} < \sigma^2 < \dfrac{9 \cdot 973.44}{2.70} \Rightarrow$
$460.618 < \sigma^2 < 3244.8 \Rightarrow 21.46 < \sigma < 56.96$ minutes.

53. Here we want to find n so that $z\hat{\sigma}_p < 5$ percent when $\bar{x} = \dfrac{24}{36} = 66.67$ percent, $z = 1.645$, and $\hat{\sigma}_p = \sqrt{\dfrac{66.67 \cdot 33.33}{n}} = \dfrac{47.139}{\sqrt{n}}$. Substituting into the inequality $z\hat{\sigma}_p < 5$ we get $1.645 \cdot \dfrac{47.139}{\sqrt{n}} < 5$ solving we get $\dfrac{1.645 \cdot 47.139}{5} < \sqrt{n} \Rightarrow 15.51 < \sqrt{n}$ so $240.5 < n$. Thus the sample size must contain at least 241 World War II POW survivors.

55. As in Exercise 53, we want to find n so that $z\hat{\sigma}_{\bar{x}} < 2$ words when $\bar{x} = 58.3$, $s = 20.5$. So $z = 1.96$ and $\hat{\sigma}_{\bar{x}} = \dfrac{s}{\sqrt{n}} = \dfrac{20.5}{\sqrt{n}}$. Substituting into the inequality $z\hat{\sigma}_{\bar{x}} < 2$ we get $1.96 \cdot \dfrac{20.5}{\sqrt{n}} < 2$ solving we get $\dfrac{196 \cdot 20.5}{2} < \sqrt{n} \Rightarrow 20.09 < \sqrt{n}$, so $403.6 < n$. Thus the sample size must contain at least 404 students.

57. We are given the following data: $n = 10$, $\bar{x} = 24.7$, $s = 3.6$. Because $n < 30$ we use the t distribution and with 9 df, $t_{0.025} = 2.262$. Thus $\hat{\sigma}_{\bar{x}} = \dfrac{s}{\sqrt{n}} = \dfrac{3.6}{\sqrt{10}} = 1.14$ kg. Hence the interval is $\bar{x} \pm t_{\alpha/2} \cdot \hat{\sigma}_{\bar{x}} = 24.7 \pm 2.262(1.14)$ or 22.12 to 27.28 kg.

59. Here we want to find n so that $z\hat{\sigma}_p < 3$ percent when $\bar{x}$ is unknown, $z = 1.96$, and σ_p is maximized when $p = 1 - p = 50$ percent, thus we use 50 percent as our point estimate. Then $\hat{\sigma}_p = \sqrt{\dfrac{50 \cdot 50}{n}} = \dfrac{50}{\sqrt{n}}$.

Substituting these values into the inequality $z\widehat{\sigma}_p < 3$ we get $1.96 \cdot \dfrac{50}{\sqrt{n}} < 3$ solving we get $\dfrac{1.96 \cdot 50}{3} < \sqrt{n}$

$\Rightarrow \quad 32.67 < \sqrt{n}$ so $1067.1 < n$. Thus the same size must contain at least $1,068$ voters.

61. Here we want to find n so that $z\sigma_{\overline{x}} < 15$ minutes when $s = 127$. Using $z = 2.575$ and $\sigma_{\overline{x}} = \dfrac{s}{\sqrt{n}} = \dfrac{127}{\sqrt{n}}$.

Substituting these values into the inequality $z\sigma_{\overline{x}} < 15$ we get $2.575 \cdot \dfrac{127}{\sqrt{n}} < 15$ solving we get

$\dfrac{2.575 \cdot 127}{15} < \sqrt{n} \quad \Rightarrow \quad 21.8 < \sqrt{n}$ so $475.3 < n$. Thus the same size must contain at least 476 six-year old children.

63. We calculate the mean using the first column, then use it to calculate the other two columns.

x	$x - \overline{x}$	$(x - \overline{x})^2$
1493	−100.7	10140.49
1737	143.3	20534.89
1319	−274.7	75460.09
1193	−400.7	160560.49
1253	−340.7	116076.49
1893	299.3	89580.49
1651	57.3	3283.29
1532	−61.7	3806.89
1812	218.3	47654.89
1717	123.3	15202.89
1682	88.3	7796.89
1842	248.3	61652.89
19124		611752.68

$\overline{x} = \dfrac{19124}{12} = 1593.7$. Then $s = \sqrt{\dfrac{\Sigma(x - \overline{x})^2}{n - 1}} = \sqrt{\dfrac{611752.68}{11}} = 235.83$.

65. Data: $n = 12$ and $s^2 = (235.83)^2 = 55615.79$. $\alpha = 0.01$ and with 11 df we have $\chi^2_{\alpha/2} = \chi^2_{0.005} = 26.8$ and $\chi^2_{1-\alpha/2} = \chi^2_{0.995} = 2.60$. Substituting

$$\dfrac{(n-1)s^2}{\chi^2_{\alpha/2}} < \sigma^2 < \dfrac{(n-1)s^2}{\chi^2_{1-\alpha/2}}$$
$$\dfrac{11 \cdot 55615.79}{26.8} < \sigma^2 < \dfrac{11 \cdot 55615.79}{2.60}$$
$$22,827.38 < \sigma^2 < 235,297.57.$$

Chapter 8 Testing Hypotheses: One-Sample Procedures

Study Aids and Practice Exercises

8-1 The Hypothesis-Testing Procedure in General

Study Objectives
You should be able to:
1. Understand the purpose of hypothesis testing.
2. Outline the seven steps in classical hypothesis testing.
3. Form appropriate null and alternative hypotheses.
4. Use terms such as level of significance, Type I and II errors, rejection or critical regions, and test statistic.

Section Overview
   In estimation, our interest was in getting a good guess for the value of a parameter. For example, we might estimate the mean income of residents of Ames, Iowa, or we might estimate the percentage of voters who favor federally mandated standards of education at the elementary school level. In hypothesis testing, we are again concerned with the value of a parameter. However, now we have an idea or hypothesis about the value of the parameter. We might think that the mean income of Ames is $33,000. Or we might think that the percentage of voters who favor federally mandated standards of education is 20%. We want to see if these values are reasonable or not, i.e., we want to put these values "to the test." So we gather a sample and base our decision on a sample statistic. For example, we might sample 50 residents of Ames and calculate the mean income of that sample. If the sample mean is $33,003 or $32,994, we will conclude that the mean salary could be $33,000 for all residents of Ames. If, however, the sample mean is $14,500 or $65,666, we would conclude that the mean for all of Ames is not $33,000. The question is when do we start to believe the mean for all Ames is not $33,000? When the sample mean is $33,100? $33,500? $34,000? To answer this question, we must look at the sampling distribution of the statistic involved. For the Ames example, the statistic is the sample mean. From the sampling distribution of means, we can decide what values of the sample mean are reasonable if the mean is really $33,000, and what values are unreasonable. We will base our decision upon the reasonableness or unreasonableness of the value of the statistic.
   We can divide the hypothesis testing procedure into a series of seven steps.
1. *State the null (H_0) and alternative (H_1) hypothesis.* The null hypothesis always involves a statement of equality. For example, H_0: $\mu = \$33,000$ or H_0: $\pi = 20\%$. The specific numbers are called the hypothesized values and are designated by μ_0 or π_0. The null hypothesis is assumed to be true and is rejected only if the sample information contradicts this assumption. The alternative hypothesis may have any of three forms, inequality,

greater than, or less than. For example, we could have H_1: $\mu \neq \$33,000$ or H_1: $\mu > \$33,000$ or H_1: $\mu < \$33,000$. The appropriate alternative hypothesis depends on the purpose of the research.

2. *Select the level of significance.* There are two mistakes possible in hypothesis testing. We make a Type I error when we reject a true null hypothesis. We make a Type II error when we fail to reject a false null hypothesis. While we should worry about the possibility of either type of error, we actually control the probability of a Type I error. This probability is designated α and is the level of significance of the test. The more harmful a Type I error is, the smaller we should choose α to be.

3. *Determine the test distribution to use.* The distribution to use matches the distribution used in obtaining confidence intervals. For example, in those situations where the sample mean is normal and σ is known, we use the normal (z); if σ is unknown and the sample is from a normal population, we use Student's t; if a test is needed about the variance of a normal population, we use the χ^2 distribution.

4. *Define the rejection or critical region.* This consists of the values of the test statistic that are unlikely to be observed if H_0 is true. If the calculated value of the test statistic falls into this rejection region, our decision will be to reject the null hypothesis in favor of the alternative. This is sometimes called a significant difference, in that there is a significant difference between the hypothesized value and the test statistic. If the calculated value of the test statistic does not fall in the rejection region, there is insufficient evidence to reject the null hypothesis, so our decision is to "fail to reject H_0". The start of the rejection or critical region is called the critical value and its value depends on the distribution of the test statistic and the level of significance of the test.

    Hint: Based on the rules of hypothesis testing, we will fail to reject H_0 unless there is strong evidence against it. So failing to reject the null hypothesis does not "prove" it is true; rather, it demonstrates that it is a possibility, but it is one of many possibilities. Consider the Ames example. If our sample mean turned out to be $\$33,011$, we would not have substantial evidence that the mean of the population was not $\$33,000$ and we would decide in favor of H_0: $\mu = \$33,000$. But maybe the population mean is $\$32,900$ or $\$33,100$ or $\$33,200$ or anything else close to $\$33,011$. So in deciding in favor of null, we are saying nothing more than that the hypothesized value is feasible.

5. *State the decision rule.* This is the formal statement indicating when H_0 should be rejected.

6. *Make the necessary calculations.* Based on the appropriate sampling distribution, we will have a test statistic. At this step we will calculate the value of this test statistic.

7. *Make a statistical decision.* If the test statistic falls in the rejection region, we reject H_0; if not, we fail to reject H_0.

Key Terms & Formulas

Hypothesis Testing The statistical process by which decisions are made about the value of a population parameter.

Classical Hypothesis Testing A seven-step procedure that gives general rules for the performance of a hypothesis test.

Null Hypothesis Designated H_0, it is a statement about the value of a parameter. This statement is assumed to be true unless the sample provides sufficient evidence that this assumption is incorrect. The null hypothesis always involves a statement of equality, i.e., a statement that a parameter is equal to a particular value. For example, H_0: $\mu = 27$ or H_0: $\pi = 66\%$.

Hypothesized Value The value specified in the null hypothesis. It is usually written by attaching a subscript of 0 to the parameter, e.g., μ_0 or π_0.

Alternative Hypothesis Designated H_1, it is the contradiction of H_0. It is the hypothesis that will be believed if the null is proven to be unreasonable.

Type I Error The error of rejecting the null hypothesis when the null hypothesis is true.

Type II Error The error of failing to reject the null hypothesis when the null hypothesis is false.

> Hint: The definitions of Type I and II errors sound silly as they are stated. Why would we reject H_0 if H_0 is true? The point is that we do not know which hypothesis is true. We are trying to develop a procedure that would make as small as possible the probability of either type of error. To do so, we need to be able to distinguish between these two potential errors.

Level of Significance Designated by α, it is the probability of committing a Type I error.

Test distribution The probability distribution of the statistic used to do the test.

Rejection Region Also called the critical region, it contains the values of the test statistic that would be unlikely to occur if the null hypothesis is true. So when we calculate the value of the test statistic, if it falls in the rejection region, we reject H_0. The critical region depends on the alternative hypothesis and the value chosen for the level of significance.

Significant Difference A difference between a test statistic and the hypothesized value that is large enough for the test statistic to fall into the rejection region and lead to a rejection of H_0. This is what has occurred if a test is declared to be "statistically significant."

Decision Rule A formal statement that states when to reject and when not to reject the null hypothesis. It says to reject H_0 if the test statistic falls in the rejection region and fail to reject H_0 otherwise.

Worked Examples

Aches and Pains

Suppose a pharmaceutical firm is considering manufacturing a new pain-relief drug. It will make this new drug if it has a better time-to-relief than a similar drug that is already available. The currently available drug has a mean time to relief of 30 minutes. What null and alternative hypothesis are appropriate to this situation?

Solution

Because the drug company will only market the new drug if the mean time-to-relief is less than 30 minutes, one of the two hypotheses must be $\mu < 30$. The null hypothesis must have equality, and since this does not, it must be the alternative hypothesis. So the hypotheses are:

H_0: $\mu = 30$
H_1: $\mu < 30$.

> Caution: Notice that the hypothesis the drug company is really interested in is the alternative. So though we assume that H_0 is true at the start of the hypothesis testing procedure, that does not imply that we prefer to decide in favor of H_0. It is more common that the reverse is true, i.e., we would like to reject the null in favor of the alternative. So the word "alternative" is a bit misleading, especially is you think of it as implying secondary or unimportant.

College Graduation

Typically, the percentage of first-year students that eventually graduate from college is 40%. A college believes that their rate is superior to this norm. What hypothesis pair would the college use to evaluate their belief? What hypothesis pair would the college use if they only wanted to see if their rate is different from the norm?

Solution

If the college is better at graduating students than the norm, their percentage would be more than 40%. Because this statement does not involve equality, this must be the alternative hypothesis. Therefore, the hypothesis pair is:

H_0: $\pi = 40\%$
H_1: $\pi > 40\%$.

If the college only wanted to learn if their students had a different graduation rate than the norm of 40%, the hypotheses would be:

H_0: $\pi = 40\%$
H_1: $\pi \neq 40\%$.

Practice Exercises 8-1

1. A recent survey of college students reported that 25% did not know that Columbus discovered America before 1500. Not trusting this survey, a historian decides to try to prove that the rate is greater than this 25%. What hypotheses are appropriate for this situation?

2. The Environmental Protection Agency has set an upper limit of 5 ppm for the amount of PCB allowed in drinking water. What pair of hypotheses should be used to find out if the average PCB amount in water coming from a well is above the EPA's limit?

3. One portion of a canning operation involves placing appropriate amounts of food in a container. Suppose the mean fill in a can is supposed to be 12 ounces. Too much in the can is wasteful, too little is false advertising. The manager would like to see if the canning operation is filling the can with the appropriate mean amount. What hypotheses should be tested?

4. If the manager of the canning operation from the previous exercise wants to see if more than 20% of the cans are under filled, what hypotheses should the manager use?

Solutions to Practice Exercises 8-1

1. Because "greater than" does not include equality, it must be the alternative hypothesis. So the two hypotheses are:

   H_0: $\pi = 25\%$
   H_1: $\pi > 25\%$.

2. The water is outside the EPA's limit if it is more than 5 ppm. Since this does not involve equality, it must be the alternative hypothesis. The hypotheses are:

   H_0: $\mu = 5$
   H_1: $\mu > 5$.

3. The appropriate mean amount equals 12 oz. Because this statement involves equality, it will be the null. Since the manager neither wants an underfill nor underfill, the alternative hypothesis will be an inequality:

   H_0: $\mu = 12$
   H_1: $\mu \neq 12$.

4. Because "more than" does not include equality, the more than statement must be the alternative hypothesis. So the two hypotheses are:

   H_0: $\pi = 20\%$
   H_1: $\pi > 20\%$.

8-2 One-Sample Hypothesis Tests of Means

Study Objectives

You should be able to:

1. Distinguish between two-tailed and one-tailed hypothesis tests.
2. Decide when the normal distribution should be used for a test and when Student's t should be used.
3. Perform a hypothesis test using the classical seven-step procedure.
4. Do a hypothesis test using the p-value approach.

Section Overview

There are many points to consider when we are planning to do a hypothesis test about a mean. One choice is whether a two-tailed or one-tailed test is more appropriate. A two-tailed test has a general alternative that says that the parameter is not equal to the hypothesized value. A one-tailed test specifies that the parameter is greater than the hypothesized value (a right-tail test) or is less than the hypothesized value (a left-tail test). A second point is whether we know the value of σ. A third is whether to use the classical approach to hypothesis testing or the p-value approach. Finally, we need to consider whether the normal (z) distribution or Student's t distribution is the sampling distribution of the test statistic.

A two-tailed hypothesis test has the form:

H_0: μ = hypothesized value

H_1: $\mu \neq$ hypothesized value.

In a two-tailed test, the null hypothesis is rejected if the sample statistic is significantly larger or smaller than the hypothesized value of the population parameter. A one-tailed hypothesis test can be either right- or left-tailed. A right-tailed has the form:

H_0: μ = hypothesized value

H_1: $\mu >$ hypothesized value.

In a right-tailed test, the null hypothesis is rejected if the sample statistic is significantly larger than the hypothesized value of the population parameter. A left-tailed test has the form:

H_0: μ = hypothesized value

H_1: $\mu <$ hypothesized value.

In a left-tailed test, the null hypothesis is rejected if the sample statistic is significantly smaller than the hypothesized value of the population parameter.

To perform a classical test when $n > 30$ or when σ is known and the population is normally distributed, the seven steps are as follows:

1. Select the hypotheses. This can result in a two-, right-, or left-tailed test.

2. Choose a level of significance. The most common values are 0.10, 0.05, and 0.01. A number as small as 0.01 is usually selected when a Type I error is more harmful than a Type II error. A number as large as 0.10 is used when a Type II error is more bothersome, since allowing a larger probability of a Type I error decreases the chance of a Type II error.

3. If $n > 30$ or σ is known and the population is normally distributed, the normal (z) distribution may be used.

4. The rejection region depends on whether we have a two-, right-, or left-tailed test.

   A. Two-tailed test, i.e., alternative hypothesis is H_1: $\mu \neq$ hypothesized value:

   $z > z_{\alpha/2}$ or $z < -z_{\alpha/2}$, where the $z_{\alpha/2}$ value is found in the normal table by finding the value associated with a probability of $0.5 - \alpha/2$. For example, to find the z value for a test with a level of significance of $\alpha = 0.05$, we look for

   $0.5 - \dfrac{0.05}{2} = 0.5 - 0.025 = 0.475$ in the probability portion of the table. The

   corresponding z value is 1.96 (which should be familiar from interval estimation). Then for $\alpha = 0.05$, the rejection region of a two-tailed test is:

   $z > 1.96$ or $z < -1.96$.

   B. Right-tailed test, i.e., alternative hypothesis is H_1: $\mu >$ hypothesized value:

   $z > z_\alpha$, where the z_α is found in the normal table by finding the value associated with a probability of $0.5 - \alpha$. For example, to find the z value for a test with $\alpha = 0.05$, we look for $0.5 - 0.05 = 0.45$ in the probability portion of the table. The corresponding z value is 1.645. Then for $\alpha = 0.05$, the rejection region of a right-tailed test is:

   $z > 1.645$.

   C. Left-tailed test, i.e., alternative hypothesis is H_1: $\mu <$ hypothesized value:

$z < -z_\alpha$, where the z_α is found in the normal table by finding the value associated with a probability of $0.5 - \alpha$, i.e., in the same fashion as with a right-tailed test. For example, for a test with a level of significance 0.05, the z value is 1.645. Then for $\alpha = 0.05$, the rejection region of a left-tailed test is: $z < -1.645$.

5. The decision rule is always the same. Reject the null hypothesis if the test statistic falls into the rejection region; otherwise, fail to reject H_0.

6. Calculate the test statistic:

$$z = \frac{\overline{x} - \mu_0}{\sigma_{\overline{x}}} = \frac{\overline{x} - \mu_0}{\sigma/\sqrt{n}}.$$

When σ is unknown but $n > 30$, we substitute s in place of σ to calculate the test statistic.

7. Make a statistical decision based on the value of the test statistic.

The p-value approach to hypothesis testing will modify some of the steps to the classical approach, <u>but will always reach the same decision as the classical approach</u>. For a p-value approach to testing a hypothesis about μ when $n > 30$ or σ is known and the population is normally distributed, the revised seven steps are as follows:

1. Select the hypotheses. This is the same as the classical approach.

2. Choose a level of significance. This is the same as the classical approach.

3. The normal (z) distribution is used. This is the same as the classical approach.

4. State the decision rule. The decision rule using p-values is to reject H_0 if the p-value is less than α, otherwise, fail to reject H_0.

5. Calculate the test statistic:

$$z = \frac{\overline{x} - \mu_0}{\sigma_{\overline{x}}} = \frac{\overline{x} - \mu_0}{\sigma/\sqrt{n}} \left(\text{or } \frac{\overline{x} - \mu_0}{s/\sqrt{n}} \text{ if } \sigma \text{ is unknown but } n > 30 \right).$$

This is the sixth step in classical hypothesis testing.

6. Compute the p-value. The computation of the p-value depends on whether the test is two-, right-, or left-tailed.

    A. Two-tailed test, i.e., alternative hypothesis is H_1: $\mu \neq$ hypothesized value:
        p-value $= P(z < -|z|) + P(z > |z|) = 2\,P(z > |z|)$
        where $|z|$ represents the absolute value of the test statistic.

    B. Right-tailed test, i.e., alternative hypothesis is H_1: $\mu >$ hypothesized value:
        p-value $= P(z > z)$.

    C. Left-tailed test, i.e., alternative hypothesis is H_1: $\mu <$ hypothesized value:
        p-value $= P(z < z)$.

7. Make a statistical decision based on the p-value.

When the sample size is small, i.e., less than 30 and σ is not known, we can still do a hypothesis test as long as it is reasonable to believe that the population is normal. Then, we use $\widehat{\sigma}_{\overline{x}} = \dfrac{s}{\sqrt{n}}$ for $\sigma_{\overline{x}} = \dfrac{\sigma}{\sqrt{n}}$ and the test statistic has a Student's t distribution with $n - 1$ degrees of freedom rather than a normal distribution. This causes us to use a different table and allows us to only place limits on p-values. Otherwise either testing procedure, the classical or p-value approach, remains basically the same.

Note: Whenever s is used in place of σ, we are really doing a t-test. However, when $n > 30$ the z and t distributions are similar. Because the z table is more extensive, we often use the z table in place of the t table when n is large.

Key Terms & Formulas

Two-Tailed Test on μ A two-tailed hypothesis test on μ has the form:
H_0: $\mu =$ hypothesized value
H_1: $\mu \neq$ hypothesized value.
In a two-tailed test, the null hypothesis is rejected if the sample statistic is significantly larger or smaller than the hypothesized value of the population parameter.

One-Tailed, Right-Tailed Test on μ A right-tailed has the form:
H_0: $\mu =$ hypothesized value
H_1: $\mu >$ hypothesized value.
In a right-tailed test, the null hypothesis is rejected if the sample statistic is significantly larger than the hypothesized value of the population parameter.

One-Tailed, Left-Tailed Test on μ A left-tailed has the form:
H_0: $\mu =$ hypothesized value
H_1: $\mu <$ hypothesized value.
In a left-tailed test, the null hypothesis is rejected if the sample statistic is significantly smaller than the hypothesized value of the population parameter.

P-value The probability, if H_0 is true, of obtaining a test statistic at least as extreme (or unlikely) as the one observed. The smaller the p-value is, the stronger the case against H_0.

Worked Examples

Aches and Pains Keep Coming Back

Back to the pharmaceutical firm that is considering manufacturing a new pain-relief drug. As a reminder, it will make this new drug if it has a better mean time-to-relief then a similar drug that is already available. They take a sample of 36 subjects and learn that the mean time-to-relief for these was 27.4, while the standard deviation was 5.6 minutes. Use the classical approach to hypothesis testing to analyze this information.

Solution

The first step is to form the hypotheses. We have already developed hypotheses for this situation in the previous section. They are:
H_0: $\mu = 30$
H_1: $\mu < 30$.
The next step is to choose a level of significance. Since the pharmaceutical company does not want to market the new drug unless they are sure it is effective, they choose $\alpha = 0.01$. The third step is to figure out the probability distribution to use for the test. Because $n = 36 > 30$, the normal (z) distribution is used. Next is the creation of the rejection region. This is a left-tailed test. We find z_α by looking for $0.5 - 0.01 = 0.49$ in the probability portion of the table. There we find 2.33. Then the rejection region is: $z < -2.33$. The decision rule is to reject the null hypothesis if $z < -2.33$; otherwise, fail to reject H_0. Next we calculate the test statistic, z.

$$z = \frac{\overline{x} - \mu_0}{\sigma_{\overline{x}}} = \frac{\overline{x} - \mu_0}{\sigma/\sqrt{n}} \approx \frac{\overline{x} - \mu_0}{s/\sqrt{n}} = \frac{27.4 - 30}{5.6/\sqrt{36}} = -2.79.$$

Because $-2.79 < -2.33$, we reject H_0 and conclude at $\alpha = 0.01$ that the mean time-to-relief for this new drug is less than 30 minutes.

Aches and Pains Keep Coming Back Again and Again

Redo the pharmaceutical firm illustration using the p-value approach to hypothesis testing.

Solution

The first three steps are the same. The hypotheses are:
H_0: $\mu = 30$
H_1: $\mu < 30$.

The level of significance can remain at $\alpha = 0.01$. The probability distribution to use for the test is still the normal distribution. The fourth step is the decision rule. With the p-value approach, the decision rule is to reject H_0 if the value of the test statistic is less than α. So our decision rule is to reject H_0 if the p-value < 0.01. The computation of the test statistic would again yield a value of -2.79. Then to calculate the p-value, we would recognize that this is a left-tailed test and would look in the normal table to find the probability of finding a z value less than -2.79. This turns out to be $0.5 - 0.4974 = 0.0026$. Since $0.0026 < 0.01$, we again reject the null hypothesis and conclude that it would be worthwhile to market the new drug.

Aches and Pains Keep Coming Back Again and Again and Again

One last time with the drug illustration. What changes would occur with the classical and p-value approach if this had been a two-tailed test, i.e., if the hypotheses had been:

H_0: $\mu = 30$
H_1: $\mu \neq 30$?

Solution

With the classical approach, the first alteration after the hypotheses would be in the rejection region. Now we are dealing with a two-tailed test. The z value involved is found by looking in the normal tables for $0.5 - \dfrac{0.01}{2} = 0.495$. We find a z value of 2.58. Then the rejection region is: $z < -2.58$ or $z > 2.58$. The decision rule is to reject the null hypothesis if $z < -2.58$ or $z > 2.58$; otherwise, fail to reject H_0. The test statistic remains -2.79. Because $-2.79 < -2.58$, we reject H_0 in favor of H_1 and conclude that the mean time-to-relief for this new drug is not 30 minutes.

With the p-value approach, the only substantial change affects the calculation of the p-value. Rather than being 0.0026, it will be $2(0.0026) = 0.0052$. We still reject H_0.

Sharks

One of the most feared predators in the ocean is the great white shark. Although it is known that the white shark grows to a mean length of 21 feet, a marine biologist believes that the great white sharks off the Bermuda coast grow much longer due to unusual feeding habits. To test this claim, full-grown great white sharks are captured off the Bermuda coast, measured, and then set free. However, because the capture of sharks is difficult, costly, and very dangerous, only three are sampled. Their lengths are 24, 20, and 22 feet. Test the appropriate hypothesis, using a level of significance of 0.10.

Solution

The first step is to form the hypotheses. With the situation as described, where we want to see if the average length of the great whites exceeds 21 feet for the Bermuda sharks, we would have:

H_0: $\mu = 21$
H_1: $\mu > 21$.

The level of significance is given as $\alpha = 0.10$. If it is reasonable to believe that the lengths of great white sharks form a normal population, we can use the Student's t distribution with $n - 1 = 3 - 1 = 2$ degrees of freedom to do the test. Next we need a rejection region for this right-tailed test. We find the t critical value in the 2 df row and the $\alpha = 0.10$ column of the t table. There we find 1.886. The rejection region is: $t > 1.886$. The decision rule is to reject the null hypothesis if the $t > 1.886$; otherwise, fail to reject H_0. Then we need to calculate the test statistic. To do so, we must first compute the sample mean and standard deviation.

$$\overline{x} = \frac{24 + 20 + 22}{3} = 22$$

$$s = \sqrt{\frac{\Sigma(x - \overline{x})^2}{n - 1}} = \sqrt{\frac{(24 - 22)^2 + (20 - 22)^2 + (22 - 22)^2}{3 - 1}} = 2.$$

The test statistic is: $t = \dfrac{\overline{x} - \mu_0}{\hat{\sigma}_{\overline{x}}} = \dfrac{\overline{x} - \mu_0}{s/\sqrt{n}} = \dfrac{22 - 21}{2/\sqrt{3}} = 0.87.$

Because 0.87 is not greater than 1.886, the null hypothesis is not rejected. We conclude that there is insufficient evidence to support the theory that Bermuda-based great whites sharks are any larger than expected.

Hint: The MINITAB output for this problem is:
```
T-Test of the Mean


Test of mu = 21.00 vs mu > 21.00

Variable      N      Mean     StDev    SE Mean       T          P
Sharks        3     22.00      2.00       1.15     0.87       0.24
```

t p-values

Suppose we do a hypothesis test using the t distribution with 10 degrees of freedom. Further suppose the value of the test statistic is 2.10. What are the p-values for a right-tailed, left-tailed, and two-tailed test?

Solution

If we look in the t table in the 10-df line, we find 1.812 in the 0.05 column and 2.228 in the 0.025 column. The test statistic, 2.10, is between these two values. For a right-tailed test, we have:
p-value $= P(t > 2.10) =$ somewhere between 0.025 and 0.05.
That is, while we cannot specify the exact p value, we can say that it is between 0.025 and 0.05.

Hint: For this example, although we do not know the exact p-value, we can still make a decision. If we had chosen $\alpha = 0.01$, the p-value is not less than this amount, so we would fail to reject H_0. If we had selected $\alpha = 0.05$ or 0.10, the p-value is less than either and we would reject H_0.

For a left-tailed test, p-value $= P(t < 2.10) =$ somewhere between $1 - 0.05$ and $1 - 0.025$, i.e., $0.95 < p$-value < 0.975.
For a two-tailed test, p-value $= 2[P(t > |2.10|)] = 2(0.025$ to $0.05) =$ somewhere between 0.05 and 0.10.

Practice Exercises 8-2

1-2. Traffic studies before the widening of Los Osos Valley Road (LOVR) indicated that the average speed of cars was 53 mph. There is concern that widening LOVR has increased the mean speed. A sample of 60 cars yielded a mean of 55.2 mph and a standard deviation of 8.6. With a level of significance of 0.05, decide if the concern is justified.

   1. Use the classical approach.
   2. Use the p-value approach.

3-4. Last year, the mean number of lottery tickets purchased by Pennsylvania state residents was 25. The director of the state lottery wanted to test to see if there had been a change since last year. He did a survey of 64 Pennsylvania residents and found out that they had purchased an average of 22.7 tickets with a standard deviation of 11.9. Use this information to do an appropriate test using a level of significance of 0.01.

   3. Use the classical approach.
   4. Use the p-value approach.

5-6. Most international data transfer goes through satellites, and the average amount of time necessary for a signal to go from station to station is 270 msec. A company claims that they have produced a telecommunication relay system employing lasers that improve on these times. A sample of 20 data transfers yielded a mean of 265 msec. and a standard deviation of 15 msec. Do the appropriate hypothesis test, using a level of significance of 0.05.

   5. Use the classical approach.
   6. Use the p-value approach.

7-8. Bolts produced by a manufacturing process are designed to have a mean diameter of 0.2000 cm. A random sample of 14 bolts is examined and found to have the following diameters:

0.1991	0.1967	0.1987	0.1992	0.2005	0.1984	0.1995
0.2019	0.1988	0.2007	0.1972	0.2006	0.1967	0.1983

For the above sample, the following were calculated: $\Sigma x = 2.78630$, $\Sigma x^2 = 0.554565$.
Use this information to decide if the process is working as it was designed to do, using $\alpha = 0.10$.

   7. Use the classical approach.
   8. Use the p-value approach.

Solutions to Practice Exercises 8-2

1. Because the concern is that the mean speed might now exceed 53 mph, the hypotheses are:

    H_0: $\mu = 53$

    H_1: $\mu > 53$.

    The level of significance has been chosen to be 0.05. Since the sample size $= 60 > 30$, we may use the normal (z) distribution for the test. The rejection region for this right-tailed test is found by looking for $0.5 - 0.05 = 0.45$ in the probability portion of the normal tables. We find a z value of 1.645. Then the rejection region is: $z > 1.645$. The decision rule is to reject the null hypothesis if $z > 1.645$; otherwise, fail to reject H_0.

    The test statistic is: $z = \dfrac{\overline{x} - \mu_0}{\sigma_{\overline{x}}} = \dfrac{\overline{x} - \mu_0}{\sigma/\sqrt{n}} \approx \dfrac{\overline{x} - \mu_0}{s/\sqrt{n}} = \dfrac{55.2 - 53}{8.6/\sqrt{60}} = 1.98$.

    Because $1.98 > 1.645$, we reject H_0 and conclude at $\alpha = 0.05$ that the mean speed on Los Osos Valley Road is now greater than 53 mph.

2. The first three steps are the same as Exercise 1. The hypotheses are:

    H_0: $\mu = 53$

    H_1: $\mu > 53$.

    The level of significance is 0.05. Since the sample size $= 60 > 30$, we use the normal (z) distribution for the test. The decision rule is to reject H_0 if the value of the test statistic is less than $\alpha = 0.05$. The computation of the test statistic would again yield a value of 1.98. To calculate the p-value, we find $P(z > 1.98) = 0.5 - 0.4761 = 0.0239$. Since $0.0239 < 0.05$, we reject the null hypothesis and conclude that the mean speed exceeds 53 mph.

3. Because the director is interested in seeing if there has been any change, a two-tailed test is appropriate. The hypotheses are:

    H_0: $\mu = 25$

    H_1: $\mu \neq 25$.

    The level of significance has been chosen to be 0.01. Since the sample size $= 64 > 30$, we will use the normal (z) distribution for the test. The rejection region for this two-tailed test is found by looking for $0.5 - 0.01/2 = 0.495$ in the probability portion of the normal tables. We find a z value of 2.58. The rejection region is: $z < -2.58$ or $z > 2.58$. The decision rule is to reject the null hypothesis if $z < -2.58$ or $z > 2.58$; otherwise, fail to reject H_0.

    The test statistic, $z = \dfrac{\overline{x} - \mu_0}{\sigma_{\overline{x}}} = \dfrac{\overline{x} - \mu_0}{\sigma/\sqrt{n}} \approx \dfrac{\overline{x} - \mu_0}{s/\sqrt{n}} = \dfrac{22.7 - 25}{11.9/\sqrt{64}} = -1.55$.

    Because -1.55 is not in the rejection region, we fail to reject H_0 and conclude at $\alpha = 0.05$ that the mean number of lottery tickets for Pennsylvania residents could still be 25.

4. The first three steps are the same as Exercise 3. The hypotheses are:

    H_0: $\mu = 25$

    H_1: $\mu \neq 25$.

    The level of significance has been chosen to be 0.01. We will use the normal (z) distribution for the test. The decision rule is to reject H_0 if the value of the test statistic is less than $\alpha = 0.01$. The computation of the test ratio would again yield a value of -1.55. To calculate the p-value, we find $2P(z > |-1.55|) = 2(5 - 0.4394) = 2(0.0606) = 0.1212$. Since $0.1212 > 0.01$, we fail to reject H_0 and conclude that the mean number of lottery tickets for Pennsylvania residents could still be 25.

5. An improvement would be a decrease in the mean time, so the hypotheses are:

    H_0: $\mu = 270$

    H_1: $\mu < 270$.

    The level of significance has been chosen to be 0.05. Since the sample size is $20 < 30$, we will use Student's t distribution for the test. This is a valid approach if the times for the data transmissions have a normal distribution. The rejection region for this left-tailed test is found by looking in the t table in the $df = 20 - 1 = 19$ row and the 0.05 column. There we find a t value of 1.729. The rejection region is:

$t < -1.729$. The decision rule is to reject the null hypothesis if $t < -1.729$; otherwise, fail to reject H_0. The test statistic is: $t = \dfrac{\overline{x} - \mu_0}{\widehat{\sigma}_{\overline{x}}} = \dfrac{\overline{x} - \mu_0}{s/\sqrt{n}} = \dfrac{265 - 270}{15\sqrt{20}} = -1.49$.

Because -1.49 is not less than -1.729, we fail to reject H_0 and conclude that there is not strong evidence that the mean is below 270 msec.

6. The first three steps are the same as Exercise 5. The hypotheses are:

H_0: $\mu = 270$

H_1: $\mu < 270$.

The level of significance has been chosen to be 0.05 and Student's t distribution will be used for the test. The decision rule is to reject H_0 if the value of the test statistic is less than $\alpha = 0.05$. The computation of the test ratio would again yield a value of -1.49. To calculate the p-value, we find $P(t < -1.49)$. From the t table we find $t_{0.10} = 1.328$ and $t_{0.05} = 1.729$. Using the symmetry of the t table, since $-1.729 < -1.49 < -1.328$, then $0.05 < p\text{-value} < 0.10$. Since $p\text{-value} > 0.05$, we fail to reject H_0 and conclude that there is not strong evidence that the mean is below 270 msec.

7. A deviation in the diameter in either direction is undesirable, so we will do a two-tailed test. The hypotheses are:

H_0: $\mu = 0.2000$

H_1: $\mu \neq 0.2000$.

The level of significance has been chosen to be 0.10. Since the sample size $= 14 < 30$, we will use Student's t distribution for the test. This is valid if the population of diameters has a normal distribution. The rejection region for this two-tailed test is found by looking in the t table in the df $= 14 - 1 = 13$ row and the $\dfrac{0.10}{2} = 0.05$ column. There we find a t value of 1.771. The rejection region is: $t < -1.771$ or $t > 1.771$. The decision rule is to reject the null hypothesis if the $t > 1.771$ or $t < -1.771$; otherwise, fail to reject H_0. Before we can calculate the test statistic, we need to compute the sample mean and standard deviation.

$$\overline{x} = \dfrac{2.78630}{14} = 0.199021 \text{ and } s = \sqrt{\dfrac{n\Sigma x^2 - (\Sigma x)^2}{n(n-1)}} = \sqrt{\dfrac{14(0.554565) - (2.78630)^2}{14(14-1)}} = 0.001559.$$

The test statistic is: $t = \dfrac{\overline{x} - \mu_0}{\widehat{\sigma}_{\overline{x}}} = \dfrac{\overline{x} - \mu_0}{s/\sqrt{n}} = \dfrac{0.199021 - 0.2000}{0.001559/\sqrt{14}} = -2.35$.

Since $-2.35 < -1.771$, we reject H_0 in favor of H_1 and conclude at $\alpha = 0.10$ that the bolts do not have a mean diameter of 0.2000.

8. The first three steps are the same as Exercise 7. The hypotheses are:

H_0: $\mu = 0.2000$

H_1: $\mu \neq 0.2000$.

The level of significance is 0.10. Student's t distribution with 13 degrees of freedom will be used for the test. The decision rule is to reject H_0 if the value of the test statistic is less than $\alpha = 0.10$. The test ratio is -2.35. To calculate the p-value, we find $2P(t > |-2.35|) = 2(0.01 \text{ and } 0.025)$. Therefore, the p-value is between 0.02 and 0.05. Since all values in this interval are less than 0.10, we reject H_0 in favor of H_1 and conclude at $\alpha = 0.10$ that the bolts do not have a mean diameter of 0.2000.

8-3 One-Sample Hypotheses Tests of Percentages

Study Objectives

You should be able to:

1. Recognize the similarities in the hypothesis testing procedures for percentages and means.
2. Perform hypothesis tests for percentages using either the classical or p-value approach.

Section Overview

The testing procedures for population percentages are similar to those for means. We still have the choice of using the seven-step classical or p-value approach. As with confidence intervals, what constitutes a large sample when dealing with percentages is different from what is a large sample when dealing with means. Rather than a large sample being a sample greater than 30, it is np and $n(100 - p)$ both being at least 500. Of course, since we are dealing with percentages rather than means, the test statistic changes. It is $z = \dfrac{p - \pi_0}{\sigma_p}$, where π_0 is the hypothesized value of the population percentage. Because we do not know the value of π, we use the hypothesized value in the calculation of the standard error: $\sigma_p = \sqrt{\dfrac{\pi_0(100 - \pi_0)}{n}}$.

Key Terms & Formulas

Test Statistic for Population Percentages $\dfrac{p - \pi_0}{\sigma_p}$, where π_0 is the hypothesized value of the population percentage

and $\sigma_p = \sqrt{\dfrac{\pi_0(100 - \pi_0)}{n}}$.

Worked Examples

Oyster Bed Patrol

An entrepreneur is considering marketing a new game called "Oyster Bed Patrol" in Pismo Beach, California. Experience tells this entrepreneur that if more than 10 percent of the population of an area this size enjoy the game, it will be profitable. To make the appropriate decision, the entrepreneur takes a random sample of 100 Pismo Beach residents and finds that 15 of them enjoy the game. Does this provide sufficient evidence for the entrepreneur to decide to market the game in Pismo Beach using a level of significance of 0.01?

Solution

Due to the interest of the entrepreneur, one hypothesis must be that the population proportion is more than 10%. Since this does not include equality, it is the alternative hypothesis. The two hypotheses are:

H_0: $\pi = 10\%$

H_1: $\pi > 10\%$.

We know that $\alpha = 0.01$. For this sample, $p = 15\%$, so $np = 1500$ and $n(100 - p) = 8500$. Because both exceed 500, we can use the normal distribution for the test. The rejection region for this right-tailed test is found by looking for $0.5 - 0.01 = 0.49$ in the normal table. There we find 2.33. The decision rule is to reject H_0 if $z > 2.33$; otherwise, fail to reject H_0. To compute the test statistic, we first calculate the standard error.

$$\sigma_p = \sqrt{\frac{\pi_0(100 - \pi_0)}{n}} = \sqrt{\frac{10(100 - 10)}{100}} = \sqrt{9} = 3. \text{ Then the test statistic is: } z = \frac{p - \pi_0}{\sigma_p} = \frac{15 - 10}{3} = 1.67.$$

Because 1.67 is not greater than 2.33, H_0 is not rejected. There is insufficient evidence to justify the entrepreneur's marketing the game.

Drug Relapse

The first-year recidivism rate for drug abuse for patients at a drug rehabilitation clinic has been 35%. A staff member has introduced a new therapy and wants to see if it has changed the recidivism rate. She takes a sample of 122 former patients who were treated by the new therapy and discovers 55 cases of repeated drug abuse in the first year after treatment. Use the p-value approach with $\alpha = 0.05$ to see if there has been any change in recidivism rate.

Solution

Because we want to see if any change has occurred, we test:

H_0: $\pi = 35\%$

H_1: $\pi \neq 35\%$.

We have $\alpha = 0.05$. For this sample $p = 45.082 \approx 45\%$, $np = 5500$ and $n(100 - p) = 6700$. Because both exceed 500, we can use the normal distribution for the test. The decision rule is to reject H_0 if the p-value < 0.05; otherwise, fail to reject H_0. To compute the test statistic, we need:

$$\sigma_p = \sqrt{\frac{\pi_0(100 - \pi_0)}{n}} = \sqrt{\frac{35(100 - 35)}{122}} = \sqrt{18.64754} \approx 4.32.$$

The test statistic $z = \dfrac{p - \pi_0}{\sigma_p} = \dfrac{45.082 - 35}{4.32} \approx 2.33.$

For this two-tailed test: p-value $= P(z < -|2.33|) + P(z > |2.33|) = 0.0099 + 0.0099 = 0.0198.$
Because $0.0198 < 0.05$, we reject H_0 and conclude at $\alpha = 0.05$ that there has been a change in the recidivism rate.

Practice Exercises 8-3

1-2. A delivery service claims it delivers at least 80% of packages within 4 hours. A consumer agency has trouble believing this claim and furtively uses this service to deliver 225 packages. Of these, 171 are delivered within this time.
   1. Using the classical method, do the appropriate test with $\alpha = 0.10$.
   2. What is the p-value of the test?

3-4. The personnel director of a company believes that a majority of the company's employees would rate work conditions as more important than salary. She samples 76 employees and finds that 41 rate work conditions higher than salary.
   3. Test the appropriate hypothesis using the classical approach with $\alpha = 0.01$.
   4. Calculate the p-value of the test.

Solutions to Practice Exercises 8-3

1. The two hypotheses are:

   H_0: $\pi = 80\%$

   H_1: $\pi < 80\%$.

   We know that $\alpha = 0.10$. For this sample, $p = 76\%$, so $np = 17,100$ and $n(100 - p) = 5,400$. Because both exceed 500, we can use the normal distribution for the test. The table value for this left-tailed test is found by looking for $0.5 - 0.10 = 0.40$ in the normal table. There we find 1.28. The decision rule is to reject H_0 if the test statistic $z < -1.28$; otherwise, fail to reject. Next we calculate

   $$\sigma_p = \sqrt{\frac{\pi_0(100 - \pi_0)}{n}} = \sqrt{\frac{80(100 - 10)}{225}} = \sqrt{7.11111} \approx 2.6667.$$

   The test statistic is: $z = \dfrac{p - \pi_0}{\sigma_p} = \dfrac{76 - 80}{2.6667} \approx -1.50.$

   Because $-1.50 < -1.28$, we reject H_0 and conclude at $\alpha = 0.10$ that the delivery service's claim is not true.

2. The p-value for this left-tailed test $= P(z < -1.50) = 0.5 - 0.4332 = 0.0668.$

3. A majority rating working condition over salary would mean that the percentage is more than 50%. Therefore, the hypotheses are:

   H_0: $\pi = 50\%$

   H_1: $\pi > 50\%$.

   We have $\alpha = 0.01$. For this sample, $p \approx 54\%$, so $np = 4,100$ and $n(100 - p) = 3,500$. Because both exceed 500, we can use the normal distribution for the test. The table value for this right-tailed test is found by looking for $0.5 - 0.01 = 0.49$ in the normal table. There we find 2.33. The decision rule is to reject H_0 if $z > 2.33$; otherwise, fail to reject. Next we calculate the standard error:

   $$\sigma_p = \sqrt{\frac{\pi_0(100 - \pi_0)}{n}} = \sqrt{\frac{50(100 - 50)}{76}} = \sqrt{32.8947} = 5.7354$$

   The test statistic is: $z = \dfrac{p - \pi_0}{\sigma_p} = \dfrac{54 - 50}{5.7354} \approx 0.69.$

   Because 0.69 is not greater than 2.33, we fail to reject H_0 and conclude that there is not substantial evidence that a majority think work conditions are more important than salary.

4. The p-value for this right-tailed test $= P(z > 0.69) = 0.5 - 0.2549 = 0.2451.$

8-4 One-Sample Hypothesis Tests of Variances and Standard Deviations

Study Objectives
You should be able to:
1. Recognize when to test a hypothesis about a variance or a standard deviation.
2. Do a hypothesis test about a variance or standard deviation.

Section Overview
An important characteristic of data is its dispersion or spread. Suppose we are manufacturing a product that should be a certain size (or weight or density or whatever). We would want the product to meet specifications as closely as possible. Not only does that mean we want the average size to be the prescribed amount, it also means we want the product to be consistently close to that size, that is, we want the variance (or standard deviation) to be small. In such cases, a hypotheses test about a population variance is important. If the population involved has a normal distribution, we can perform such a test based on the χ^2 distribution. The test statistic is $\chi^2 = \dfrac{(n-1)s^2}{\sigma_0^2}$, where σ_0^2 is the hypothesized variance. We can use either the classical or p-value approach to do this test.

Key Terms & Formulas
Test Ratio for Variances/Standard Deviations If a sample is taken from a normal population, we can test a hypothesis that the population variance is σ^2 by using the test statistic: $\chi^2 = \dfrac{(n-1)s^2}{\sigma_0^2}$. This test statistic has a χ^2 distribution with $n-1$ degrees of freedom.

Worked Examples
Peter's Pepper Poi Potpourri
Suppose Peter, the maker of a Hawaiian hot sauce, bottles his sauce with a machine that has a variance of 0.080 oz.^{2} in the amount of fill. He is considering replacing the machine with a new model. He wants to make the switch only if there is strong evidence that the new model will produce a more consistent fill, i.e., has a smaller variance. Peter takes a sample of 19 bottles and finds that the variance is 0.038 oz.^{2}. Is this sufficient evidence to justify switching to the new machine using a test with a level of significance of 0.05?

Solution
Peter wants to switch only if the variance of the new machine is less than the old. So the hypotheses are:
H_0: $\sigma^2 = 0.080$
H_1: $\sigma^2 < 0.080$.
We want $\alpha = 0.05$. If it is reasonable to believe that the fill amounts are normal, we can use the χ^2 distribution to run this test. The rejection region of this left-tailed test would consist of values of the test statistic below a χ^2 table value in the df $= 19 - 1 = 18$ row and the $1 - \alpha = 1 - 0.05 = 0.95$ column. That value is 9.39. The decision rule is to reject H_0 if $\chi^2 < 9.39$; otherwise, fail to reject H_0.

The test statistic is: $\chi^2 = \dfrac{(n-1)s^2}{\sigma_0^2} = \dfrac{(19-1)(0.038)}{0.080} \approx 8.55$.

Because $8.55 < 9.39$, H_0 is rejected. There is sufficient evidence at $\alpha = 0.05$ to justify switching to the new model.

Peter's Pepper Poi Potpourri p-value
For Peter's problem, what is the p-value of the test? Additionally, what would the p-value be if this had been a two-tailed test?

Solution

To find the p-value of this test, we would look in the χ^2 table in the df $= 18$ row. We would look for two values that bracket the value of the test ratio, 8.55. We find 8.23 and 9.39 in the 0.975 and 0.95 columns respectively. Since this is a lower-tailed test, this means that the p-value is somewhere between $1 - 0.975$ and $1 - 0.95$, i.e., between 0.025 and 0.05. If this had been a two-tailed test, the p-value would be between twice these amounts, i.e., the p-value would be between 0.05 and 0.10.

Practice Exercises 8-4

1-2. One theory of designing exams says that, on a 100-point exam, the optimal standard deviation for the test is 15 points. Less than that, the questions are too easy; more than that, the questions are too hard. A statistics instructor decides to evaluate her exams by testing to see if the standard deviation is 15. She takes a sample of 26 test grades and finds that the standard deviation is 19.3.

   1.   Use the classical approach to test the appropriate hypothesis at $\alpha = 0.01$.
   2.   What is the p-value of the test?

3-4. The diameter of ball bearings used in bicycles should have a variance no more than 0.03 mm^2. The maker of Teton Tough bikes takes a sample of 16 ball bearings from a shipment and finds that the variance is 0.056 mm^2. Is this sufficient evidence to reject the shipment at $\alpha = 0.10$?

   3.   Use the classical approach to test the appropriate hypothesis.
   4.   What is the p-value of the test?

Solutions to Practice Exercises 8-4

   1.   The hypotheses are:
        H_0: $\sigma = 15$
        H_1: $\sigma \neq 15$.
        We want $\alpha = 0.01$. If the test grades are normal, we can use the χ^2 distribution to run this test. The critical values of this two-tailed test are found in the χ^2 table in the df $= 26 - 1 = 25$ row and the $1 - \alpha/2 = 1 - 0.01/2 = 0.995$ and the $\alpha/2 = 0.01/2 = 0.005$ columns. Those values are 10.52 and 46.9. The decision rule is to reject H_0 if $\chi^2 < 10.52$ or if $\chi^2 > 46.9$; otherwise, fail to reject H_0. The test statistic is: $\chi^2 = \dfrac{(n-1)s^2}{\sigma_0^2} = \dfrac{(26-1)19.3^2}{15^2} = 41.39$. Since the χ^2 does not fall in the rejection region, we would conclude that the test could have a standard deviation of 15.

   2.   The calculated value of the test statistic of 41.39 is found in the 25-df row between 40.6 (in the 0.025 column) and 44.3 (in the 0.01 column). Because this is a two-tailed test, this means that the p-value is between $2(0.01)$ and $2(0.025)$, i.e., $0.02 < p\text{-value} < 0.05$.

   3.   The hypotheses are:
        H_0: $\sigma^2 = 0.03$
        H_1: $\sigma^2 > 0.03$.
        We want $\alpha = 0.10$. If the diameters are normal, we can use the χ^2 distribution to run this test. The critical value of this right-tailed test is found in the χ^2 table in the df $= 16 - 1 = 15$ row and the $\alpha = 0.10$ column. This value is 22.3. The decision rule is to reject H_0 if $\chi^2 > 22.3$; otherwise, fail to reject H_0. The test statistic is: $\chi^2 = \dfrac{(n-1)s^2}{\sigma_0^2} = \dfrac{(16-1)0.056^2}{0.03} = 28.00$. Since $28.0 > 22.3$, the null hypothesis is rejected and we conclude at $\alpha = 0.10$ that the shipment should be rejected.

   4.   The calculated value of the test statistic of 28.0 is found in the 15 df row between 27.5 (in the 0.025 column) and 30.6 (in the 0.01 column). Because this is a right-tailed test, the p-value is between the column headings, i.e., $0.01 < p\text{-value} < 0.025$.

Solutions to Odd-Numbered Exercises

1. Step 1: State the Null and Alternative Hypotheses. H_0: $\mu = 7$ and H_1: $\mu > 7$.

Step 2: Select the Level of Significance. $\alpha = 0.05$ (given).

Step 3: Determine the Test Distribution to Use. Since $n > 30$, we use the z distribution.

Step 4: Define the Rejection or Critical Regions. This is a one-tailed test since H_1 states that $\mu > 7$. With $\alpha = 0.05$, the remaining area is $0.5000 - 0.0500 = 0.4500$ which corresponds to a z value of 1.645.

Step 5: State the Decision Rule. Reject H_0 in favor of H_1 if $z > 1.645$. Otherwise, fail to reject H_0.

Step 6: Make the Necessary Computations. We are given the following data:

$$\bar{x} = 8.1; \quad s = 2.3; n = 84. \text{ Since } \sigma \text{ is unknown, we use } \hat{\sigma}_{\bar{x}} = \frac{s}{\sqrt{n}} = \frac{2.3}{\sqrt{84}} = 0.2509.$$

$$z = \frac{\bar{x} - \mu_0}{\hat{\sigma}_{\bar{x}}} = \frac{8.1 - 7}{0.2509} = 4.38.$$

Step 7: Make a Statistical Decision. Since $z = 4.38 > 1.645$, we reject H_0.

3. Step 1: State the Null and Alternative Hypotheses. H_0: $\pi = 75$ and H_1: $\pi < 75$.

Step 2: Select the Level of Significance. $\alpha = 0.01$ (given).

Step 3: Determine the Test Distribution to Use. Since $np > 500$ and $n(100 - p) > 500$, we use the z distribution.

Step 4: Define the Rejection or Critical Regions. This is a one-tailed test since H_1 states that $\pi < 75$. With $\alpha = 0.01$, the remaining area is $0.5000 - 0.0100 = 0.4900$ which corresponds to a z value of 2.33.

Step 5: State the Decision Rule. Reject H_0 in favor of H_1 if $z < -2.33$. Otherwise, fail to reject H_0.

Step 6: Make the Necessary Computations. We are given the following data: 452 successful out of 639.

$$p = \frac{452}{639} \times (100 \text{ percent}) = 70.7355\% \text{ and } \sigma_p = \sqrt{\frac{\pi_0(100 - \pi_0)}{n}} = \sqrt{\frac{75 \cdot 25}{639}} = 1.713. \text{ Then}$$

$$z = \frac{p - \pi_0}{\sigma_p} = \frac{70.7355 - 75}{1.713} = -2.4895.$$

Step 7: Make a Statistical Decision. Since $z = 2.4895 < -2.33$, we reject H_0.

5. Step 1: H_0: $\mu = 20.9$ and H_1: $\mu \neq 20.9$.

Step 2: $\alpha = 0.01$.

Step 3: Since $n > 30$, we use the z distribution.

Step 4: This is a two-tailed test. With $\alpha = 0.01$ the risk of error in each tail is 0.005, the remaining area is $0.5000 - 0.0050 = 0.4950$ which corresponds to a z value of 2.575.

Step 5: Reject H_0 in favor of H_1 if $z < -2.575$ or if $z > 2.575$. Otherwise, fail to reject H_0.

Step 6: Data: $\bar{x} = 12.8$; $s = 6.98$; $n = 32$. Since σ is unknown, we use $\hat{\sigma}_{\bar{x}} = \frac{s}{\sqrt{n}} = \frac{6.98}{\sqrt{32}} = 1.234$. Then

$$z = \frac{\bar{x} - \mu_0}{\hat{\sigma}_{\bar{x}}} = \frac{12.8 - 20.9}{1.234} = -6.565.$$

Step 7: Since $z = -6.565 < -2.575$, we reject H_0.

7. Step 1: H_0: $\pi = 48$ and H_1: $\pi < 48$.

Step 2: $\alpha = 0.01$.

Step 3: Since $np > 500$ and $n(100 - p) > 500$, we use the z distribution.

Step 4: This is a one-tailed test. With $\alpha = 0.01$, the remaining area is $0.5000 - 0.0100 = 0.4900$ which corresponds to a z value of -2.33. Otherwise, fail to reject H_0.

Step 5: Reject H_0 in favor of H_1 if $z < -2.33$.

Step 6: Data: 1,392 out of 6,629. $p = \dfrac{1,392}{6,629} \times (100 \text{ percent}) = 20.9986\%$ and

$$\sigma_p = \sqrt{\frac{\pi_0(100 - \pi_0)}{n}} = \sqrt{\frac{48 \cdot 52}{6,629}} = 0.6136. \text{ Then } z = \frac{p - \pi_0}{\sigma_p} = \frac{20.9986 - 48}{0.6136} = -44.0035.$$

Step 7: Since $z = -44.0035 < -2.33$, we reject H_0.

9. Step 1: H_0: $\mu = 9.64$ and H_1: $\mu > 9.64$.
    Step 2: $\alpha = 0.05$.
    Step 3: Since $n > 30$, we use the z distribution.
    Step 4: This is a one-tailed test. With $\alpha = 0.05$, the remaining area is $0.5000 - 0.0500 = 0.4500$ which corresponds to a z value of 1.645.
    Step 5: Reject H_0 in favor of H_1 if $z > 1.645$. Otherwise, fail to reject H_0.
    Step 6: Data: $\bar{x} = 11.52$; $s = 7.06$; $n = 41$. Since σ is unknown, we use $\hat{\sigma}_{\bar{x}} = \dfrac{s}{\sqrt{n}} = \dfrac{7.06}{\sqrt{41}} = 1.1026$.

    Then $z = \dfrac{\bar{x} - \mu_0}{\hat{\sigma}_{\bar{x}}} = \dfrac{11.52 - 9.64}{1.1026} = 1.7051$.
    Step 7: Since $z = 1.7051 > 1.645$, we reject H_0.

11. Step 1: H_0: $\pi = 30$ and H_1: $\pi \neq 30$.
    Step 2: $\alpha = 0.05$.
    Step 3: Since $np > 500$ and $n(100 - p) > 500$, we use the z distribution.
    Step 4: This is a two-tailed test. With $\alpha = 0.05$ the risk of error in each tail is 0.025, the remaining area is $0.5000 - 0.0250 = 0.4750$ which corresponds to a z value of 1.96.
    Step 5: Reject H_0 in favor of H_1 if $z < -1.96$ or if $z > 1.96$. Otherwise, fail to reject H_0.
    Step 6: Data: 357 out of 1,192. $p = \dfrac{357}{1,192} \times (100 \text{ percent}) = 29.9497\%$ and

    $$\sigma_p = \sqrt{\frac{\pi_0(100 - \pi_0)}{n}} = \sqrt{\frac{30 \cdot 70}{1,192}} = 1.3273. \text{ Then } z = \frac{p - \pi_0}{\sigma_p} = \frac{29.9497 - 30}{1.3273} = -0.0379.$$

    Step 7: Since $z = -0.0379$ is between ± 1.96, we fail to reject H_0.

13. Step 1: H_0: $\mu = 14.91$ and H_1: $\mu < 14.91$.
    Step 2: $\alpha = 0.05$.
    Step 3: Since $n < 30$ and we assume that population values are normally distributed, we use the t distribution with 25 df.
    Step 4: This is a one-tailed test, with $\alpha = 0.05$, $t_\alpha = t_{0.01} = -1.708$.
    Step 5: Reject H_0 in favor of H_1 if $t < -1.708$. Otherwise, fail to reject H_0.
    Step 6: Data: $\bar{x} = 13.75$; $s = 3.26$; $n = 26$. $\hat{\sigma}_{\bar{x}} = \dfrac{s}{\sqrt{n}} = \dfrac{3.26}{\sqrt{26}} = 0.6393$. Then

    $t = \dfrac{\bar{x} - \mu_0}{\hat{\sigma}_{\bar{x}}} = \dfrac{13.75 - 14.91}{0.6393} = -1.8144$
    Step 7: Since $t = -1.8144 < -1.708$, we reject H_0.

15. Step 1: H_0: $\sigma^2 = 16$ and H_1: $\sigma^2 < 16$.
    Step 2: $\alpha = 0.01$.
    Step 3: Use χ^2 distribution on variances with 28 df.
    Step 4: This is a one-tailed test, the rejection region falls below $\chi^2_{1-\alpha}$. With $\alpha = 0.01$ and 28 df and $\chi^2_{1-\alpha} = \chi^2_{.99} = 13.56$.
    Step 5: Reject H_0 in favor of H_1 if $\chi^2 < 13.56$. Otherwise, fail to reject H_0.
    Step 6: Data: $s^2 = 15.62$ and $\sigma^2 = 16$. Then $\chi^2 = \dfrac{(n - 1)s^2}{\sigma^2} = \dfrac{28 \cdot 15.62}{16} = 27.3350$.

Step 7: Since $\chi^2 > 13.56$, we fail to reject H_0.

17. Step 1: H_0: $\pi = 68$ and H_1: $\pi < 68$.
Step 2: $\alpha = 0.05$.
Step 3: Since $np > 500$ and $n(100 - p) > 500$, we use the z distribution.
Step 4: This is a one-tailed test. With $\alpha = 0.05$, the remaining area is $0.5000 - 0.0500 = 0.4500$ which corresponds to a z value of -1.645.
Step 5: Reject H_0 in favor of H_1 if $z < -1.645$. Otherwise, fail to reject H_0.
Step 6: Data: 164 out of 338. $p = \dfrac{164}{338} \times (100 \text{ percent}) = 48.5207\%$ and

$$\sigma_p = \sqrt{\frac{\pi_0(100 - \pi_0)}{n}} = \sqrt{\frac{68 \cdot 32}{338}} = 2.5373. \text{ Then } z = \frac{p - \pi_0}{\sigma_p} = \frac{48.5207 - 68}{2.5373} = -7.6772.$$

Step 7: Since $z < -1.645$, we reject H_0.

19. Step 1: H_0: $\mu = 4.5$ and H_1: $\mu \neq 4.5$.
Step 2: $\alpha = 0.05$.
Step 3: Since $n < 30$ and we assume that population values are normally distributed, we use the t distribution.
Step 4: This is a two-tailed test. With $\alpha = 0.05$ and $n = 6$, we have 5 df so we use $t_{\alpha/2} = t_{.025} = 2.571$.
Step 5: Reject H_0 in favor of H_1 if $t < -2.571$ or if $t > 2.571$. Otherwise, fail to reject H_0.
Step 6: Data: $\bar{x} = 4.17$; $s = 0.32$; $n = 6$. $\hat{\sigma}_{\bar{x}} = \dfrac{s}{\sqrt{n}} = \dfrac{0.32}{\sqrt{6}} = 0.1306$. Then

$$t = \frac{\bar{x} - \mu_0}{\hat{\sigma}_{\bar{x}}} = \frac{4.17 - 4.5}{0.1306} = -2.5626.$$

Step 7: Since $t = -2.526$ is within ± 2.571, we fail to reject H_0.

21. Step 1: H_0: $\sigma^2 = 1.5$ and H_1: $\sigma^2 > 1.5$.
Step 2: $\alpha = 0.01$.
Step 3: Use χ^2 distribution on variances with 14 df.
Step 4: This is a one-tailed test, the acceptance region falls below χ^2_α. With $\alpha = 0.01$ and 14 df and $\chi^2_\alpha = \chi^2_{.01} = 29.1$.
Step 5: Reject H_0 in favor of H_1 if $\chi^2 > 29.1$. Otherwise, fail to reject H_0.
Step 6: Data: $s = 1.7$ (note you are given sample standard deviation and a hypothesized population variance) and $n = 15$. Then $\chi^2 = \dfrac{(n - 1)s^2}{\sigma^2} = \dfrac{14 \cdot (1.7)^2}{1.5} = 26.9733$.
Step 7: Since $\chi^2 < 26.9733$, we fail to reject H_0.

23. Step 1: H_0: $\mu = 55$ and H_1: $\mu \neq 55$.
Step 2: $\alpha = 0.01$.
Step 3: Since $n > 30$, we use the z distribution.
Step 4: This is a two-tailed test. With $\alpha = 0.01$, the remaining area is $0.5000 - 0.0050 = 0.4950$ which corresponds to a z value of 2.575.
Step 5: Reject H_0 in favor of H_1 if $z < -2.575$ or if $z > 2.575$. Otherwise, fail to reject H_0.
Step 6: Data: $\bar{x} = 58.3$; $s = 20.5$; $n = 281$. $\hat{\sigma}_{\bar{x}} = \dfrac{s}{\sqrt{n}} = \dfrac{20.5}{\sqrt{281}} = 1.2229$. Then

$$z = \frac{\bar{x} - \mu_0}{\hat{\sigma}_{\bar{x}}} = \frac{58.3 - 55}{1.2229} = 2.6984$$

Step 7: Since $z = 2.6984 > 2.575$, we reject H_0.

25. Step 1: H_0: $\pi = 33$ and H_1: $\pi > 33$.
Step 2: $\alpha = 0.01$.

Step 3: Since $np > 500$ and $n(100 - p) > 500$, we use the z distribution.

Step 4: This is a one-tailed test. With $\alpha = 0.01$, the remaining area is $0.5000 - 0.0100 = 0.4900$ which corresponds to a z value of 2.33.

Step 5: Reject H_0 in favor of H_1 if $z > 2.33$. Otherwise, fail to reject H_0.

Step 6: Data: 17 out of 49. $p = \dfrac{17}{49} \times (100 \text{ percent}) = 34.6939\%$ and

$$\sigma_p = \sqrt{\frac{\pi_0(100 - \pi_0)}{n}} = \sqrt{\frac{33 \cdot 67}{49}} = 6.7173. \text{ Then } z = \frac{p - \pi_0}{\sigma_p} = \frac{34.6939 - 33}{6.7173} = 0.2522.$$

Step 7: Since $z = 0.2522$ is below 2.33, we fail to reject H_0.

27. Step 1: H_0: $\mu = 430$ and H_1: $\mu < 430$.

Step 2: $\alpha = 0.05$.

Step 3: Since $n < 30$ and we assume that population values are normally distributed, we use the t distribution.

Step 4: This is a one-tailed test. With $\alpha = 0.05$ and $n = 26$, we have 25 df so we use $t_\alpha = t_{0.05} = -1.708$.

Step 5: Reject H_0 in favor of H_1 if $t < -1.708$. Otherwise, fail to reject H_0.

Step 6: Data: $\bar{x} = 427$; $s = 5$; $n = 26$.

$$\hat{\sigma}_{\bar{x}} = \frac{s}{\sqrt{n}} = \frac{5}{\sqrt{26}} = 0.9806. \text{ Then } t = \frac{\bar{x} - \mu_0}{\hat{\sigma}_{\bar{x}}} = \frac{427 - 430}{0.9806} = -3.0594.$$

Step 7: Since $t = -3.0594 < -1.708$, we reject H_0.

29. Step 1: H_0: $\mu = 50$ and H_1: $\mu < 50$.

Step 2: $\alpha = 0.05$.

Step 3: Although $n < 30$, we still use the z distribution since σ is known.

Step 4: This is a one-tailed test. With $\alpha = 0.05$, the remaining area is $0.5000 - 0.0500 = 0.4500$ which corresponds to a z value of -1.645.

Step 5: Reject H_0 in favor of H_1 if $z < -1.645$. Otherwise, fail to reject H_0.

Step 6: Data: $\bar{x} = 49.65$; $s = 0.60$; $n = 4$. So $\hat{\sigma}_{\bar{x}} = \dfrac{s}{\sqrt{n}} = \dfrac{0.60}{\sqrt{4}} = 0.300$. Then

$$z = \frac{\bar{x} - \mu_0}{\hat{\sigma}_{\bar{x}}} = \frac{49.65 - 50}{0.300} = -1.1667.$$

Step 7: Since $z > -1.645$, we fail to reject H_0.

31. Step 1: H_0: $\mu = 2.6$ and H_1: $\mu \neq 2.6$.

Step 2: $\alpha = 0.10$.

Step 3: Since $n < 30$ and we assume that population values are normally distributed, we use the t distribution.

Step 4: This is a two-tailed test. With $\alpha = 0.10$ and $n = 12$, we have 11 df so we use $t_{\alpha/2} = t_{0.05} = 1.796$.

Step 5: Reject H_0 in favor of H_1 if $t < -1.796$ or if $t > 1.796$. Otherwise, fail to reject H_0.

Step 6: Data: $\bar{x} = 2.58$; $s = 0.035$; $n = 12$. So $\hat{\sigma}_{\bar{x}} = \dfrac{s}{\sqrt{n}} = \dfrac{0.035}{\sqrt{12}} = 0.0101$. Then

$$t = \frac{\bar{x} - \mu_0}{\hat{\sigma}_{\bar{x}}} = \frac{2.58 - 2.6}{0.0101} = -1.9795.$$

Step 7: Since $t < -1.796$, we reject H_0.

33. Step 1: H_0: $\pi = 60$ and H_1: $\pi < 60$.

Step 2: $\alpha = 0.01$.

Step 3: Since $np > 500$ and $n(100 - p) > 500$, we use the z distribution.

Step 4: This is a one-tailed test. With $\alpha = 0.01$, the remaining area in each tail is $0.5000 - 0.0100 = 0.4900$ which corresponds to a z value of 2.33.

Step 5: Reject H_0 in favor of H_1 if $z < -2.33$. Otherwise, fail to reject H_0.

Step 6: Data: 116 out of 200. $p = \dfrac{116}{200} \times (100 \text{ percent}) = 58\%$ and $\sigma_p = \sqrt{\dfrac{\pi_0(100 - \pi_0)}{n}} = \sqrt{\dfrac{60 \cdot 40}{200}} = 3.4641$. $z = \dfrac{p - \pi_0}{\sigma_p} = \dfrac{58 - 60}{3.4641} = -0.5774$.

Step 7: Since $z > -2.33$, we fail to reject H_0. Senator Wilson is more likely right.

35. Step 1: H_0: $\mu = 750$ and H_1: $\mu < 750$.

Step 2: $\alpha = 0.10$.

Step 3: Since $n < 30$ and we assume that population values are normally distributed, we use the t distribution.

Step 4: This is a one-tailed test. With $\alpha = 0.10$ and $n = 10$, we have 9 df so we use $t_\alpha = t_{0.01} = -1.383$.

Step 5: Reject H_0 in favor of H_1 if $t < -1.383$. Otherwise, fail to reject H_0.

Step 6: Data: $\bar{x} = 710$; $s = 40$; $n = 10$. $\hat{\sigma}_{\bar{x}} = \dfrac{s}{\sqrt{n}} = \dfrac{40}{\sqrt{10}} = 12.6491$. So

$$t = \dfrac{\bar{x} - \mu_0}{\hat{\sigma}_{\bar{x}}} = \dfrac{710 - 750}{12.6491} = -3.1623.$$

Step 7: Since $t = -3.1623 < -1.383$, we reject H_0.

37. Step 1: H_0: $\sigma^2 = 10{,}500$ and H_1: $\sigma^2 < 10{,}500$.

Step 2: $\alpha = 0.05$.

Step 3: Use χ^2 distribution on variances with 12 df.

Step 4: This is a one-tailed test, the rejection region falls below $\chi^2_{1-\alpha}$. With $\alpha = 0.05$, 12 df, and $\chi^2_{1-\alpha} = \chi^2_{0.95} = 5.23$.

Step 5: Reject H_0 in favor of H_1 if $\chi^2 < 5.23$. Otherwise, fail to reject H_0.

Step 6: Data: $s = 100$ (note: you are given sample standard deviation and a hypothesized population variance) and $n = 13$. $\chi^2 = \dfrac{(n-1)s^2}{\sigma^2} = \dfrac{12 \cdot (100)^2}{10}$

Step 7: Since $\chi^2 > 5.23$, we fail to reject H_0.

39. Step 1: H_0: $\pi = 50$ and H_1: $\pi \neq 50$.

Step 2: $\alpha = 0.01$.

Step 3: Since $np > 500$ and $n(100 - p) > 500$, we use the z distribution.

Step 4: This is a two-tailed test. With $\alpha = 0.01$, the remaining area in each tail is $0.5000 - 0.0050 = 0.4950$ which corresponds to a z value of 2.575.

Step 5: Reject H_0 in favor of H_1 if $z < -2.575$ or if $z > 2.575$. Otherwise, fail to reject H_0.

Step 6: Data: 696 out of 1293. $p = \dfrac{696}{1{,}293} \times (100 \text{ percent}) = 48.41\%$ and

$$\sigma_p = \sqrt{\dfrac{\pi_0(100 - \pi_0)}{n}} = \sqrt{\dfrac{50 \cdot 50}{1{,}293}} = 1.3905. \quad z = \dfrac{p - \pi_0}{\sigma_p} = \dfrac{48.41 - 50}{1.3905} = -1.14.$$

Step 7: Since $-2.575 < z < 2.575$, we fail to reject H_0.

41. Step 1: H_0: $\sigma = 875$ and H_1: $\sigma < 875$.

Step 2: $\alpha = 0.05$.

Step 3: Use χ^2 test on variances.

Step 4: One-tailed test, the rejection region falls below $\chi^2_{1-\alpha}$. So with $\alpha = 0.05$ and 5 df we have $\chi^2_{1-\alpha} = \chi^2_{0.95} = 1.145$.

Step 5: Reject H_0 in favor of H_1 if $\chi^2 < 1.145$. Otherwise, fail to reject H_0.

Step 6: Data: $n = 6$. Calculate $\bar{x}$ and s^2. From Exercise 40, $\bar{x} = 72137.5$.

x	$x - \bar{x}$	$(x - \bar{x})^2$
7512	298.5	89102.25
7620	406.5	165242.25
6982	−231.5	53592.25
7256	42.5	1806.25
6766	−447.5	200256.25
7145	−68.5	4692.25
		514691.50

$$s^2 = \frac{\Sigma(x - \bar{x})^2}{n - 1} = \frac{514691.5}{5} = 102,938.3. \text{ Then } \chi^2 = \frac{(n-1)s^2}{\sigma^2} = \frac{5 \cdot 102,938.3}{(875)^2} = 0.672.$$

Step 7: Since $\chi^2 < 0.672$ we reject H_0, and conclude that the standard deviation is less that 875.

43. Step 1: H_0: $\mu = 50,000$ and H_1: $\mu > 50,000$.

Step 2: $\alpha = 0.01$.

Step 3: t distribution.

Step 4: One-tailed test with $\alpha = 0.05$ and 21 df we use $t_\alpha = t_{0.01} = 2.831$.

Step 5: Reject H_0 in favor of H_1 if $t > 2.831$. Otherwise, fail to reject H_0.

Step 6: Data: (from MINITAB) $\bar{x} = 41,354$ and $s = 7912$; $n = 22$. Then $\hat{\sigma}_{\bar{x}} = \dfrac{s}{\sqrt{n}} = \dfrac{7912}{\sqrt{22}} = 1686.84$.

Then $t = \dfrac{41,354 - 50,000}{1686.84} = -5.126$.

Step 7: Since $t < 2.831$, we fail to reject H_0.

45. Step 1: H_0: $\mu = 100$ mg/dl and H_1: $\mu \neq 100$ mg/dl.

Step 2: $\alpha = 0.05$.

Step 3: t distribution.

Step 4: Two-tailed test with $\alpha = 0.05$ and 18 df we use $t_{\alpha/2} = t_{0.025} = 2.101$.

Step 5: Reject H_0 in favor of H_1 if $t > 2.101$ or if $t < -2.101$. Otherwise, fail to reject H_0.

Step 6: Data: (from MINITAB) $\bar{x} = 120.8$, $\sigma = 30$; $n = 19$. So $\sigma_{\bar{x}} = \dfrac{\sigma}{\sqrt{n}} = \dfrac{30}{\sqrt{19}} = 6.882$. Then

$t = \dfrac{120.8 - 100}{6.882} = 3.922$.

Step 7: Since $t > 2.101$, we reject H_0.

47. Step 1: H_0: $\pi = 20$ and H_1: $\pi < 20$.

Step 2: $\alpha = 0.10$.

Step 3: Since $np > 500$ and $n(100 - p) > 500$, we use the z distribution.

Step 4: This is a one-tailed test. With $\alpha = 0.10$, the remaining area in each tail is $0.5000 - 0.1000 = 0.4000$ which corresponds to a z value of 1.28.

Step 5: Reject H_0 in favor of H_1 if $z < -1.28$. Otherwise, fail to reject H_0.

Step 6: Data: 29 out of 200. $p = \dfrac{29}{200} \times (100 \text{ percent}) = 14.5\%$ and

$-\sigma_p = \sqrt{\dfrac{\pi_0(100 - \pi_0)}{n}} = \sqrt{\dfrac{20 \cdot 80}{200}} = 2.828. \quad z = \dfrac{p - \pi_0}{\sigma_p} = \dfrac{14.5 - 20}{2.828} = -1.945.$

Step 7: Since $z < -1.28$, we reject H_0, and conclude that the percent of Mexican Americans at Stanislaus State has decreased.

49. Step 1: H_0: $\pi = 26.7$ and H_1: $\pi > 26.7$.

Step 2: $\alpha = 0.10$.

Step 3: Since $np > 500$ and $n(100 - p) > 500$, we use the z distribution.

Step 4: This is a one-tailed test. With $\alpha = 0.10$, the remaining area in each tail is $0.5000 - 0.1000 = 0.4000$ which corresponds to a z value of 1.28.

Step 5: Reject H_0 in favor of H_1 if $z > 1.28$. Otherwise, fail to reject H_0.

Step 6: Data: 144 out of 450. $p = \dfrac{144}{450} \times (100 \text{ percent}) = 32\%$ and

$$\sigma_p = \sqrt{\frac{\pi_0(100 - \pi_0)}{n}} = \sqrt{\frac{26.7 \cdot 73.3}{450}} = 2.085. \quad z = \frac{p - \pi_0}{\sigma_p} = \frac{32 - 26.7}{2.085} = 2.542.$$

Step 7: Since $z > 1.28$, we reject H_0, and conclude that the percent of Asian Americans at San Francisco State has increased.

51. Step 1: H_0: $\mu = 2,100$, and H_1: $\mu > 2,100$.

Step 2: $\alpha = 0.05$.

Step 3: Use the t distribution with 8 df.

Step 4: The values for a right-tailed test at the 0.05 level, $t = 1.860$.

Step 5: Reject H_0 in favor of H_1 if $z > 1.860$. Otherwise, fail to reject H_0.

Step 6: From MINITAB, $\bar{x} = 2,493$ and $s = 500$. Since σ is unknown we use $\hat{\sigma}_{\bar{x}} = \dfrac{s}{\sqrt{n}} = \dfrac{500}{\sqrt{9}} = 166.67$.

Then $z = \dfrac{2,493 - 2,100}{166.67} = 2.358$.

Step 7: Since the z value of 2.358 fall in the rejection region, we reject H_0.

Chapter 9 Inference: Two-Sample Procedures

Study Aids and Practice Exercises

9-1 Hypothesis Tests of Two Variances

Study Objectives
You should be able to:
1. Use the F table.
2. Do a hypothesis test on the equality of two variances.

Section Overview
   Rather than comparing a parameter to a hypothesized value (as we did in the previous chapter), we often try to compare the parameters of two different populations. We might want to see if the mean salary offer to males is identical to the mean salary offer to females. Or we might want to see if the percentage of cases of cancer among smokers is equal to the percentage among nonsmokers. Or, as we do in this section, we might want to see if the variances of two populations are equal.

   To test a hypothesis about the equality of two population variances, we need to work with a new probability distribution. If we take two random samples independently from normal populations with equal variances, then the test statistic $\dfrac{s_1^2}{s_2^2}$ has an F distribution with numerator degrees of freedom $n_1 - 1$ and denominator degrees of freedom $n_2 - 1$. This is the test ratio for testing the equality of two variances. The critical values of the test are found in the F table.

Key Terms & Formulas

F Distribution The probability distribution of $\dfrac{s_1^2}{s_2^2}$ when two independent random samples are taken from normal populations with equal variances.

Worked Examples
Scuba
The manager of a scuba shop is trying to choose between two brands of air compressors. Both brands fill air tanks with appropriate mean amounts of air. One compressor is cheaper than the other. The manager will purchase the cheaper brand unless he thinks that the expensive brand is less variable in the fill amounts. Both manufacturers were willing to demonstrate their products. The cheaper brand filled six tanks, and the fill amounts had a variance of 492 lb/in^2. The more expensive brand filled seven tanks and had a variance of 329 lb/in^2. Perform the appropriate test, using a level of significance of 0.05.

Solution

This is a one-tailed hypothesis test. The hypotheses are:

H_0: $\sigma_1^2 = \sigma_2^2$
H_1: $\sigma_1^2 > \sigma_2^2$

where $\sigma_1^2 =$ the variance of the cheaper brand and $\sigma_2^2 =$ the variance of the expensive compressor.

Hint: Because of the way the F tables are constructed, the only one-tailed alternative we can test is H_1: $\sigma_1^2 > \sigma_2^2$. This is not a problem as the numbers 1 and 2 used to identify the samples are arbitrary numbers. If we want to test what is originally written as H_1: $\sigma_1^2 < \sigma_2^2$, we can switch the meaning of 1 and 2 and test H_1: $\sigma_1^2 > \sigma_2^2$.

The level of significance has been chosen to be 0.05. If the amount of fill is normal, the F distribution may be used to perform the test. The test statistic is $\dfrac{s_1^2}{s_2^2}$. This test statistic has an F distribution with 5 numerator degrees of freedom and 6 denominator degrees of freedom. The critical value is in the 0.05 F table in the 5 degrees of freedom column and the 6 degrees of freedom row. There we find 4.39. So the decision rule is to reject H_0 if the test statistic exceeds 4.39. The calculated value of the test statistic is: $\dfrac{492}{392} = 1.50$.

Since this is not in the rejection region, we fail to reject H_0 and conclude that there is no evidence of a significant difference in the variances between the two air compressors.

Hint: If this had been a two-tailed test, i.e., if the alternative hypothesis had been H_1: $\sigma_1^2 \neq \sigma_2^2$, then the test statistic is the larger sample variance over the smaller sample variance. In this example, the test statistic and the associated degrees of freedom remain the same. If, however, the sample variance of the more expensive brand was greater than that of the cheaper brand, the test statistic would be the variance of the more expensive brand divided by the variance of the cheaper brand. This would interchange the degrees of freedom (as the first set of degrees of freedom is always the degrees of freedom associated with the numerator). No matter which ratio is used, the F table is entered at the $\alpha/2$ level. For this example, we would use the $0.05/2 = 0.025$ F table.

Electric Life

The data that follows represents the life (in months) of components of an electrical system supplied by two different manufacturers. Use this data to test to the hypothesis that the variances in the component lives are identical for the two different suppliers, with $\alpha = 0.05$.

Supplier 1	Supplier 2
60 53 53 69 56 54	57 58 51 55
64 65 47 38 40	47 57 55 58

Solution

This is a two-tailed hypothesis test. The hypotheses are:

H_0: $\sigma_1^2 = \sigma_2^2$
H_1: $\sigma_1^2 \neq \sigma_2^2$.

The level of significance has been chosen to be 0.05. If the component life is normal, the F distribution may be used to perform the test. The test statistic is $\dfrac{s_1^2}{s_2^2}$, where s_1^2 is the larger of the two sample variances. So we need to calculate the variances before we know the test statistic and the degrees of freedom for the test. The following table supplies most of the calculations necessary to obtain the sample variance for the first sample.

x	$x - \overline{x}_1$	$(x - \overline{x}_1)^2$
60	5.5455	30.752
53	−1.4545	2.116
53	−1.4545	2.116
69	14.5455	211.57
56	1.5455	2.388
54	−0.4545	0.207
64	9.5455	91.116
65	10.5455	111.207
47	−7.4545	55.57
38	−16.4545	270.752
40	−14.4545	208.934
599		986.728

We find that the sample mean of this first sample is $\bar{x}_1 = \dfrac{599}{11} = 54.4545$. The sample variance is

$s_1^2 = \dfrac{\Sigma(x - \bar{x}_1)^2}{n_1 - 1} = \dfrac{986.728}{11 - 1} = 98.6728$. A similar table helps to calculate the second sample variance.

x	$x - \bar{x}_2$	$(x - \bar{x}_2)^2$
57	2.25	5.0625
58	3.25	10.5625
51	3.75	14.0625
55	0.25	0.0625
47	7.75	60.0625
57	2.25	5.0625
55	0.25	0.0625
58	3.25	10.5625
438		105.5

We find that the second sample mean is $\bar{x}_2 = \dfrac{438}{8} = 54.75$. The sample variance is

$s_2^2 = \dfrac{\Sigma(x - \bar{x}_2)^2}{n_2 - 1} = \dfrac{105.5}{8 - 1} = 15.0714$.

Since the first sample variance is larger, the F distribution will have 10 numerator degrees of freedom and 7 denominator degrees of freedom. The critical F value is found by looking in the 0.025 table in the 10 degrees of freedom column and the 7 degrees of freedom row. There we find 4.76. So the decision rule is to reject H_0 if the test statistic exceeds 4.76. The calculated value of the test statistic is: $\dfrac{98.6728}{15.0714} = 6.547$.

Since $6.547 > 4.76$, we reject H_0 and conclude at $\alpha = 0.05$ that there is a difference in the two variances.

Practice Exercises 9-1

1. A consumer group is comparing the amounts of pyridoxine being placed in the multivitamin tablets of 2 companies, Vitowin and Hugo's. While both satisfy FDA rules concerning the average amount of pyridoxine, there is some concern that there is more variability in the amounts of pyridoxine in the Vitowin tablets. A sample of 16 Vitowin tablets resulted in a variance of 0.56, while 10 Hugo tablets had a variance of 0.29. Perform the appropriate test to check the validity of this concern using a level of significance of 0.05.

2. A community is worried about the amount of nitrates in the water it gets from its two wells. The average is under the safe limit; but, if the amount varies, individuals could be receiving nitrates in amounts over the safe limits. The county engineer is told to compare the variances in the nitrate amounts in the two wells. He takes water samples from both wells and determines the amount of nitrates with the following results.
    Well 1: 5 9 11 12 7 6 6 9 8 17 7
    Well 2: 13 12 11 10 12 13 12 11 10 12 11 12 11.
    Perform the appropriate test using $\alpha = 0.05$.

Solutions to Practice Exercises 9-1

1. This is a one-tailed hypothesis test. The hypotheses are:
    H_0: $\sigma_1^2 = \sigma_2^2$
    H_1: $\sigma_1^2 > \sigma_2^2$.
    The level of significance is 0.05. If the amount of pyridoxine is normal, the F distribution may be used to perform the test. The test statistic is $\dfrac{s_1^2}{s_2^2}$. This test statistic has an F distribution with 15 numerator degrees of freedom and 9 denominator degrees of freedom. The critical F value is found by looking in the 0.05 table in the 15 degrees of freedom column and the 9 degrees of freedom row. There we find 3.77. So the decision rule is to reject H_0 if the test statistic exceeds 3.77. The calculated value of the test statistic is $\dfrac{0.56}{0.29} = 1.93$.
    Since this is not in the rejection region, we fail to reject H_0 and conclude that there is not a significant difference in the variances in the amounts of pyridoxine.

2. This is a two-tailed hypothesis test. The hypotheses are:
H_0: $\sigma_1^2 = \sigma_2^2$
H_1: $\sigma_1^2 \neq \sigma_2^2$.

The level of significance is 0.05. If the amount of nitrates is normal, the F distribution may be used to perform the test. Since this is a two-tailed test, the test statistic is the ratio of the larger sample variance divided by the smaller sample variance. So to continue the test, we need to calculate the sample variances. For well 1, the following table contains most of the calculations.

x	$x - \overline{x}_1$	$(x - \overline{x}_1)^2$
5	−3.81818	14.5785
9	0.18182	0.0331
11	2.18182	4.7603
12	3.18182	10.1240
7	−1.81818	3.3058
6	−2.81818	7.9421
6	−2.81818	7.9421
9	0.18182	0.0331
8	−0.81818	0.6694
17	8.18182	66.9421
7	−1.81818	3.3058
97		119.6363

From the first column, we can get the mean for well 1: $\overline{x}_1 = \dfrac{97}{11} = 8.81818$. With this, we complete the rest of the table and calculate the sample variance for well 1: $s_1^2 = \dfrac{\Sigma(x - \overline{x}_1)^2}{n_1 - 1} = \dfrac{119.6363}{11 - 1} = 11.964$.

Going through a similar process for well 2, we have:

x	$x - \overline{x}_2$	$(x - \overline{x}_2)^2$
5	−3.81818	14.5785
12	0.46154	0.21302
11	−0.53846	0.28994
10	−1.53846	2.36686
12	0.46154	0.21302
9	0.18182	0.0331
11	2.18182	4.7603
12	3.18182	10.124
7	−1.81818	3.3058
11	2.18182	4.7603
12	3.18182	10.124
9	0.18182	0.0331
8	−0.81818	0.6694
150		11.2308

The sample mean and variance for the second well are $\overline{x}_2 = \dfrac{150}{13} = 11.53846$ and

$s_2^2 = \dfrac{\Sigma(x - \overline{x}_2)^2}{n_2 - 1} = \dfrac{11.23077}{13 - 1} = 0.9359$.

Because the sample variance of the first well is greater than that of the second, the test statistic is: $\dfrac{s_1^2}{s_2^2}$. This test statistic has an F distribution with 10 numerator degrees of freedom and 12 denominator degrees of freedom. The critical F value is found by looking in the 0.025 table in the 10 degrees of freedom column and the 12 degrees of freedom row. There we find 3.37. So the decision rule is to reject H_0 if the test statistic exceeds 3.37. The calculated value of the test statistic is $\dfrac{11.964}{0.9359} = 12.78$. Since $12.78 > 3.37$, we reject H_0 and conclude that there is a difference in the variance of the nitrates between the two wells.

9-2 Inference About Two Means

Study Objectives

You should be able to do a hypothesis test for the equality of two population means and calculate a confidence interval for their difference based on:

1. dependent samples from normal populations.
2. two large independent samples.
3. two small independent samples from normal populations where $\sigma_1^2 \neq \sigma_2^2$.
4. two small independent samples from normal populations where $\sigma_1^2 = \sigma_2^2$.

Section Overview

Often we are interested in comparing the means of two populations. For example, we might want to compare the mean recovery times of patients under two different therapies or the mean salary offers to males and females. There is not one hypothesis test that is appropriate to use. Instead, there are several procedures that apply to different sets of conditions. In this section, we consider four procedures.

The first procedure involves dependent samples. Two samples are dependent when there is an obvious connection between pairs of observations. One example would be the same people doing two tasks, perhaps using two techniques to assemble a product. Or we might compare peoples' test scores of reading ability before and after training in decoding. In such situations, we work with the differences between pairs and do a paired t test. This paired t test is nothing more than a one-sample t test applied to the set of differences.

The other three techniques all apply to independently obtained random samples. If both of the samples are over 30 or if both populations are normal with known variances, we can use a z test. The basis of this z test is the sampling distribution of the difference between sample means. First, the mean of the difference between the sample means is the corresponding difference of the population means, i.e.: $\mu_{\overline{x}_1 - \overline{x}_2} = \mu_1 - \mu_2$. Second, the standard error of the difference of the sample means is:

$$\sigma_{\overline{x}_1 - \overline{x}_2} = \sqrt{\frac{\sigma_1^2}{n_1} + \frac{\sigma_2^2}{n_2}}.$$

Finally, if we have samples from normal populations with known variances or if we have large samples, the sampling distribution of the differences between means is normal in the former case and approximately normal in the latter case. This says that we can use the standardized difference of the sample means

$$z = \frac{\overline{x}_1 - \overline{x}_2}{\sigma_{\overline{x}_1 - \overline{x}_2}}$$

as a test statistic. This test statistic will have a standard normal distribution when the null hypothesis of no difference between the population means is true. Additionally, if the population variances are unknown but the sample sizes are both over 30, we can substitute the sample variances for the population variances and perform an approximate test with no other changes.

The remaining two procedures involve Student's t distribution. They are valid when small samples are taken independently from two normal populations. One is appropriate if the population variances are different, and one is appropriate if the population variances are equal. The decision about whether we think the two variances are equal can be based on the F test in the previous section. If we conclude that the variances are not equal, we use as the test statistic:

$$t = \frac{\overline{x}_1 - \overline{x}_2}{\sqrt{\dfrac{s_1^2}{n_1} + \dfrac{s_2^2}{n_2}}}.$$

This test statistic has a t distribution with degrees of freedom equal to the smaller of $n_1 - 1$ and $n_2 - 1$. If we conclude that the variances are equal, the test statistic is:

$$t = \frac{\overline{x}_1 - \overline{x}_2}{s_p\sqrt{\dfrac{1}{n_1} + \dfrac{1}{n_2}}} = \frac{\overline{x}_1 - \overline{x}_2}{\sqrt{\dfrac{s_1^2(n_1 - 1) + s_2^2(n_2 - 1)}{n_1 + n_2 - 2}}\sqrt{\dfrac{1}{n_1} + \dfrac{1}{n_2}}}.$$

This has a t distribution with $n_1 + n_2 - 2$ degrees of freedom.

Each of these four procedures has a corresponding confidence interval. In each case, the formula for the confidence interval is:

estimate of the difference $\pm$ (a z or t table value)(denominator of the test statistic).

Key Terms & Formulas

<u>Dependent Samples</u> Two samples taken so that the results consist of pairs of observations from each population.

<u>Paired t Test on μ_d</u> The appropriate procedure for paired or dependent samples. The test statistic is a one-sample t test statistic applied to the differences between the pairs of observations. Using d to represent differences, the test statistic is: $t = \dfrac{\overline{d} - \mu_d}{s_d / \sqrt{n}}$. This has a t distribution with $n - 1$ degrees of freedom, where n is the number of pairs.

<u>Paired t Confidence Interval for μ_d</u> Under the same condition for which it is valid to perform a hypotheses test, a confidence interval for μ_d is computed using: $\overline{d} \pm t_{\alpha/2}\dfrac{s_d}{\sqrt{n}}$

<u>Z Inference for the Difference of Two Means from Independent Populations</u> The procedure to use when two independent random samples are taken from normal populations with known variances or when both sample sizes exceed 30. To test H_0: $\mu_1 = \mu_2$, the test statistic is: $z = \dfrac{\overline{x}_1 - \overline{x}_2}{\sigma_{\overline{x}_1 - \overline{x}_2}}$, where $\sigma_{\overline{x}_1 - \overline{x}_2} = \sqrt{\dfrac{\sigma_1^2}{n_1} + \dfrac{\sigma_2^2}{n_2}}$.

If the population variances are unknown, we can substitute the sample variances for the population variances as long as the sample sizes are larger than 30. The formula for a conference interval in this situation for $\mu_1 - \mu_2$ is: $\overline{x} \pm z_{\alpha/2}\,\sigma_{\overline{x}_1 - \overline{x}_2}$.

<u>t Inferences for the Difference of Two Means from Independent Populations when $\sigma_1^2 \neq \sigma_2^2$</u> The procedure to use if two independent random samples, at least one of which has a sample size less than 30, are taken from normal populations with unknown variances when these variances are unequal. To test H_0: $\mu_1 = \mu_2$, the test statistic is $t = \dfrac{\overline{x}_1 - \overline{x}_2}{\sqrt{\dfrac{s_1^2}{n_1} + \dfrac{s_2^2}{n_2}}}$. This test statistic has a t distribution with degrees of freedom equal to the smaller of $n_1 - 1$ and $n_2 - 1$. The formula for a corresponding conference interval for $\mu_1 - \mu_2$ is: $\overline{x}_1 - \overline{x}_2 \pm t_{\alpha/2}\sqrt{\dfrac{s_1^2}{n_1} + \dfrac{s_2^2}{n_2}}$.

<u>t Inference for the Difference of Two Means from Independent Populations when $\sigma_1^2 = \sigma_2^2$</u> The test procedure to use if two independent random samples, at least one of which has a sample size less than 30, are taken from normal populations with unknown but equal variances. To test H_0: $\mu_1 = \mu_2$, the test statistic is:

$$t = \frac{\overline{x}_1 - \overline{x}_2}{s_p\sqrt{\dfrac{1}{n_1} + \dfrac{1}{n_2}}} = \frac{\overline{x}_1 - \overline{x}_2}{\sqrt{\dfrac{s_1^2(n_1 - 1) + s_2^2(n_2 - 1)}{n_1 + n_2 - 2}}\sqrt{\dfrac{1}{n_1} + \dfrac{1}{n_2}}}.$$

This has a t distribution with $n_1 + n_2 - 2$ degrees of freedom. The formula for a conference interval for $\mu_1 - \mu_2$ in this case is: $\bar{x}_1 - \bar{x}_2 \pm t_{\alpha/2} s_p \sqrt{\dfrac{1}{n_1} + \dfrac{1}{n_2}}$.

Hint: This last test is usually called a "pooled t test." This is based on s_p being a weighted average of the two sample variances.

Worked Examples

Honest?

A buyer for a manufacturing plant suspects that his primary supplier of raw materials is overcharging. To decide if his suspicion is correct, he contacts a second supplier known for honesty, and asks for the prices of various materials. He wants to compare these prices with those of his primary supplier for the same materials. The data collected is presented in the table below, with some summary statistics presented (all these might not be necessary). Does it appear that the primary supplier is overcharging? Use a level of significance of 0.05. Also obtain a 95% confidence interval for the difference in the mean amounts charged.

Material	Primary Supplier	Secondary Supplier	Difference
1	$55	$45	$10
2	$48	$47	$1
3	$31	$32	-$1
4	$83	$77	$6
5	$37	$37	$0
6	$55	$54	$1
$\Sigma X_i =$	$309	$292	$\Sigma d_i = \$17$
$\Sigma X_i^2 =$	17573	15472	$\Sigma d_i^2 = 139$

Solution

The prices are matched or paired on the identical materials, so this is a situation where a paired t test should be run. The differences we chose to use were Primary - Secondary, so a positive number indicates a material for which the primary supplier is higher than the secondary supplier. The hypotheses are:

H_0: $\mu_d = 0$

H_0: $\mu_d > 0$.

We use a right-tailed test because the buyer's concern only involves the primary supplier's possible overcharging.

Because of the paired nature of the sample, the test statistic will be $t = \dfrac{\bar{d} - \mu_d}{s_d/\sqrt{n}}$, which has a t distribution with $6 - 1 = 5$ degrees of freedom. With $\alpha = 0.05$, the value that bounds the rejection region is found in the t table by looking in the five degrees of freedom row and the 0.05 column. This t value is 2.015. Thus the decision rule is to reject H_0 if $t > 2.015$, otherwise, fail to reject H_0. To obtain the test statistic, we must first calculate the mean and standard deviation of the differences.

Hint: The values of ΣX and ΣX^2 for the primary and secondary supplier are irrelevant for this test. The only relevant summary information from the table is Σd and Σd^2 for the differences.

$$\bar{d} = \frac{\Sigma d}{n} = \frac{17}{6} = 2.8333 \text{ and } s_d = \sqrt{\frac{n\Sigma d^2 - (\Sigma d)^2}{n(n-1)}} = \sqrt{\frac{6(139) - 17^2}{6(6-1)}} = 4.2622.$$

Then $t = \dfrac{\bar{d} - \mu_d}{s_d/\sqrt{n}} = \dfrac{2.8333 - 0}{4.2622/\sqrt{6}} = 1.628.$

Because 1.628 is not greater than 2.015, we fail to reject the null hypothesis and conclude that there is insufficient evidence at $\alpha = 0.05$ to conclude that the mean price charged by the primary supplier exceeds that of the secondary supplier.

To obtain a 95% confidence interval for μ_d, we need $t_{0.025}$ with five degrees of freedom $= 2.571$. Then the confidence interval is: $2.8333 \pm (2.571) \dfrac{4.2662}{\sqrt{6}} = 2.8333 \pm 4.4736$ or from -1.64 to 7.31. This interval indicates with 95% confidence, that μ_1 might be greater than μ_2 by as much as 7.31, or μ_1 might be less than μ_2 by as much as 1.64. It is also possible, since 0 is in the interval, that there is no difference between the two means.

Abalone

A marine biologist suspects that the warmth of the water near power plants would decrease the growth of nearby marine life. She is particularly concerned with abalone. She samples 50 abalone near a power plant and another 100 in a cold water location. The average diameter of the warm water abalone is 19.2 cm. with a standard deviation of 2.2 cm. The cold water abalone have a mean of 20.5 cm. with a standard deviation of 4.3 cm. Using a test with a level of significance of 0.01, determine if the marine biologist's suspicion is justified. Afterwards, compute a 99% confidence interval for $\mu_1 - \mu_2$.

Solution

Because the sample sizes are both large, we will not need to worry about the assumption of normality. Additionally, it is not necessary that we know the population variances--we can substitute their sample counterparts. If we let μ_1 represent the mean diameter of warm water abalone and μ_2 the mean for the warm water abalone, the hypotheses are:
H_0: $\mu_1 = \mu_2$
H_1: $\mu_1 < \mu_2$.
We use a left-tailed test because the marine biologist's suspicion is that the mean for the warm water abalone is smaller than the cold water abalone. With $\alpha = 0.01$, the z value that bounds the rejection region is found from the normal tables by looking for $0.5 - 0.01 = 0.49$ in the probability portion of the table. This z value is 2.33. Thus the decision rule is to reject H_0 if the test statistic $z < -2.33$; otherwise, fail to reject H_0. The test statistic is:

$$z = \frac{\bar{x}_1 - \bar{x}_2}{\sigma_{\bar{x}_1 - \bar{x}_2}} = \frac{\bar{x}_1 - \bar{x}_2}{\sqrt{\frac{\sigma_1^2}{n_1} + \frac{\sigma_2^2}{n_2}}} = \frac{19.2 - 20.5}{\sqrt{\frac{2.2^2}{50} + \frac{4.3^2}{100}}} = \frac{-1.3}{\sqrt{0.0968 + 0.1849}} = -2.45.$$

Because $-2.45 < -2.33$, we reject the null hypothesis and conclude with 99% confidence that the suspicion of the marine biologist is justified.

To obtain a 99% confidence interval for $\mu_1 - \mu_2$, we need $z_{\alpha/2} = z_{0.005} = 2.58$. Then we compute:

$$(19.2 - 20.5) \pm (2.58)\sqrt{\frac{2.2^2}{50} + \frac{4.3^2}{100}} = -1.30 \pm 1.37 \text{ or from } -2.67 \text{ to } +0.07.$$

Electric Life Plugged in Again

In the previous section we looked at the data below that represents the life of components of an electrical system from two different manufacturers. We tested the hypothesis that the variances were identical and concluded at the 0.05 level that there is a difference between the two means. Again using a level of 0.05, test the hypothesis that the mean lifetimes of the components are equal. Then obtain a 95% confidence interval for $\mu_1 - \mu_2$.

Supplier 1	Supplier 2
60 53 53 69 56 54	57 58 51 55
64 65 47 38 40	47 57 55 58

Solution

This is a two-tailed hypothesis test. The hypotheses are:
H_0: $\mu_1 = \mu_2$
H_1: $\mu_1 \neq \mu_2$.
The level of significance has been chosen to be 0.05. If the component life is normal, the t distribution may be used to perform the test. Since we have rejected the hypothesis of equal variances, we will use the test statistic:

$$t = \frac{\bar{x}_1 - \bar{x}_2}{\sqrt{\frac{s_1^2}{n_1} + \frac{s_2^2}{n_2}}}.$$

This test statistic has a t distribution with degrees of freedom equal to the smaller of $n_1 - 1$ and $n_2 - 1$. Here $n_1 = 11$ and $n_2 = 8$, so the number of degrees of freedom is $8 - 1 = 7$. The rejection region is bounded by a critical value found in the t table in the 7 degrees of freedom row and the $0.05/2 = 0.025$ column. This value is 2.365. The decision rule is to reject H_0 if $t < -2.365$ or if $t > 2.365$; otherwise, fail to reject H_0. In the previous section, we calculated the following statistics:

$$\bar{x}_1 = \frac{599}{11} = 54.4545, s_1^2 = 98.6728, \bar{x}_2 = \frac{438}{8} = 54.75, \text{ and } s_2^2 = 15.0714.$$

Using this information, we can calculate the test statistic:

$$t = \frac{\bar{x}_1 - \bar{x}_2}{\sqrt{\dfrac{s_1^2}{n_1} + \dfrac{s_2^2}{n_2}}} = \frac{54.4545 - 54.75}{\sqrt{\dfrac{98.6728}{11} + \dfrac{15.0714}{8}}} = \frac{-0.2955}{3.2546} = -0.09.$$

Since -0.09 is not in the rejection region, we fail to reject H_0 and conclude that there is no significant evidence of a difference in the two means.

Most of the calculations for the confidence interval have been done. Additionally, the same t value of $t_{0.025}$ with 7 degrees of freedom is appropriate. So the confidence interval is: $-0.2955 \pm (2.365)(3.2546) = -0.2955 \pm 7.6971$ or from -7.99 to 7.40.

Wheelpeople

A member of "Wheelpeople," a bicycle club, is interested in comparing the effects of two winter conditioning programs she designed. One involved the use of free weights and jogging, while the other involved dietary control plus exercises involving various stationary machines. She involved ten people in her experiment, five using the weights/jogging regimen and five using the diet/machine program. At the end of the winter she asked them to cycle up Cuesta Grade (a 7% incline) and timed their rides. The average for the first group was 41 minutes with a variance of 21 while the second group had a mean of 35 and a variance of 24. Does this data show, using a level of significance of 0.01, that one plan is superior to the other? After making the decision, calculate a 99% confidence interval for the difference in the mean ride times.

Solution

Since there is no reason to favor one program over the other, this is a two-tailed hypothesis test. The hypotheses are:
H_0: $\mu_1 = \mu_2$
H_1: $\mu_1 \neq \mu_2$.
The level of significance has been chosen to be 0.01. If the times are normal, the t distribution may be used to perform the test. The result of the F test (calculations not given here) for this data indicates that there is not a significant difference in the variances. Therefore, we will use the test statistic:

$$t = \frac{\bar{x}_1 - \bar{x}_2}{s_p\sqrt{\dfrac{1}{n_1} + \dfrac{1}{n_2}}} = \frac{\bar{x}_1 - \bar{x}_2}{\sqrt{\dfrac{s_1^2(n_1 - 1) + s_2^2(n_2 - 1)}{n_1 + n_2 - 2}}\sqrt{\dfrac{1}{n_1} + \dfrac{1}{n_2}}}.$$

The test statistic has a t distribution with degrees of freedom $n_1 + n_2 - 2$. Here $n_1 = n_2 = 5$, so the number of degrees of freedom is $5 + 5 - 2 = 8$. The rejection region is bounded by a critical value found in the t table in the 8 degrees of freedom row and the $\frac{0.01}{2} = 0.005$ column. This value is 3.355. The decision rule is to reject H_0 if $t < -3.355$ or if $t > 3.355$; otherwise, fail to reject H_0. Calculating the test statistic, we get:

$$t = \frac{\bar{x}_1 - \bar{x}_2}{\sqrt{\dfrac{s_1^2(n_1 - 1) + s_2^2(n_2 - 1)}{n_1 + n_2 - 2}}\sqrt{\dfrac{1}{n_1} + \dfrac{1}{n_2}}} = \frac{41 - 35}{\sqrt{\dfrac{21(5 - 1) + 24(5 - 1)}{5 + 5 - 2}}\sqrt{\dfrac{1}{5} + \dfrac{1}{5}}} = \frac{6.0}{3.0} = 2.00.$$

Since 2.00 is not in the rejection region, we fail to reject H_0 and conclude that there is insignificant evidence of a difference between the two conditioning program means.

The 99% confidence interval is: $(\bar{x}_1 - \bar{x}_2) \pm t_{\alpha/2}\, s_p\sqrt{\dfrac{1}{n_1} + \dfrac{1}{n_2}} = 6 \pm (3.355)(3.0) = 6 \pm 10.1$ or from -4.1 to 16.1.

Practice Exercises 9-2

1. A reading specialist wants to see if teaching children how to decode words improves reading skills. He gives a test of reading achievement to 13 students before and after he teaches them how to decode. The results are in the table below (Note: The higher the score, the greater the reading skill.) Use this information to test if there has been an improvement in the mean score using $\alpha = 0.05$.

Pretest	Posttest
57	59
45	46
48	55
35	33
35	44
67	68
48	61
44	49
29	39
49	64
51	55
61	60
55	55

2. A woodworker wants to compare the quality of two wood preservatives, Tyrone and Strolak. He takes six pieces of wood and treats half with Tyrone and half with Strolak. After weathering the wood, he evaluates the amount of discoloration. The table below contains these amounts. Calculate a 95% confidence interval for the difference in the mean amount of discoloration.

Tyrone	Strolak
55	59
32	50
87	97
67	71
63	76
23	25

3. The captain of a fishing boat wants to compare the number of rock cod caught by sports fishers off Cayucos with the number caught off Cambria. She has her hands take a count of the number of rock cod caught by individuals at both locations. She finds that the mean number caught off Cayucos by a sample of 120 fishers was 4.167 with a standard deviation of 1.621. The corresponding numbers for a sample of 113 fishers off Cambria were 5.717 and 1.398. Compute a 99 confidence interval for $\mu_1 - \mu_2$.

4. Education officials in the Dakotas agreed to use two different programs at the secondary level to teach students Logic and Critical Thinking. South Dakota used the traditional classroom setting while North Dakota used a computer based method of instruction. After the programs had run for a year, random samples of students from both states were given a multiple-choice exam of Logic and Critical Thinking. The scores on the test are the number of incorrect answers. They wanted to see if the computer-based program would lower the mean number of incorrect answers. The mean and standard deviation for a sample of 56 South Dakota students was 4.29 and 1.81, while for 68 North Dakota students these statistics were 4.44 and 2.40. Perform the appropriate test with a level of significance of 0.05.

5. An animal trainer wants to compare the mean weights of adult Australian and Hungarian Shepherds. From his records he finds that the mean weight of 6 Australian Shepherds is 67.33 lbs. with a standard deviation of 10.29. He finds the records of 4 Hungarian Shepherds and finds that their mean and standard deviation are 56.00 and 2.94. Use this information to compute a 99% confidence interval on the difference in the mean weights for the two populations. Use $\alpha = 0.01$.

6. An international corporation has begun a test of reading, mathematical, and reasoning skills that employees must pass to have any chance of promotion or advancement. The minimum passing grade is 22. Due to the high failure rate, the corporation has hired a firm to help the employees develop the skills necessary to pass this exam. After training 14 employees, the firm compares their mean score to another 24 employees that took the exam at the same time. The mean score of the employees who took the class was 24.210 with a standard deviation of 2.170. These statistics were 21.643 and 9.460 for those who did not take the class. Does this information prove that the firm is doing a good job, i.e., that the training is increasing the mean score? Use $\alpha = 0.05$ in performing this test.

7. A member of a statistics class is upset that he has to compete against Math majors for a grade. He takes a random sample of Math majors and a second random sample of non-Math majors to try to learn if he is worrying needlessly. Of 8 Math majors sampled, the mean grade was 67.37 with a standard deviation of 18.61. For 5 non-Math majors, the mean was 71.00 and the standard deviation was 14.92. Does this data show, using a level of significance of 0.10, that the Math majors have a significantly higher mean score than the non-Math majors?

8. A forestry major is interested in comparing the number of mites found in trees exposed to substantial amounts of sunlight versus trees that are primarily shaded. He samples five of each type of tree and counts the number of mites found on a total of 20 leaves per tree. The data follows with some summary values. Use this information to calculate a 90% confidence interval for $\mu_1 - \mu_2$.

	Sun	Shade
	27	42
	33	58
	43	48
	29	44
	37	41
$\Sigma x =$	169	233
$\Sigma x^2 =$	5877	11049

Solutions to Practice Exercises 9-2

1. The pre- and postscores are matched or paired because they come from the same individuals, therefore a paired t test should be run. The differences can be taken in either direction. However, it seems to make more sense to subtract the pretest score from the posttest score so that the differences would represent reading gains. The hypotheses are:

H_0: $\mu_d = 0$

H_0: $\mu_d > 0$.

The test statistic is $t = \dfrac{\bar{d} - \mu_d}{s_d/\sqrt{n}}$, which has a t distribution with $13 - 1 = 12$ degrees of freedom. With $\alpha = 0.05$, the t value that bounds the rejection region is found in the t table by looking in the 12 degree of freedom row and the 0.05 column. This t value is 1.782. Thus the decision rule is to reject H_0 if $t > 1.782$; otherwise, fail to reject H_0. To obtain the test statistic, we must first calculate the mean and standard deviation of the differences.

Pretest	Posttest	Difference	Diff^2
57	59	2	4
45	46	1	1
48	55	7	49
35	33	−2	4
35	44	9	81
67	68	1	1
48	61	13	169
44	49	5	25
29	39	10	100
49	64	15	225
51	55	4	16
61	60	−1	1
55	55	0	0
		64	676

From the table, we can calculate: $\bar{d} = \dfrac{\Sigma d}{n} = \dfrac{64}{13} = 4.9231$ and

$$s_d = \sqrt{\dfrac{n\Sigma d^2 - (\Sigma d)^2}{n(n-1)}} = \sqrt{\dfrac{13(676) - 64^2}{13(13-1)}} = 5.4842. \text{ So } t = \dfrac{\bar{d} - \mu_d}{s_d/\sqrt{n}} = \dfrac{4.9231 - 0}{5.4842/\sqrt{13}} = 3.237.$$

Because $3.237 > 1.782$, we reject the null hypothesis and conclude that the mean score on reading has gone up.

2. The measurements taken upon the same piece of wood leads to an obvious pairing. The formula for a confidence interval is: $\overline{d} \pm t_{\alpha/2}\dfrac{s_d}{\sqrt{n}}$, which has a t distribution with $6 - 1 = 5$ degrees of freedom. With $\alpha = 0.05$, the t value for the confidence interval is found in the t table by looking in the 5 degree of freedom row and the $0.05/2 = 0.025$ column. This t value is 2.571. Before computing the confidence interval, we first calculate the mean and standard deviation of the differences.

Tyrone	Strolak	Difference	Diff^2
55	59	-4	16
32	50	-18	324
87	97	-10	100
67	71	-4	16
63	76	-13	169
23	25	-2	4
		-51	629

Then $\overline{d} = \dfrac{\Sigma d}{n} = \dfrac{-51}{6} = -8.5$, $s_d = \sqrt{\dfrac{n\Sigma d^2 - (\Sigma d)^2}{n(n-1)}} = \sqrt{\dfrac{6(629) - 51^2}{6(6-1)}} = 6.2530$ and the confidence interval is: $\overline{d} \pm t_{\alpha/2}\dfrac{s_d}{\sqrt{n}} = -8.5 \pm (2.571)\dfrac{6.2530}{\sqrt{6}} = -8.5 \pm 6.6$. We conclude that the mean discoloration is greater for Strolak by between 1.9 and 15.1.

3. Let μ_1 represent the mean number of rock cod caught off Cayucos and μ_2 the mean for Cambria. Both sample sizes are over 30. With large samples, we do not need the assumption of normality nor do we need to know the population variances to use the normal distribution to compute a confidence interval. With $\alpha = 0.01$, the z value is found from the normal tables by finding $0.5 - \frac{0.01}{2} = 0.495$ in the probability portion of the table. This z value is 2.58. In calculating the confidence interval, we will substitute the values of the sample variances for the population variances. The confidence interval is:

$$(\overline{x}_1 - \overline{x}_2) \pm z_{\alpha/2}\sqrt{\dfrac{\sigma_1^2}{n_1} + \dfrac{\sigma_2^2}{n_2}} = (4.167 - 5.717) \pm (2.575)\sqrt{\dfrac{1.621^2}{120} + \dfrac{1.398^2}{113}}$$

$$= -1.55 \pm (2.575)\sqrt{0.0219 + 0.0173} = -1.55 \pm 0.51.$$

So we conclude with 99% confidence that there is a difference in the mean number of rock cod caught at the two locations with the mean being greater at Cambria by between 1.04 and 2.06 fish.

4. If we let μ_1 represent the mean number of errors for SD and μ_2 the mean for ND, the hypotheses are:
H_0: $\mu_1 = \mu_2$
H_1: $\mu_1 > \mu_2$.
We use a right-tailed test because if there is a positive effect associated with the computer-based program, the mean number of errors for North Dakota, μ_2, would be the smaller mean. Both samples sizes are over 30. With large samples, we do not need the assumption of normality or knowledge of the population variances to use the normal distribution for the test. With $\alpha = 0.05$, the critical z value that bounds the rejection region is found by locating $0.5 - 0.05 = 0.45$ in the probability portion of the normal table. This z value is 1.645. Thus the decision rule is to reject H_0 if $z > 1.645$; otherwise, fail to reject H_0. In calculating the test statistic, we will substitute the values of the sample variances for the population variances. The test statistic is:

$$z = \dfrac{\overline{x}_1 - \overline{x}_2}{\sigma_{\overline{x}_1 - \overline{x}_2}} = \dfrac{\overline{x}_1 - \overline{x}_2}{\sqrt{\dfrac{\sigma_1^2}{n_1} + \dfrac{\sigma_2^2}{n_2}}} = \dfrac{4.286 - 4.441}{\sqrt{\dfrac{1.806^2}{56} + \dfrac{2.396^2}{68}}} = \dfrac{-0.155}{\sqrt{0.0582 + 0.0844}} = -0.41.$$

Because -0.41 is not in the rejection region, we fail to reject the null hypothesis and conclude that there is a not enough evidence to show an improvement in the mean number of errors associated with the computer-based program.

Hint: Most of the above arithmetic in calculating the test statistic was unnecessary. The rejection region for this test statistic consisted of values of the test statistic greater than 1.645, i.e., relatively large positive values. When it became apparent that the test statistic was negative, it was not necessary to finish the calculations.

5. If the weights of the two types of shepherds are normal, the t distribution may be used compute the confidence interval. The question is which type of t interval to calculate. That would depend on whether we believe that the two population variances are equal. While we are not going to show the details here, if we used the F test to test for equal variances, we would reject that null hypothesis in favor of the alternative that they differ. Because of that decision, we will compute $(\overline{x}_1 - \overline{x}_2) \pm t_{\alpha/2}\sqrt{\dfrac{s_1^2}{n_1} + \dfrac{s_2^2}{n_2}}$. This is based on a t distribution with degrees of freedom the smaller of $n_1 - 1$ and $n_2 - 1$. Here $n_1 = 6$ and $n_2 = 4$, so the number of degrees of freedom is $4 - 1 = 3$. Thus we need the value found in the t table in the 3 degrees of freedom row and the $0.01/2 = 0.005$ column. This value is 5.841. Then the confidence interval is:

$$(\overline{x}_1 - \overline{x}_2) \pm t_{\alpha/2}\sqrt{\frac{s_1^2}{n_1} + \frac{s_2^2}{n_2}} = (67.33 - 56.00) \pm (5.841)\sqrt{\frac{10.29^2}{6} + \frac{2.94^2}{4}}$$

$$= 11.33 \pm (5.841)(4.4506) = 11.33 \pm 26.00.$$

The 99% confidence interval for $\mu_1 - \mu_2$ is from -14.67 to 37.33.

6. We will let μ_1 represent the mean of those who took the training, μ_2 the mean of those who did not. This is a right-tailed test with the hypotheses:
H_0: $\mu_1 = \mu_2$
H_1: $\mu_1 > \mu_2$.
The level of significance has been chosen to be 0.05. If the test scores are normal, the t distribution may be used to perform the test. Doing an F test for equal variances, we find that we would reject that null hypothesis in favor of the alternative that they differ. Therefore, the test statistic is

$t = \dfrac{\overline{x}_1 - \overline{x}_2}{\sqrt{\dfrac{s_1^2}{n_1} + \dfrac{s_2^2}{n_2}}}$. The test statistic has a t distribution with degrees of freedom equal to the smaller of $n_1 - 1$

and $n_2 - 1$. Here $n_1 = 14$ and $n_2 = 24$, so the number of degrees of freedom is $14 - 1 = 13$. The rejection region is bounded by a critical value found in the t table in the 13 degrees of freedom row and the 0.05 column. This value is 1.771. Then the decision rule is to reject H_0 if $t > 1.771$; otherwise, fail to reject H_0. The test statistic is:

$$t = \frac{\overline{x}_1 - \overline{x}_2}{\sqrt{\dfrac{s_1^2}{n_1} + \dfrac{s_2^2}{n_2}}} = \frac{24.210 - 21.643}{\sqrt{\dfrac{2.170^2}{14} + \dfrac{9.460^2}{24}}} = \frac{2.567}{2.0162} = 1.273.$$

Since 1.273 is not in the rejection region, we fail to reject H_0 and conclude that there is not strong evidence that the training is effective in increasing the mean.

Hint: While the firm did not increase the mean, the standard deviation was significantly smaller. If we were to compare the percentage who pass the exam in the two groups, the consistency of getting slightly above average grades probably implies that the training improves the pass rate.

7. If the Math majors are group 1 and non-Math group 2, the hypotheses are:
H_0: $\mu_1 = \mu_2$
H_1: $\mu_1 > \mu_2$.
The level of significance has been chosen to be 0.10. If the test grades are normal, the t distribution may be used to perform the test. The result of the F test (calculations not given here) for this data suggests that there is not a significant difference in the variances. Therefore, we will use the test statistic:

$$t = \frac{\overline{x}_1 - \overline{x}_2}{s_{po}\sqrt{\dfrac{1}{n_1} + \dfrac{1}{n_2}}} = \frac{\overline{x}_1 - \overline{x}_2}{\sqrt{\dfrac{s_1^2(n_1 - 1) + s_2^2(n_2 - 1)}{n_1 + n_2 - 2}}\sqrt{\dfrac{1}{n_1} + \dfrac{1}{n_2}}}.$$

This ratio has a t distribution with degrees of freedom $n_1 + n_2 - 2$. Here $n_1 = 8$ and $n_2 = 5$, so the number of degrees of freedom is $8 + 5 - 2 = 11$. The rejection region is bounded by a value found in the t table in the 11 degrees of freedom row and the 0.10 column. This value is 1.363. The decision rule is to reject H_0 if $t > 1.363$; otherwise, fail to reject H_0. The test statistic equals:

$$t = \frac{\overline{x}_1 - \overline{x}_2}{\sqrt{\frac{s_1^2(n_1 - 1) + s_2^2(n_2 - 1)}{n_1 + n_2 - 2}} \sqrt{\frac{1}{n_1} + \frac{1}{n_2}}}$$

$$= \frac{67.37 - 71.00}{\sqrt{\frac{18.61^2(8 - 1) + 14.92^2(5 - 1)}{8 + 5 - 2}} \sqrt{\frac{1}{8} + \frac{1}{5}}} = -0.21.$$

Since -0.21 is not in the rejection region, we fail to reject H_0 and conclude that there is not significant evidence that Math majors have a higher mean grade than non-Math majors.

> Hint: Again, it is not really necessary to fully calculate the test statistic--it is negative, and the rejection region only involves positive values of the test statistic.

8. Let $\mu_1 =$ the mean number of mites on sun lit plants and $\mu_2 =$ the mean number of mites on shaded plants. If the mite counts are normal, the t distribution may be used to obtain the interval. Before we decide which t interval to use, we would first need to calculate the sample variances and do an F test for equal population variances. For convenience, we obtain the sample means simultaneously.

$$\overline{x}_1 = \frac{169}{5} = 33.8 \qquad s_1^2 = \frac{5(5877) - 169^2}{5(5 - 1)} = 41.20$$

$$\overline{x}_2 = \frac{23.3}{5} = 46.6 \qquad s_2^2 = \frac{5(11049) - 233^2}{(5 - 1)} = 47.80$$

The result of the F test (calculations not given here) for this data suggests that there is not a significant difference in the variances. Therefore, we will compute the confidence interval using:

$$(\overline{x}_1 - \overline{x}_2) \pm t_{\alpha/2} s_p \sqrt{\frac{1}{n_1} + \frac{1}{n_2}} = (\overline{x}_1 - \overline{x}_2) \pm t_{\alpha/2} \sqrt{\frac{s_1^2(n_1 - 1) + s_2^2(n_2 - 1)}{n_1 + n_2 - 2}} \sqrt{\frac{1}{n_1} + \frac{1}{n_2}}.$$

This interval is based on a t distribution with degrees of freedom $n_1 + n_2 - 2$. Here $n_1 = n_2 = 5$, so the number of degrees of freedom is $5 + 5 - 2 = 8$. The value found in the t table in the 8 degrees of freedom row and the 0.05 column, is 1.860. The confidence interval is:

$$(33.80 - 46.60) \pm (1.860)\sqrt{\frac{41.20(5 - 1) + 47.80(5 - 1)}{5 + 5 - 2}} \sqrt{\frac{1}{5} + \frac{1}{5}} = -12.8 \pm 7.8.$$

So we conclude with 90% confidence that the mean number of mites is greater in the shaded trees by between 5.0 and 20.6.

9-3 Inference About Two Percentages

Study Objectives

You should be able to:

> Do a hypothesis test on the equality of two population percentages and compute a confidence interval for their difference based on large, independently taken, random samples.

Section Overview

When comparing two populations, sometimes we will be interested in the occurrence or nonoccurrence of an attribute rather than the average of a variable. If so, we will work with the percentage of occurrence of the attribute for both groups. For example, we might want to compare the percentage of cancer cases for smokers versus nonsmokers or the percentage of cases of AIDS for homosexuals versus heterosexuals. In this section, we consider how to perform a hypothesis test on these percentages and how to compute a confidence interval for their difference.

We can do a hypothesis test on the difference of two percentages if we have large independent random samples from both populations. (Remember that a large sample for percentages is not 30, but when np and $n(100 - p)$ are both at least 500.) This test is based on the normal distribution. The details of this z test rely on an understanding of the sampling distribution of the difference between sample percentages. First, the mean of the difference between the sample percentages is the corresponding difference of the population percentages, i.e.: $\mu_{p_1-p_2} = \pi_1 - \pi_2$. If the null hypothesis

H_0: $\pi_1 = \pi_2$

is true, then the mean of the sampling distribution of $p_1 - p_2$ is zero. Second, the standard error of the difference of the sample percentages is:

$$\sigma_{p_1-p_2} = \sqrt{\frac{\pi_1(100 - \pi_1)}{n_1} + \frac{\pi_2(100 - \pi_2)}{n_2}}.$$

The values of the population percentages are unknown (or why do a test about their values?). So in doing the test, we will estimate the standard error with:

$$\widehat{\sigma}_{p_1-p_2} = \sqrt{\frac{p_1(100 - p_1)}{n_1} + \frac{p_2(100 - p_2)}{n_2}}$$

Finally, if we have large samples, the sampling distribution of the differences between percentages is approximately normal. This means that we can use the standardized difference of the sample percentages

$$z = \frac{p_1 - p_2}{\widehat{\sigma}_{p_1-p_2}}$$

as a test statistic. The test statistic will have a standard normal distribution when the null hypothesis of no difference between the population percentages is true.

Similarly, a confidence interval for $\pi_1 - \pi_2$ is obtained by calculating $(p_1 - p_2) \pm z_{\alpha/2}\widehat{\sigma}_{p_1-p_2}$.

Key Terms & Formulas

Mean of the Sampling Distribution of the Difference between Sample Percentages This mean is the corresponding difference of the population percentages, $\mu_{p_1-p_2} = \pi_1 - \pi_2$.

Standard Error of the Difference of Sample Percentages $\sigma_{p_1-p_2} = \sqrt{\dfrac{\pi_1(100 - \pi_1)}{n_1} + \dfrac{\pi_2(100 - \pi_2)}{n_2}}.$

Estimated Standard Error of the Difference of Sample Percentages $\widehat{\sigma}_{p_1-p_2} = \sqrt{\dfrac{p_1(100 - p_1)}{n_1} + \dfrac{p_2(100 - p_2)}{n_2}}.$

Hypothesis test for H_0: $\pi_1 = \pi_2$ The test statistic for the equality of two population percentages based on large, independent, random samples is $z = \dfrac{p_1 - p_2}{\widehat{\sigma}_{p_1-p_2}}.$

Confidence Interval for $\pi_1 - \pi_2$ The confidence interval for the difference of two population percentages is:

$(p_1 - p_2) \pm z_{\alpha/2}\widehat{\sigma}_{p_1-p_2}.$

Worked Examples

Insurance Costs

Traditionally males have paid more for auto insurance than females. The major reason for this is the belief that males are in more accidents than females. An actuary is interested in learning if this is true. To decide, she pulls the records on 200 males and 200 females and discovers that the males had 106 accidents while the females had 94 accidents. Perform the appropriate hypothesis test. Use a level of significance of 0.10. Then compute a 90% confidence interval for $\pi_1 - \pi_2$.

Solution

Since the actuary specifically is interested in finding out if males have a higher incidence of accidents, the hypotheses are:

H_0: $\pi_1 = \pi_2$

H_1: $\pi_1 > \pi_2$

where π_1 is the percentage of accidents for males and π_2 is the percentage of accidents for females. The level of significance has been chosen to be 0.10. For the first sample, $p_1 = \dfrac{106}{200} \times (100 \text{ percent})$ or 53 percent. For the second sample, $p_2 = \dfrac{94}{200} \times (100 \text{ percent})$ or 47 percent. To ensure that we may use the normal distribution to do the test, we check np and $n(100 - p)$ for both samples. For the first sample these are 10,600 and 9,400; for the second they are 9,400 and 10,600. Because all four of these exceed 500, we have large samples. The rejection region for this right-tailed test is bounded by a value found in the z table by looking in the probability portion for $0.5 - 0.10 = 0.40$. This critical value is 1.28. The decision rule is to reject H_0 if $z > 1.28$; otherwise, fail to reject H_0. The test statistic is:

$$z = \frac{p_1 - p_2}{\hat{\sigma}_{p_1 - p_2}} = \frac{53 - 47}{\sqrt{\dfrac{53(47)}{200} + \dfrac{47(53)}{200}}} = 1.20.$$

Since 1.20 is not in the rejection region, we fail to reject H_0 and conclude that there is not evidence of a significantly greater percentage of accidents for males.

The confidence interval uses a z value of $z_{0.05}$ found by searching for a table probability of $0.5 - (0.10/2) = 0.45$. This z value is 1.645. Then the confidence interval is: $(53 - 47) \pm (1.645)\sqrt{\dfrac{53(47)}{200} + \dfrac{47(53)}{200}} = 6 \pm 8.2$ or from -2.2 to 14.2.

Practice Exercises 9-3

1. Baseball fans' opinions concerning the designated hitter rule are elicited in Saint Louis and in Detroit. The purpose is to determine if the rate of approval is less in Saint Louis, which hosts a National League team (National League teams do not use the designated hitter rule), than in Detroit, an American League city. The results are below. Use this data to do the appropriate test at $\alpha = 0.05$.

Saint Louis	Detroit
400 surveyed	300 surveyed
184 approved	159 approved

2. A quality control engineer is interested in comparing the percentages of defective computer boards being produced by two different manufacturing processes. He samples 80 of the computer boards manufactured by the first method and 75 of the computer boards manufactured by the second method. He finds that 8 and 12 of them, respectively, are defective. Calculate a 99% confidence interval for $\pi_1 - \pi_2$.

Solutions to Practice Exercises 9-3

1. Since we are interested in finding out if the approval rate is less in Saint Louis, the hypotheses are:

    H_0: $\pi_1 = \pi_2$

    H_1: $\pi_1 < \pi_2$

    where π_1 is the percentage of people who approve the designated hitter rule in Saint Louis and π_2 is the percentage for Detroit. The level of significance has been chosen to be 0.05. For the first sample, $p_1 = \dfrac{184}{400} \times (100 \text{ percent})$ or 46 percent. For the second sample, $p_2 = \dfrac{159}{300} \times (100 \text{ percent})$ or 53 percent. To ensure that we may use the normal distribution to perform the test, we check np and $n(100 - p)$ for both samples. For the first sample these are $18,400$ and $21,600$, for the second they are $15,900$ and $14,100$. Because all four of these exceed 500, we have large samples. The rejection region for this left-tailed test is bounded by a z critical value of $-z_{0.05}$ found in the z table by looking in the probability portion of the table for $0.5 - 0.05 = 0.45$. This value is 1.645. Our decision rule is to reject H_0 if $z < -1.645$; otherwise, fail to reject H_0. Calculating the test statistic, we get:

$$z = \frac{p_1 - p_2}{\hat{\sigma}_{p_1 - p_2}} = \frac{46 - 53}{\sqrt{\frac{46(54)}{400} + \frac{53(47)}{300}}} = -1.84.$$ Because $-1.84 < -1.645$, we reject H_0 and conclude that

the percentage who approve the designated hitter rule is less in Saint Louis than in Detroit.

2. For the first sample, $p_1 = \frac{8}{80} \times (100$ percent$)$ or 10 percent. For the second sample, $p_2 = \frac{12}{75} \times (100$ percent$)$ or 16 percent. To ensure that we may use the normal distribution to calculate the confidence interval, we check np and $n(100 - p)$ for both samples. For the first sample these are 800 and 7,200; for the second they are 1,200 and 6,300. Because all four of these exceed 500, we have large samples. The value for a 99% confidence interval is found in the z table by looking in the probability portion of the table for $0.5 - \frac{0.01}{2} = 0.495$. This value is 2.575. Then the confidence interval is:

$$(p_1 - p_2) \pm z_{\alpha/2}\hat{\sigma}_{p_1 - p_2} = (10 - 16) \pm (2.575)\sqrt{\frac{10(90)}{80} + \frac{16(84)}{75}} = -6 \pm 13.9 \text{ or } -19.9 < \pi_1 - \pi_2 < 7.9.$$

Solutions to Odd-Numbered Exercises

1. Step 1: State the Null and Alternative Hypotheses. H_0: $\sigma_1^2 = \sigma_2^2$ and H_1: $\sigma_1^2 \neq \sigma_2^2$.
   Step 2: Select the Level of Significance. $\alpha = 0.05$ (given).
   Step 3: Determine the Test Distribution to Use. Because this is a test of variance we use an F distribution.
   Step 4: Define the Rejection or Critical Regions. Since $\alpha = 0.05$ and this is a two-tailed test, we use the table for $\alpha = 0.025$. There are 16 in the group with the larger variance and 17 in the group with the smaller variance, so we have 15 degrees of freedom for the numerator and 16 degrees of freedom for the denominator. The critical F value is 2.79.
   Step 5: State the Decision Rule. Reject H_0 in favor of H_1 if $F > 2.79$. Otherwise, fail to reject H_0.
   Step 6: Make the Necessary Computations. Then $F = \frac{s_1^2}{s_2^2} = \frac{5.67^2}{4.74^2} = \frac{32.1489}{22.4676} = 1.4309.$
   Step 7: Make a Statistical Decision. Since $F = 1.4309$, we fail to reject H_0 that the population variances are equal.

3. In this Exercise we have independent samples, large sample sizes, and we know both σ_1 and σ_2, so we use procedure 2 with a z distribution.
   Step 1: H_0: $\mu_1 = \mu_2$ and H_1: $\mu_1 < \mu_2$.
   Step 2: $\alpha = 0.01$.
   Step 3: z distribution.
   Step 4: This is a one-tail test with $\alpha = 0.01$, thus the critical z value is -2.33.
   Step 5: Reject H_0 in favor of H_1 if $z < -2.33$. Otherwise, fail to reject H_0.

   Step 6: The standard error of the difference between means $\hat{\sigma}_{\bar{x}_1 - \bar{x}_2} = \sqrt{\frac{\sigma_1^2}{n_1} + \frac{\sigma_2^2}{n_2}} =$

   $\sqrt{\frac{2.10^2}{110} + \frac{3.15^2}{73}} = \sqrt{0.1760} = 0.4195.$ Then $z = \frac{\bar{x}_1 - \bar{x}_2}{\hat{\sigma}_{\bar{x}_1 - \bar{x}_2}} = \frac{4.2 - 4.6}{0.4195} = \frac{-0.4}{0.4195} = -0.9535.$
   Step 7: Since $z = -0.9535$ does not fall in the rejection region, we fail to reject H_0 that the mean number of drugs needed in each population is the same.

5. Since the samples are independent and the sample sizes are large, we conduct a two sample test on percentages with a z distribution.
   Step 1: H_0: $\pi_1 = \pi_2$ and H_1: $\pi_1 \neq \pi_2$.
   Step 2: $\alpha = 0.01$.

Step 3: We use the z distribution.

Step 4: This is a two-tailed test, with $\alpha = 0.01$ the critical z values are ± 2.575.

Step 5: Reject H_0 in favor of H_1 if $z < -2.575$ or if $z > 2.575$. Otherwise, fail to reject H_0.

Step 6: Calculate the percentages: $p_1 = \dfrac{60}{186} \times (100 \text{ percent}) = 32.3\%$ and

$p_2 = \dfrac{20}{97} \times (100 \text{ percent}) = 20.6\%$. Calculate the standard error of the difference between percentages:

$$\widehat{\sigma}_{p_1 - p_2} = \sqrt{\frac{p_1(100 - p_1)}{n_1} + \frac{p_2(100 - p_2)}{n_2}} = \sqrt{\frac{32.3 \cdot 67.7}{186} + \frac{20.6 \cdot 79.4}{97}} =$$

$\sqrt{28.619} = 5.350$. Thus $z = \dfrac{p_1 - p_2}{\widehat{\sigma}_{p_1 - p_2}} = \dfrac{32.3 - 20.6}{5.350} = 2.187$.

Step 7: Since z does not fall in the rejection region, we fail to reject H_0 that the announcement had no effect.

7. The 95% confidence interval for μ_d is completed using: $\overline{d} \pm t_{\alpha/2} \dfrac{s_d}{\sqrt{n}}$. We use the t distribution with

df $= 7$, so $t_{\alpha/2} = t_{0.025} = 2.365$. Now $\overline{d} = \dfrac{-38}{8} = -4.75$ and $s_d = \sqrt{\dfrac{883.499}{7}} = \sqrt{126.2141} = 11.2345$.

So the confidence interval is: $-4.75 \pm (2.365)\dfrac{11.2345}{\sqrt{8}} = -4.75 \pm 9.39$ or -4.64 to 14.14.

9. Since the samples are independent and the sample sizes are large, so we conduct a two sample test on percentages with a z distribution.

Step 1: H_0: $\pi_1 = \pi_2$ and H_1: $\pi_1 \neq \pi_2$.

Step 2: $\alpha = 0.05$.

Step 3: We use the z distribution.

Step 4: This is a two-tail test, with $\alpha = 0.05$ the critical z values are ± 1.96.

Step 5: Reject H_0 in favor of H_1 if $z < -1.96$ or if $z > 1.96$. Otherwise, fail to reject H_0.

Step 6: Calculate the percentages: $p_1 = \dfrac{2,519}{5,834} \times (100 \text{ percent}) = 43.178\%$ and

$p_2 = \dfrac{2,983}{4,110} \times (100 \text{ percent}) = 72.580\%$. Calculate the standard error of the difference between

percentages: $\widehat{\sigma}_{p_1 - p_2} = \sqrt{\dfrac{p_1(100 - p_1)}{n_1} + \dfrac{p_2(100 - p_2)}{n_2}} = \sqrt{\dfrac{43.178 \cdot 56.822}{5,834} + \dfrac{72.580 \cdot 27.420}{4,110}} =$

$\sqrt{0.9048} = 0.951$. Thus $z = \dfrac{p_1 - p_2}{\widehat{\sigma}_{p_1 - p_2}} = \dfrac{43.178 - 72.580}{0.951} = \dfrac{-29.402}{0.951} = -30.9093$.

Step 7: Since $z = -30.9093$ falls in the rejection region, we reject H_0. The percent of restrained patients who show signs of cognitive impairment is different that the percent of unrestrained patients who show such impairment.

11. Step 1: H_0: $\mu_1 = \mu_2$ and H_1: $\mu_1 \neq \mu_2$.

Step 2: $\alpha = 0.01$.

Step 3: We showed in Exercise 10 Review that the variance are equal, so we use procedure 4 with a t distribution.

Step 4: This is a two-tailed test. With $\alpha = 0.01$ and df $= n_1 + n_2 - 2 = 30$, so we use $t_{0.005} = \pm 2.750$.

Step 5: Reject H_0 in favor of H_1 if $t < -2.750$ or if $t > 2.750$. Otherwise, fail to reject H_0.

Step 6: Calculate the pooled standard deviation, $s_p = \sqrt{\dfrac{s_1^2(n_1 - 1) + s_2^2(n_2 - 1)}{n_1 + n_2 - 2}}$

$= \sqrt{\dfrac{(0.81)^2(16 - 1) + (0.99)^2(16 - 1)}{16 + 16 - 2}} = \sqrt{\dfrac{24.543}{30}} = 0.904$. The estimated standard error is

$\hat{\sigma}_{\bar{x}_1-\bar{x}_2} = s_p\sqrt{\dfrac{1}{n_1}+\dfrac{1}{n_2}} = 0.904\sqrt{\dfrac{1}{16}+\dfrac{1}{16}} = 0.904\cdot 0.354 = 0.320.$ Computing we get

$t = \dfrac{\bar{x}_1-\bar{x}_2}{\hat{\sigma}_{\bar{x}_1-\bar{x}_2}} = \dfrac{25.58-25.80}{0.320} = -0.688.$

Step 7: Since $t = -0.688$, we fail to reject H_0.

13. The 95% confidence interval for μ_d is completed using: $\bar{d}\pm t_{\alpha/2}\dfrac{s_d}{\sqrt{n}}.$ We use the t distribution with

df $= 3$, so $t_{\alpha/2} = t_{0.025} = 3.182$. Now $\bar{d} = 4$ and $s_d = \sqrt{\dfrac{4}{3}} = \sqrt{1.3333} = 1.155$. So the confidence

interval is: $4\pm(3.182)\dfrac{1.155}{\sqrt{4}} = 4\pm 1.84$ or 2.16 to 5.84.

15. Sample sizes are over 30 so use procedure 2 with a z distribution.
Step 1: H_0: $\mu_1 = \mu_2$ and H_1: $\mu_1 \neq \mu_2$.
Step 2: $\alpha = 0.01$.
Step 3: z distribution.
Step 4: This is a two-tailed test with $\alpha = 0.01$, thus the critical z values are ± 2.575.
Step 5: Reject H_0 in favor of H_1 if $z < -2.575$ or if $z > 2.575$. Otherwise, fail to reject H_0.
Step 6: The standard error of the difference between means

$\hat{\sigma}_{\bar{x}_1-\bar{x}_2} = \sqrt{\dfrac{\sigma_1^2}{n_1}+\dfrac{\sigma_2^2}{n_2}} = \sqrt{\dfrac{(7.65)^2}{222}+\dfrac{(6.83)^2}{64}} = \sqrt{0.9925} = 0.9962.$ Then

$z = \dfrac{\bar{x}_1-\bar{x}_2}{\hat{\sigma}_{\bar{x}_1-\bar{x}_2}} = \dfrac{30.47-29.64}{0.9962} = \dfrac{0.83}{0.9962} = 0.8332.$

Step 7: Since $z = 0.8332$ does not fall in the rejection region, we fail to reject H_0 that the mean age of the husband is equal for both groups.

17. Since the samples are independent and the sample sizes ≥ 30 we conduct a two sample test on percentages with a z distribution.
Step 1: H_0: $\pi_1 = \pi_2$ and H_1: $\pi_1 \neq \pi_2$.
Step 2: $\alpha = 0.01$.
Step 3: We use the z distribution.
Step 4: This is a two-tailed test with $\alpha = 0.01$ the critical z values are ± 2.575.
Step 5: Reject H_0 and in favor of H_1 if $z < -2.575$ or if $z > 2.575$. Otherwise, fail to reject H_0.
Step 6: Calculate the percentages: $p_1 = \dfrac{22}{30}\times(100 \text{ percent}) = 73.333\%$ and

$p_2 = \dfrac{32}{54}\times(100 \text{ percent}) = 59.259\%$. Then

$z = \dfrac{p_1-p_2}{\sqrt{\dfrac{p_1(100-p_1)}{n_1}+\dfrac{p_2(100-p_2)}{n_2}}} = \dfrac{73.333-59.259}{\sqrt{\dfrac{73.333\cdot 26.667}{30}+\dfrac{59.259\cdot 40.741}{54}}} = \dfrac{14.074}{\sqrt{109.894}} = 1.3426.$

Step 7: Since $z = 1.3426$ does not fall in the rejection region, we fail to reject H_0 that the population success rates for the two techniques are the same.

19. To obtain a 99% confidence interval for $\mu_1 - \mu_2$, we need $z_{\alpha/2} = z_{0.005} = 2.575$. Then we compute:

$\bar{x}_1 - \bar{x}_2 \pm z_{\alpha/2}\sqrt{\dfrac{\sigma_1^2}{n_1}+\dfrac{\sigma_2^2}{n_2}} = (4.02-2.93)\pm(2.575)\sqrt{\dfrac{(0.83)^2}{84}+\dfrac{(0.97)^2}{32}}$

$= 1.09\pm(2.575)(0.1939) = 0.83\pm 0.50$ or 0.33 to 1.33.

21. Since the sample sizes are large than 30, we use procedure 2 with a z distribution.

Step 1: H_0: $\mu_1 = \mu_2$ and H_1: $\mu_1 < \mu_2$.

Step 2: $\alpha = 0.05$.

Step 3: z distribution.

Step 4: This is a one-tail test with $\alpha = 0.05$, thus the critical z value is -1.645.

Step 5: Reject H_0 in favor of H_1 if $z < -1.645$. Otherwise, fail to reject H_0.

Step 6: $\hat{\sigma}_{\bar{x}_1 - \bar{x}_2} = \sqrt{\dfrac{\sigma_1^2}{n_1} + \dfrac{\sigma_2^2}{n_2}} = \sqrt{\dfrac{(5.21)^2}{46} + \dfrac{(7.07)^2}{41}} = \sqrt{1.8092} = 1.3451$. Then

$$z = \frac{\bar{x}_1 - \bar{x}_2}{\hat{\sigma}_{\bar{x}_1 - \bar{x}_2}} = \frac{9.64 - 11.52}{1.3451} = \frac{-1.88}{1.3451} = -1.398.$$

Step 7: Since $z = -1.398$ does not fall in the rejection region, we fail to reject H_0 that there's no difference in final exam mean scores for the two groups.

23. Step 1: H_0: $\sigma_1^2 = \sigma_2^2$ and H_1: $\sigma_1^2 \neq \sigma_2^2$.

Step 2: $\alpha = 0.05$.

Step 3: Because this is a test of variance we use an F distribution.

Step 4: Since $\alpha = 0.05$ and this is a two-tailed test, we use the table for $\alpha = 0.025$. Since there are 10 members in each group, we use 9 degrees of freedom for both the numerator and denominator. The critical F value is 4.03.

Step 5: Reject H_0 in favor of H_1 if $F > 4.03$. Otherwise, fail to reject H_0.

Step 6: Then $F = \dfrac{s_1^2}{s_2^2} = \dfrac{(0.0011)^2}{(0.00054)^2} = \dfrac{0.00000121}{0.0000002916} = 4.1495.$

Step 7: Since $F = 4.1495 > 4.03$, we reject H_0 that the population variances are equal.

25. Since the samples are independent and the sample sizes are large, so we conduct a two sample test on percentages with a z distribution.

Step 1: H_0: $\pi_1 = \pi_2$ and H_1: $\pi_1 < \pi_2$.

Step 2: $\alpha = 0.05$.

Step 3: We use the z distribution.

Step 4: This is a one-tail test, with $\alpha = 0.05$ the critical z value is -1.645.

Step 5: Reject H_0 in favor of H_1 if $z < -1.645$. Otherwise, fail to reject H_0.

Step 6: Calculate the percentages: $p_1 = \dfrac{46}{96} \times (100 \text{ percent}) = 47.92\%$ and

$p_2 = \dfrac{43}{69} \times (100 \text{ percent}) = 62.32\%$. Calculate the standard error of the difference between percentages:

$$\hat{\sigma}_{p_1 - p_2} = \sqrt{\frac{p_1(100 - p_1)}{n_1} + \frac{p_2(100 - p_2)}{n_2}} = \sqrt{\frac{47.92 \cdot 52.08}{96} + \frac{62.32 \cdot 37.68}{69}} = \sqrt{60.029} = 7.747.$$

Thus $z = \dfrac{p_1 - p_2}{\hat{\sigma}_{p_1 - p_2}} = \dfrac{47.92 - 62.32}{7.747} = \dfrac{-14.4}{7.747} = -1.859.$

Step 7: Since $z = -1.8589$ falls in the rejection region, we reject H_0. The percent of robberies related to drugs or alcohol is lower than that for assaults.

27. Since the sample sizes are large than 30, we use procedure 2 with a z distribution.

Step 1: H_0: $\mu_1 = \mu_2$ and H_1: $\mu_1 \neq \mu_2$.

Step 2: $\alpha = 0.01$.

Step 3: z distribution.

Step 4: This is a two-tailed test with $\alpha = 0.01$, thus the critical z values are ± 2.575.

Step 5: Reject H_0 in favor of H_1 if $z < -2.575$ or if $z > 2.575$. Otherwise, fail to reject H_0.

Step 6: $\widehat{\sigma}_{\overline{x}_1-\overline{x}_2} = \sqrt{\dfrac{\sigma_1^2}{n_1} + \dfrac{\sigma_2^2}{n_2}} = \sqrt{\dfrac{(2.7)^2}{31} + \dfrac{(2.4)^2}{31}} = \sqrt{0.42097} = 0.6488$. Thus

$z = \dfrac{\overline{x}_1 - \overline{x}_2}{\widehat{\sigma}_{\overline{x}_1-\overline{x}_2}} = \dfrac{75.2 - 77.4}{0.6488} = \dfrac{-2.2}{0.6488} = -3.3909$.

Step 7: Since $z = -3.3909$ falls in the rejection region, we reject H_0. There is a difference in heart rate for the two populations.

29. Since the samples are dependent, we use procedure 1 with a t test for paired differences.
Step 1: H_0: $\mu_d = 0$ and H_1: $\mu_d \neq 0$.
Step 2: $\alpha = 0.01$.
Step 3: We use the t distribution with $n = 5$.
Step 4: This is a two-tailed test. With $\alpha = 0.01$ and df $= 4$, use $t_{\alpha/2} = t_{.005} = \pm 4.604$.
Step 5: Reject H_0 in favor of H_1 if $t < -4.604$ or if $t > 4.604$. Otherwise, fail to reject H_0.
Step 6: Using MINITAB to calculate $\overline{d}$ and s_d.

Activity	Men	Women	d_i	$d_i - \overline{d}$	$(d_i - \overline{d})^2$
Studying	34.34	34.85	−0.51000	0.476002	0.226578
Working	43.95	44.31	−0.36000	0.625999	0.391875
Community	30.58	31.92	−1.34000	−0.354000	0.125316
Home/Family	44.80	46.55	−1.75000	−0.764000	0.583696
Leisure	39.42	40.39	−0.97000	0.015999	0.000256
			−4.93000		1.327721

$\overline{d} = \dfrac{-4.93000}{5} = -0.986$ and $s_d = \sqrt{\dfrac{1.327721}{4}} = \sqrt{0.33193025} = 0.5761$. Using Formula 9.2,

$t = \dfrac{\overline{d} - \mu_d}{s_d/\sqrt{n}} = \dfrac{-4.93000 - 0}{0.5761/\sqrt{5}} = -3.8268$.

Step 7: Since t is not in the rejection region, we fail to reject H_0 that there's no difference between men and women in their evaluation of the various activities in their lives.

31. Since the samples are independent and the sample sizes are large, so we conduct a two sample test on percentages with a z distribution.
Step 1: H_0: $\pi_1 = \pi_2$ and H_1: $\pi_1 > \pi_2$.
Step 2: $\alpha = 0.05$.
Step 3: We use the z distribution.
Step 4: This is a one-tail test, with $\alpha = 0.05$ the critical z value is 1.645.
Step 5: Reject H_0 in favor of H_1 if $z > 1.645$. Otherwise, fail to reject H_0.
Step 6: Calculate the percentages: $p_1 = \dfrac{108}{160} \times (100 \text{ percent}) = 67.50\%$ and

$p_2 = \dfrac{164}{338} \times (100 \text{ percent}) = 48.52\%$. Calculate the standard error of the difference between percentages:

$\widehat{\sigma}_{p_1-p_2} = \sqrt{\dfrac{p_1(100-p_1)}{n_1} + \dfrac{p_2(100-p_2)}{n_2}} = \sqrt{\dfrac{67.50 \cdot 32.50}{160} + \dfrac{48.52 \cdot 51.48}{338}} = \sqrt{21.101} = 4.594$.

Thus $z = \dfrac{p_1 - p_2}{\widehat{\sigma}_{p_1-p_2}} = \dfrac{67.50 - 48.52}{4.594} = 4.131$.

Step 7: Since $z = 4.131$ falls in the rejection region, we reject H_0. A smaller percentage of needles were infected with the HIV virus after the needle exchange program.

33. Step 1: H_0: $\sigma_1^2 = \sigma_2^2$ and H_1: $\sigma_1^2 \neq \sigma_2^2$.
Step 2: $\alpha = 0.05$.
Step 3: Because this is a test of variance we use an F distribution.

Step 4: Since $\alpha = 0.05$ and this is a two-tailed test, we use the table for $\alpha = 0.025$. There are 21 in each group, so we use 20 degrees of freedom for both the numerator and denominator. The critical F value is 2.46.

Step 5: Reject H_0 in favor of H_1 if $F > 2.46$. Otherwise, fail to reject H_0.

Step 6: $F = \dfrac{s_1^2}{s_2^2} = \dfrac{(0.792)^2}{(0.689)^2} = \dfrac{0.627264}{0.474721} = 1.3213$.

Step 7: Since $F = 1.3213$ does not fall in the rejection region, we fail to reject H_0 that the population variances are equal.

35. Since the sample sizes are larger than 30 and are independent, we use procedure 2 with a z distribution.

Step 1: H_0: $\mu_1 = \mu_2$ and H_1: $\mu_1 < \mu_2$.

Step 2: $\alpha = 0.01$.

Step 3: z distribution.

Step 4: This is a one-tail test with $\alpha = 0.01$, thus the critical z value is -2.33.

Step 5: Reject H_0 in favor of H_1 if $z < -2.33$. Otherwise, fail to reject H_0.

Step 6: $\hat{\sigma}_{\bar{x}_1 - \bar{x}_2} = \sqrt{\dfrac{s_1^2}{n_1} + \dfrac{s_2^2}{n_2}} = \sqrt{\dfrac{(13)^2}{60} + \dfrac{(17)^2}{58}} = \sqrt{7.7994} = 2.7927$. Thus

$z = \dfrac{\bar{x}_1 - \bar{x}_2}{\hat{\sigma}_{\bar{x}_1 - \bar{x}_2}} = \dfrac{22 - 62}{2.7927} = \dfrac{-40}{2.7927} = -14.3228$.

Step 7: Since $z = -14.3228$ falls in the rejection region, we reject H_0. NORVASC significantly improves exercise time.

37. Step 1: H_0: $\sigma_1^2 = \sigma_2^2$ and H_1: $\sigma_1^2 \neq \sigma_2^2$.

Step 2: $\alpha = 0.10$.

Step 3: Because this is a test of variance we use an F distribution.

Step 4: Since $\alpha = 0.10$ and this is a two-tailed test, we use the table for $\alpha = 0.025$. Since there are 7 members in each group we use 6 degrees of freedom for both the numerator and denominator. The critical F value is 4.28.

Step 5: Reject H_0 in favor of H_1 if $F > 4.28$. Otherwise, fail to reject H_0.

Step 6: Then $F = \dfrac{s_1^2}{s_2^2} = \dfrac{(0.1464)^2}{(0.0488)^2} = \dfrac{0.021433}{0.002381} = 9.002$.

Step 7: Since $F = 9.002 > 4.28$, we reject H_0, the population variances are not equal for mercury.

39. Step 1: $\sigma_1^2 = \sigma_2^2$ and H_1: $\sigma_1^2 > \sigma_2^2$.

Step 2: Use $\alpha = 0.05$.

Step 3: We use an F distribution. Since this is a right-tailed test, we use the table with 0.05 in the right tail.

Step 4: We use the table for $\alpha = 0.05$. Since there are 7 members in each group we use 6 degrees of freedom for both the numerator and denominator. The critical F value is 4.28.

Step 5: Reject H_0 in favor of H_1 if $F > 4.28$. Otherwise, fail to reject H_0.

Step 6: The $F = \dfrac{1.657^2}{0.1952^2} = \dfrac{2.74565}{0.03810} = 72.062$.

Step 7: Since $F = 72.062 > 4.28$, we reject H_0, the population variance for silver is greater at lab 4 than at lab 3..

41. In Exercise 40 we failed to reject the null hypotheses that $\sigma_1^2 = \sigma_2^2$ so in this exercise we use procedure 4 with a t distribution.

Step 1: H_0: $\mu_1 = \mu_2$ and H_1: $\mu_1 \neq \mu_2$.

Step 2: $\alpha = 0.05$.

Step 3: We use the t distribution.

Step 4: This is a two-tailed test. With $\alpha = 0.05$ and 12 df we use $t_{\alpha/2} = t_{0.025} = 2.179$.

Step 5: Reject H_0 in favor of H_1 if $t < -2.179$ or if $t > 2.179$. Otherwise, fail to reject H_0.

Step 6: Calculate the pooled standard deviation, $s_p = \sqrt{\dfrac{s_1^2(n_1 - 1) + s_2^2(n_2 - 1)}{n_1 + n_2 - 2}}$

$= \sqrt{\dfrac{(0.1113)^2(7-1) + (0.0535)^2(7-1)}{7+7-2}} = \sqrt{0.007625} = 0.0873$. The estimated standard error is

$\hat{\sigma}_{\bar{x}_1 - \bar{x}_2} = s_p \sqrt{\dfrac{1}{n_1} + \dfrac{1}{n_2}} = 0.0873 \cdot \sqrt{\dfrac{1}{7} + \dfrac{1}{7}} = 0.0873 \cdot 0.5345 = 0.0467$. Thus

$t = \dfrac{\bar{x}_1 - \bar{x}_2}{\hat{\sigma}_{\bar{x}_1 - \bar{x}_2}} = \dfrac{0.7286 - 0.7571}{0.0467} = \dfrac{-0.0285}{0.0467} = -0.610.$

Step 7: Since $t = -0.610$ does not fall within the rejection region, we fail to reject H_0, that is there is not a significant difference in the population means for cadmium.

43. In Exercise 42 we rejected the null hypotheses that $\sigma_1^2 = \sigma_2^2$ so we use procedure 3 with a t distribution.

Step 1: H_0: $\mu_1 = \mu_2$ and H_1: $\mu_1 \neq \mu_2$.

Step 2: $\alpha = 0.02$.

Step 3: t distribution.

Step 4: This is a two-tailed test with $\alpha = 0.02$. The degree of freedom is 6 since both groups have 7 members. The critical values are $t_{0.01} = \pm 3.143$.

Step 5: Reject H_0 in favor of H_1 if $t < -3.143$ or if $t > 3.143$. Otherwise, fail to reject H_0.

Step 6: Then $t = \dfrac{\bar{x}_1 - \bar{x}_2}{\sqrt{\dfrac{s_1^2}{n_1} + \dfrac{s_2^2}{n_2}}} = \dfrac{7.143 - 3.814}{\sqrt{\dfrac{(1.676)^2}{7} + \dfrac{(0.308)^2}{7}}} = \dfrac{3.329}{\sqrt{0.4148}} = \dfrac{3.329}{0.6440} = 5.169.$

Step 7: Since $t = 5.169$, falls within the rejection region, we reject H_0, there is a significant difference in the population means for lead.

45. Step 1: H_0: $\sigma_1^2 = \sigma_2^2$ and H_1: $\sigma_1^2 \neq \sigma_2^2$.

Step 2: $\alpha = 0.10$.

Step 3: Because this is a test of variance we use an F distribution.

Step 4: Since $\alpha = 0.10$ and this is a two-tailed test, we use the table for $\alpha = 0.05$. Since there are 10 members in each group we use 9 degrees of freedom for both the numerator and denominator. The critical F value is 3.18.

Step 5: Reject H_0 in favor of H_1 if $F > 5.82$. Otherwise, fail to reject H_0.

Step 6: Then $F = \dfrac{s_1^2}{s_2^2} = \dfrac{(1.602)^2}{(0.551)^2} = \dfrac{2.566404}{0.303601} = 8.453.$

Step 7: Since $F = 8.453 > 3.18$, we reject H_0 that the population variances are not equal.

47. Step 1: H_0: $\pi_1 = \pi_2$ and H_1: $\pi_1 \neq \pi_2$.

Step 2: $\alpha = 0.01$.

Step 3: We use the z distribution.

Step 4: This is a two-tailed test, with $\alpha = 0.01$ the critical z value is ± 2.575.

Step 5: Reject H_0 in favor of H_1 if $z < -2.575$ of if $z > 2.575$. Otherwise, fail to reject H_0.

Step 6: Calculate the percentages: $p_1 = \dfrac{49}{167} \times (100 \text{ percent}) = 29.34\%$ and

$p_2 = \dfrac{22}{143} \times (100 \text{ percent}) = 15.38\%$. Calculate the standard error of the difference between percentages:

$\hat{\sigma}_{p_1 - p_2} = \sqrt{\dfrac{p_1(100 - p_1)}{n_1} + \dfrac{p_2(100 - p_2)}{n_2}} = \sqrt{\dfrac{29.34 \cdot 70.66}{167} + \dfrac{15.38 \cdot 84.62}{143}} = \sqrt{21.5152} = 4.638.$

Thus $z = \dfrac{p_1 - p_2}{\hat{\sigma}_{p_1 - p_2}} = \dfrac{29.34 - 15.38}{4.638} = \dfrac{13.96}{4.638} = 3.010.$

Step 7: Since $z = 3.010$ falls in the rejection region, we reject H_0, thus there is a difference in the percentage of wells that test positive for pesticides.

49. Step 1: H_0: $\sigma_1^2 = \sigma_2^2$ and H_1: $\sigma_1^2 \neq \sigma_2^2$.

Step 2: Use $\alpha = 0.05$.

Step 3: We use an F distribution.

Step 4: Since this is a two-tailed test, we use the table with 0.025 in the right tail. The df for the numerator is $20 - 1 = 19$. And the df for the denominator is $30 - 1 = 29$. (We will use the table value for 20 df in the numerator.) The critical F value is 2.21.

Step 5: Reject H_0 in favor of H_1 if $F > 2.21$. Otherwise, fail to reject H_0.

Step 6: Then $F = \dfrac{(6.32)^2}{(5.88)^2} = \dfrac{39.9424}{34.5744} = 1.155$.

Step 7: Since $F = 1.155$, we fail to reject H_0, the variances are equal.

51. The formula for a conference interval for $\mu_1 - \mu_2$ is: $\overline{x}_1 - \overline{x}_2 \pm t_{\alpha/2} s_p \sqrt{\dfrac{1}{n_1} + \dfrac{1}{n_2}}$. With $\alpha = 0.05$ and $df = n_1 + n_2 - 2 = 48$, so we use $t_{\alpha/2} = t_{0.025} = 2.010$ (we took the average of the 40 df and the 60 df values). Then the conference interval for $\mu_1 - \mu_2$ is:

$(26.28 - 27.56) \pm (2.010) \sqrt{\dfrac{(6.32)^2(20-1) + (5.88)^2(30-1)}{20 + 30 - 2}} \sqrt{\dfrac{1}{20} + \dfrac{1}{30}} = -1.28 \pm 3.52$ or -4.8 to 2.24.

53. The 95% confidence interval for the difference of two population percentages is: $(p_1 - p_2) \pm z_{\alpha/2} \widehat{\sigma}_{p_1-p_2}$, with $z_{\alpha/2} = 1.96$. So all five questions:

Question	$(p_1 - p_2) \pm z_{\alpha/2} \widehat{\sigma}_{p_1-p_2}$
1	$(40 - 50) \pm (1.96)(9.8995) = -10 \pm 19.4030$ or -19.403 to -9.403.
2	$(78 - 64) \pm (1.96)(8.9666) = 14 \pm 17.575$ or 3.575 to 31.575.
3	$(64 - 64) \pm (1.96)(9.6000) = 0 \pm 18.816$ or -18.816 to 18.816.
4	$(32 - 46) \pm (1.96)(9.6540) = -14 \pm 18.922$ or -32.922 to 4.922.
5	$(56 - 62) \pm (1.96)(9.8184) = -6 \pm 19.244$ or -25.244 to 13.244.

55. For each of the five variables we construct a 95 percent confidence interval for the difference in the population means.

If we rejected H_0 in the first hypotheses tests on variance, then $\sigma_1^2 \neq \sigma_2^2$, so we use Procedure 3 to produce the 95% confidence interval, $\overline{x}_1 - \overline{x}_2 \pm t_{\alpha/2} \sqrt{\dfrac{s_1^2}{n_1} + \dfrac{s_2^2}{n_2}}$. We use a t distribution with $12 - 1 = 11$, $t_{\alpha/2} = t_{0.025} = 2.021$.

If we failed to reject H_0, then $\sigma_1^2 = \sigma_2^2$, so we use Procedure 4 to produce the 95% confidence interval,

$\overline{x}_1 - \overline{x}_2 \pm t_{\alpha/2} \sqrt{\dfrac{s_1^2(n_1 - 1) + s_2^2(n_2 - 1)}{n_1 + n_2 - 2}} \sqrt{\dfrac{1}{n_1} + \dfrac{1}{n_2}}$. We use a t distribution with $12 + 12 - 2 = 22$, (we use the conservative value for 40 df) $t_{\alpha/2} = t_{0.025} = 2.074$.

Variable	Procedure	95 percent confidence interval
SBP	4	$(154.83 - 134.50) \pm (2.074)(7.309) = 20.33 \pm 15.16$ or 5.17 to 35.49.
DBP	4	$(90.50 - 82.17) \pm (2.074)(3.905) = 8.33 \pm 8.10$ or 0.23 to 16.43.
SNa	4	$(144.33 - 141.92) \pm (2.074)(0.869) = 2.41 \pm 1.80$ or 0.61 to 4.21.
SK	4	$(3.38 - 4.53) \pm (2.074)(0.270) = -1.15 \pm 0.56$ or -1.71 to -0.59.
PAC	3	$(1194 - 159.2) \pm (2.021)(320.74) = 1034.8 \pm 648.2$ or 386.6 to 1683.

57. To obtain a 90% confidence interval for $\pi_1 - \pi_2$, we need $z_{\alpha/2} = z_{0.05} = 1.645$. Then we compute:

$$(p_1 - p_2) \pm z_{\alpha/2}\sqrt{\frac{\sigma_1^2}{n_1} + \frac{\sigma_2^2}{n_2}} = (38.46 - 41.67) \pm (1.645)\sqrt{\frac{38.46 \cdot 61.54}{26} + \frac{41.67 \cdot 58.33}{12}}$$

$$= -3.21 \pm (1.645)(17.134) = -3.21 \pm 28.19 \text{ or } -31.40 \text{ to } 24.98.$$

59. To obtain a 95% confidence interval for $\pi_1 - \pi_2$, we need $z_{\alpha/2} = z_{0.005} = 1.96$. Then we compute:

$$(p_1 - p_2) \pm z_{\alpha/2}\sqrt{\frac{\sigma_1^2}{n_1} + \frac{\sigma_2^2}{n_2}} = (30.77 - 91.67) \pm (1.96)\sqrt{\frac{30.77 \cdot 69.23}{26} + \frac{91.67 \cdot 8.33}{12}}$$

$$= -60.90 \pm (1.96)(12.066) = -60.90 \pm 23.65 \text{ or } -84.55 \text{ to } -37.25.$$

61. In Exercise 60 we rejected the null hypotheses that $\sigma_1^2 = \sigma_2^2$ so we use procedure 3 with a t distribution.
Step 1: H_0: $\mu_1 = \mu_2$ and H_1: $\mu_1 \neq \mu_2$.
Step 2: $\alpha = 0.05$.
Step 3: t distribution.
Step 4: This is a two-tailed test with $\alpha = 0.05$. The degree of freedom is 5. The critical values are $t_{0.025} = \pm 2.571$.
Step 5: Reject H_0 in favor of H_1 if $t < -2.571$ or if $t > 2.571$. Otherwise, fail to reject H_0.
Step 6: Then $t = \dfrac{\bar{x}_1 - \bar{x}_2}{\sqrt{\dfrac{s_1^2}{n_1} + \dfrac{s_2^2}{n_2}}} = \dfrac{4.343 - 1.433}{\sqrt{\dfrac{(1.144)^2}{7} + \dfrac{(0.151)^2}{6}}} = \dfrac{2.91}{\sqrt{0.190762}} = \dfrac{2.91}{0.437} = 6.659.$

Step 7: Since $t = 6.659$, we fall to reject H_0, the population means for pH are equal.

63. The 99% confidence interval for μ_d is completed using: $\bar{d} \pm t_{\alpha/2}\dfrac{s_d}{\sqrt{n}}$. We use the t distribution with

df $= 7$, so $t_{\alpha/2} = t_{0.005} = 5.841$. Now $\bar{d} = \dfrac{-38}{8} = -4.75$ and $s_d = 24.1$. So the confidence interval is:

$27.8 \pm (5.841)\dfrac{24.1}{\sqrt{4}} = 27.8 \pm 70.4$ or -42.6 to 98.2.

65. The 95% confidence interval for μ_d is completed using: $\bar{d} \pm t_{\alpha/2}\dfrac{s_d}{\sqrt{n}}$. We use the t distribution with

df $= 7$, so $t_{\alpha/2} = t_{0.025} = 2.306$. Now $\bar{d} = \dfrac{-38}{8} = -4.75$ and $s_d = 0.499$. So the confidence interval is:

$-0.086 \pm (2.306)\dfrac{0.499}{\sqrt{8}} = -0.086 \pm 0.407$ or -0.493 to 0.321.

67. Step 1: H_0: $\sigma_1^2 = \sigma_2^2$ and H_1: $\sigma_1^2 \neq \sigma_2^2$.
Step 2: $\alpha = 0.05$.
Step 3: Because this is a test of variance we use an F distribution. The df for the numerator is $8 - 1$ or 7. And the df for the denominator is $8 - 1$ or 7.

Step 4: Since $\alpha = 0.05$ and this is a two-tailed test, we use the table for $\alpha = 0.025$. The critical F value is 4.99.

Step 5: Reject H_0 in favor of H_1 if $F > 4.99$. Otherwise, fail to reject H_0.

Step 6: Then $F = \dfrac{s_1^2}{s_2^2} = \dfrac{(0.499)^2}{(0.299)^2} = \dfrac{0.249001}{0.089401} = 2.785$.

Step 7: Since $F = 2.785 < 4.99$, we fall to reject H_0 that the population variances are equal.

Chapter 10 Analysis of Variance

Study Aids and Practice Exercises

10-1 Analysis of Variance: Purpose and Procedure

Study Objectives
You should be able to:
1. Understand the purpose of an analysis of variance.
2. Decipher the symbols used for a one-way ANOVA.
3. Understand the types of calculations required for an ANOVA.

Section Overview
Analysis of variance (ANOVA) is a statistical technique we use to test the hypothesis that a set of two or more population means are equal. For example, we might want to compare the mean growth of a crop using four irrigation methods or the mean mileage for five brands of gasoline. When only one factor defines the populations, the analysis is a one-way analysis of variance, or a one-way ANOVA. To validly perform an analysis of variance, the following assumptions must be true:
1. The populations are normal
2. The samples are randomly and independently taken from the populations.
3. The populations have equal variances, i.e., $\sigma_1^2 = \sigma_2^2 = \cdots = \sigma_k^2$.

The hypotheses of interest in a one-way ANOVA are:

H_0: $\mu_1 = \mu_2 = \cdots = \mu_k$
H_1: At least one mean differs from the rest,

where k is the number of populations compared.

Caution: It is a common mistake to say that the alternative hypothesis is:
H_1: $\mu_1 \neq \mu_2 \neq \cdots \neq \mu_k$.

To define the test, we need several new symbols.

n_i = number of items in sample i

T = total of all samples = $n_1 + n_2 + \cdots + n_k$

$\overline{x}_i$ = mean of ith sample

$\overline{X}$ = grand mean = $\dfrac{n_1\overline{x}_1 + n_2\overline{x}_2 + \cdots + n_k\overline{x}_k}{n_1 + n_2 + \cdots + n_k}$

$\Sigma d_i^2 = \Sigma(x_i - \overline{x}_i)^2$

$\widehat{\sigma}_{\text{within}}^2 = \dfrac{\Sigma d_1^2 + \Sigma d_2^2 + \cdots + \Sigma d_k^2}{T - k}$

$\widehat{\sigma}_{\text{between}}^2 = \dfrac{n_1(\overline{x}_1 - \overline{\overline{X}})^2 + n_2(\overline{x}_2 - \overline{\overline{X}})^2 + \cdots + n_k(\overline{x}_k - \overline{\overline{X}})^2}{k - 1}$

With all the assumptions satisfied, the test is based on the F distribution with $df_{num} = k - 1$ and $df_{den} = T - k$, where T = total of all samples = $n_1 + n_2 + \cdots + n_k$. The test statistic for a one-way ANOVA is:

$$F = \frac{\hat{\sigma}^2_{between}}{\hat{\sigma}^2_{within}}.$$

H_0 is rejected if $F > F_{k-1, T-k, \alpha}$.

> Hint: Although this is a hypothesis test on the equality of a set of means, it is called analysis of variance because the test statistic involves two different estimates of the variance.

Key Terms & Formulas

Analysis of Variance (ANOVA) A statistical technique that we use to test the hypothesis that a set of two or more population means are equal, i.e., to test:

H_0: $\mu_1 = \mu_2 = \cdots = \mu_k$
H_1: At least one mean differs from the rest.

One-Way Analysis of Variance (One-Way ANOVA) An analysis of variance on populations that differ based on the values of a single factor.

ANOVA Test Statistic The test statistic for a one-way ANOVA is $F = \dfrac{\hat{\sigma}^2_{between}}{\hat{\sigma}^2_{within}}$, where

$$\hat{\sigma}^2_{within} = \frac{\Sigma d_1^2 + \Sigma d_2^2 + \cdots + \Sigma d_k^2}{T - k} \text{ and } \hat{\sigma}^2_{between} = \frac{n_1(\bar{x}_1 - \overline{\overline{X}})^2 + n_2(\bar{x}_2 - \overline{\overline{X}})^2 + \cdots + n_k(\bar{x}_k - \overline{\overline{X}})^2}{k - 1}.$$

10-2 An ANOVA Example

Study Objectives

You should be able to:
1. Calculate an analysis of variance test statistic.
2. Make a decision based on the F test statistic.

Section Overview

We work through a complete analysis of variance using the formulae introduced in the previous section.

Worked Examples

Hot Computers

The designer of a super computer is evaluating three different cooling systems for the latest model. He installs each cooling system into prototypes and measures the rise in temperature in ten minutes of operation. He repeats this a total of five times for each cooling system. The data is below.

System 1	System 2	System 3
44	26	27
41	24	33
26	21	19
34	11	25
35	15	31

Using $\alpha = 0.05$, use analysis of variance to decide if there is a significant difference in the mean increases in temperature.

Solution

The hypotheses of interest for this situation are:
H_0: $\mu_1 = \mu_2 = \mu_3$
H_1: At least one mean differs from the rest.

The level of significance is 0.05. For the F test, we need the degrees of freedom for the numerator and denominator. These involve $k = $ the number of populations $= 3$ and $T = $ total number of all items in all samples $= 5 + 5 + 5 = 15$. Then $df_{num} = k - 1 = 3 - 1 = 2$ and $df_{den} = T - k = 15 - 3 = 12$. The critical F value that starts the rejection region is $F_{2, 12, \alpha=0.05} = 3.89$. The decision rule is to reject H_0 if $F > 3.89$; otherwise, fail to reject H_0. To start the calculation process, we compute the sample means:

$$\bar{x}_1 = \frac{\Sigma x_1}{n_1} = \frac{180}{5} = 36.0, \qquad \bar{x}_2 = \frac{\Sigma x_2}{n_2} = \frac{97}{5} = 19.4, \qquad \bar{x}_3 = \frac{\Sigma x_3}{n_3} = \frac{135}{5} = 27.0.$$

Next we calculate the grand mean:

$$\overline{\overline{X}} = \frac{n_1\bar{x}_1 + n_2\bar{x}_2 + n_3\bar{x}_3}{n_1 + n_2 + n_3} = \frac{5(36.0) + 5(19.4) + 5(27.0)}{5 + 5 + 5} = \frac{412}{15} = 27.4667.$$

Now we can obtain the numerator of the test statistic:

$$\hat{\sigma}^2_{between} = \frac{n_1(\bar{x}_1 - \overline{\overline{X}})^2 + n_2(\bar{x}_2 - \overline{\overline{X}})^2 + n_3(\bar{x}_3 - \overline{\overline{X}})^2}{3 - 1}$$

$$= \frac{5(36.0 - 27.4667)^2 + 5(19.4 - 27.4667)^2 + 5(27.0 - 27.4667)^2}{2} = 345.2667$$

The calculations for $\hat{\sigma}^2_{within} = \dfrac{\Sigma d_1^2 + \Sigma d_2^2 + \cdots + \Sigma d_k^2}{T - k}$ are done in the following tables.

x_1	$x_1 - \bar{x}_1$	$(x_1 - \bar{x}_1)^2$		x_2	$x_2 - \bar{x}_2$	$(x_2 - \bar{x}_2)^2$
44	8	64		26	6.6	43.56
41	5	25		24	4.6	21.16
26	−10	100		21	1.6	2.56
34	−2	4		11	−8.4	70.56
35	−1	1		15	−4.4	19.36
		$\Sigma d_1^2 = 194$				$\Sigma d_2^2 = 157.20$

x_3	$x_3 - \bar{x}_3$	$(x_3 - \bar{x}_3)^2$
27	0	0
33	6	36
19	−8	64
25	−2	4
31	4	16
		$\Sigma d_3^2 = 120$

Then $\hat{\sigma}^2_{within} = \dfrac{\Sigma d_1^2 + \Sigma d_2^2 + \Sigma d_3^2}{T - k} = \dfrac{194 + 157.2 + 120}{15 - 3} = 39.2667.$

Finally we can calculate the test statistic: $F = \dfrac{\hat{\sigma}^2_{between}}{\hat{\sigma}^2_{within}} = \dfrac{345.2667}{39.2667} = 8.79.$

Because $8.79 > 3.89$, we reject H_0 and conclude that at least one of the cooling systems has a different mean rise in temperature.

Ropes

In an experiment to evaluate the resiliency of climbing ropes, a mountaineering shop attached 250-pound weights to four brands of ropes and dropped them 150 feet. After the drops, they measured the increase in the length of the rope. The measurements are in the table below. Do an ANOVA on this data using $\alpha = 0.01$.

Rope 1	Rope 2	Rope 3	Rope 4
13	12	7	17
17	14	9	15
18	15	11	14
15	15		
	20		
	13		

Solution

The hypotheses of interest are:

H_0: $\mu_1 = \mu_2 = \mu_3 = \mu_4$

H_1: At least one mean differs from the rest.

The level of significance is 0.01. For the F test, we need the degrees of freedom for the numerator and denominator. These involve k = the number of populations = 4 and T = total number of all items in all samples = $4 + 6 + 3 + 3 = 16$. Then $df_{num} = k - 1 = 4 - 1 = 3$ and $df_{den} = T - k = 16 - 4 = 12$, and the critical F value is $F_{3,\,12,\,\alpha=0.01} = 5.95$. The decision rule is to reject H_0 if $F > 5.95$; otherwise, fail to reject H_0. We start by computing the sample means.

$$\overline{x}_1 = \frac{\Sigma x_1}{n_1} = \frac{63}{4} = 15.75 \qquad \overline{x}_2 = \frac{\Sigma x_2}{n_2} = \frac{89}{6} = 14.8333$$

$$\overline{x}_3 = \frac{\Sigma x_3}{n_3} = \frac{27}{3} = 9.0 \qquad \overline{x}_4 = \frac{\Sigma x_4}{n_4} = \frac{46}{3} = 15.3333$$

Next the grand mean:

$$\overline{\overline{X}} = \frac{n_1 \overline{x}_1 + n_2 \overline{x}_2 + n_3 \overline{x}_3 + n_4 \overline{x}_4}{n_1 + n_2 + n_3 + n_4} = \frac{4(15.75) + 6(14.8333) + 3(9.0) + 3(15.3333)}{4 + 6 + 3 + 3} = \frac{225}{16} = 14.0625. \text{ Now we}$$

can obtain the numerator of the test statistic:

$$\hat{\sigma}^2_{between} = \frac{n_1(\overline{x}_1 - \overline{\overline{X}})^2 + n_2(\overline{x}_2 - \overline{\overline{X}})^2 + n_3(\overline{x}_3 - \overline{\overline{X}})^2 + n_4(\overline{x}_4 - \overline{\overline{X}})^2}{4 - 1}$$

$$= \frac{4(15.75 - 14.06)^2 + 6(14.83 - 14.06)^2 + 3(9.0 - 14.06)^2 + 3(15.33 - 14.06)^2}{3}$$

$$= \frac{96.6875}{3} = 32.2292.$$

The calculations for $\hat{\sigma}^2_{within} = \dfrac{\Sigma d_1^2 + \Sigma d_2^2 + \cdots + \Sigma d_k^2}{T - k}$ are presented in the following tables.

x_1	$x_1 - \overline{x}_1$	$(x_1 - \overline{x}_1)^2$
13	-2.75	7.5625
17	1.25	1.5625
18	2.25	5.0625
15	-0.75	0.5625
		$\Sigma d_1^2 = 14.75$

x_2	$x_2 - \overline{x}_2$	$(x_2 - \overline{x}_2)^2$
12	-2.833	8.0278
14	-0.833	0.6944
15	0.167	0.0278
15	0.167	0.0278
20	5.167	26.6944
13	-1.833	3.3611
		$\Sigma d_2^2 = 38.8333$

x_3	$x_3 - \overline{x}_3$	$(x_3 - \overline{x}_3)^2$
7	-2	4
9	0	0
11	2	4
		$\Sigma d_3^2 = 8$

x_4	$x_1 - \overline{x}_4$	$(x_4 - \overline{x}_4)^2$
17	1.667	2.778
15	-0.333	0.111
14	-1.333	1.778
		$\Sigma d_4^2 = 4.667$

Then $\hat{\sigma}^2_{within} = \dfrac{\Sigma d_1^2 + \Sigma d_2^2 + \Sigma d_3^2 + \Sigma d_4^2}{T - k} = \dfrac{14.75 + 38.83 + 8.0 + 4.67}{16 - 4} = 5.5208.$

Finally we can calculate the test statistic $F = \dfrac{\hat{\sigma}^2_{between}}{\hat{\sigma}^2_{within}} = \dfrac{32.2292}{5.5208} = 5.84.$

Because 5.84 is not greater than 5.95, we fail to reject H_0. We conclude that there is not enough evidence at $\alpha = 0.01$ to show any difference in the mean increase of the rope lengths.

Practice Exercises 10-2

1. An office manager is evaluating three grammar checking programs. She wants to compare the number of errors each would find on a sample of the manager's writing. Running each program four times, they detect the number of errors listed in the following table. Perform an ANOVA on this data using $\alpha = 0.05$.

Grammar1	Grammar2	Grammar3
10	11	13
8	12	10
9	8	14
9	9	15

2. Currently in the major baseball leagues, there are four schools of thought about the best method of batting instruction. The San Francisco Giants decide to find out if there is any difference in the results of the instruction. They take 20 comparable players in the instructional league and try each of the four methods on five different players. Their batting averages (with the decimals omitted) are in the table that follows. Perform an ANOVA to decide if there is a significant difference in these four methods of instruction using $\alpha = 0.05$.

HitMeth1	HitMeth2	HitMeth3	HitMeth4
311	320	325	265
283	280	299	278
302	298	305	291
290	295	297	284
295	281	304	284

3. Three different books are on the market that discuss "fuzzy logic," a concept emerging in engineering and computer systems. A publisher assigns 13 editors to each read one book and rate the book on 10 characteristics on a scale of 0 to 10. Then each editor totals his or her points and gives the book a total score. The publisher wants to know if there is a significant difference in the mean ratings. Do an ANOVA for the publisher using $\alpha = 0.01$.

Book 1	Book 2	Book 3
57	90	65
70	81	78
79	69	97
60	96	86
57		

Solutions to Practice Exercises 10-2

1. The hypotheses of interest are:

H_0: $\mu_1 = \mu_2 = \mu_3$

H_1: At least one mean differs from the rest.

The level of significance is 0.05. We have $k = 3$ and $T = 4 + 4 + 4 = 12$. Then $df_{num} = k - 1 = 3 - 1 = 2$ and $df_{den} = T - k = 12 - 3 = 9$. The critical value is $F_{2,9,\alpha=0.05} = 4.26$. The decision rule is to reject H_0 if $F > 4.26$; otherwise, fail to reject H_0. The sample means:

$$\bar{x}_1 = \frac{\Sigma x_1}{n_1} = \frac{36}{4} = 9 \qquad \bar{x}_2 = \frac{\Sigma x_2}{n_2} = \frac{40}{4} = 10 \qquad \bar{x}_3 = \frac{\Sigma x_3}{n_3} = \frac{52}{4} = 13$$

The grand mean: $\bar{\bar{X}} = \dfrac{n_1\bar{x}_1 + n_2\bar{x}_2 + n_3\bar{x}_3}{n_1 + n_2 + n_3} = \dfrac{4(9) + 4(10) + 4(13)}{4 + 4 + 4} \approx 10.67$.

We can now calculate the numerator of the test statistic:

$$\hat{\sigma}^2_{between} = \frac{n_1(\bar{x}_1 - \bar{\bar{X}})^2 + n_2(\bar{x}_2 - \bar{\bar{X}})^2 + n_3(\bar{x}_3 - \bar{\bar{X}})^2}{3 - 1}$$

$$= \frac{4(9 - 10.67)^2 + 4(10 - 10.67)^2 + 4(13 - 10.67)^2}{2} = \frac{34.6667}{2} \approx 17.3333$$

Much of the work in calculating $\hat{\sigma}^2_{within}$ is done in the following tables.

x_1	$x_1 - \overline{x}_1$	$(x_1 - \overline{x}_1)^2$
10	1	1
8	-1	1
9	0	0
9	0	0
		$\Sigma d_1^2 = 2$

x_2	$x_2 - \overline{x}_2$	$(x_2 - \overline{x}_2)^2$
11	1	1
12	2	4
8	-2	4
9	-1	1
		$\Sigma d_2^2 = 10$

x_3	$x_3 - \overline{x}_3$	$(x_3 - \overline{x}_3)^2$
13	0	0
10	-3	9
14	1	1
15	2	4
		$\Sigma d_3^2 = 14$

Then $\widehat{\sigma}^2_{within} = \dfrac{\Sigma d_1^2 + \Sigma d_2^2 + \Sigma d_3^2}{T - k} = \dfrac{2 + 10 + 14}{12 - 3} = 2.8889$. Finally we can calculate the test statistic.

$F = \dfrac{\widehat{\sigma}^2_{between}}{\widehat{\sigma}^2_{within}} = \dfrac{17.3333}{2.8889} = 6.00$. Because $6.00 > 4.26$, we reject H_0 and conclude that at least one grammar checking program has a mean number of corrections different from the others.

2. The hypotheses of interest are:

H_0: $\mu_1 = \mu_2 = \mu_3 = \mu_4$

H_1: At least one mean differs from the rest.

The level of significance is 0.05. We have $k = 4$ and $T = 5 + 5 + 5 + 5 = 20$. Then $df_{num} = k - 1 = 4 - 1 = 3$ and $df_{den} = T - k = 20 - 4 = 16$. The critical value is $F_{3,16,\alpha=0.05} = 3.24$. The decision rule is to reject H_0 if $F > 3.24$; otherwise, fail to reject H_0. The sample means are:

$$\overline{x}_1 = \frac{\Sigma x_1}{n_1} = \frac{1481}{5} = 296.2 \qquad \overline{x}_2 = \frac{\Sigma x_2}{n_2} = \frac{1474}{5} = 294.8$$

$$\overline{x}_3 = \frac{\Sigma x_3}{n_3} = \frac{1530}{5} = 306.0 \qquad \overline{x}_4 = \frac{\Sigma x_4}{n_4} = \frac{1402}{5} = 280.4.$$

The grand mean:

$$\overline{\overline{X}} = \frac{n_1\overline{x}_1 + n_2\overline{x}_2 + n_3\overline{x}_3 + n_4\overline{x}_4}{n_1 + n_2 + n_3 + n_4} = \frac{5(296.2) + 5(294.8) + 5(306.0) + 5(280.4)}{5 + 5 + 5 + 5} = 294.35.$$

Now we can obtain the numerator of the test statistic:

$$\widehat{\sigma}^2_{between} = \frac{n_1(\overline{x}_1 - \overline{\overline{X}})^2 + n_2(\overline{x}_2 - \overline{\overline{X}})^2 + n_3(\overline{x}_3 - \overline{\overline{X}})^2 + n_4(\overline{x}_4 - \overline{\overline{X}})^2}{4 - 1}$$

$$= \frac{5(296.2 - 294.35)^2 + 5(294.8 - 294.35)^2 + 5(306.0 - 294.35)^2 + 5(280.4 - 294.35)^2}{3}$$

$$= \frac{1669.75}{3} = 556.5833.$$

The following tables aid in calculating $\widehat{\sigma}^2_{within}$.

x_1	$x_1 - \overline{x}_1$	$(x_1 - \overline{x}_1)^2$
311	14.80	219.04
283	-13.20	174.24
302	5.80	33.64
290	-6.20	38.44
295	-1.20	1.44
		$\Sigma d_1^2 = 466.80$

x_2	$x_2 - \overline{x}_2$	$(x_2 - \overline{x}_2)^2$
320	25.20	635.04
280	-14.80	219.04
298	3.20	10.24
295	0.20	0.04
281	-13.80	190.44
		$\Sigma d_2^2 = 1054.80$

x_3	$x_3 - \overline{x}_3$	$(x_3 - \overline{x}_3)^2$		x_4	$x_4 - \overline{x}_4$	$(x_4 - \overline{x}_4)^2$
325	19	361		265	−15.4	237.16
299	−7	49		278	−2.4	5.76
305	−1	1		291	10.6	112.36
297	−9	81		284	3.6	12.96
304	−2	4		284	3.6	12.96
		$\Sigma d_3^2 = 496$				$\Sigma d_4^2 = 496.0$

Then $\widehat{\sigma}^2_{within} = \dfrac{\Sigma d_1^2 + \Sigma d_2^2 + \Sigma d_3^2 + \Sigma d_4^2}{T - k} = \dfrac{466.8 + 1054.8 + 496 + 381.2}{20 - 4} = 149.925$. Finally we can

calculate the test statistic. $F = \dfrac{\widehat{\sigma}^2_{between}}{\widehat{\sigma}^2_{within}} = \dfrac{556.583}{149.925} = 3.71$. Because $3.71 > 3.24$ we reject H_0 and

conclude that at least one of the mean batting averages is different from the others.

3. The hypotheses of interest are:

H_0: $\mu_1 = \mu_2 = \mu_3$

H_1: At least one mean differs from the rest.

The level of significance is 0.01. We have $k = 3$ and $T = 5 + 4 + 4 = 12$. Then

$df_{num} = k - 1 = 3 - 1 = 2$ and $df_{den} = T - k = 13 - 3 = 10$. The critical value is $F_{2, 10, \alpha=0.01} = 7.56$.

The decision rule is to reject H_0 if $F > 7.56$; otherwise, fail to reject H_0. The sample means:

$$\overline{x}_1 = \frac{\Sigma x_1}{n_1} = \frac{323}{5} = 64.6 \qquad \overline{x}_2 = \frac{\Sigma x_2}{n_2} = \frac{336}{4} = 84.0 \qquad \overline{x}_3 = \frac{\Sigma x_3}{n_3} = \frac{326}{4} = 81.5$$

The grand mean: $\overline{\overline{X}} = \dfrac{n_1\overline{x}_1 + n_2\overline{x}_2 + n_3\overline{x}_3}{n_1 + n_2 + n_3} = \dfrac{5(64.6) + 4(84.0) + 4(81.5)}{5 + 4 + 4} = 75.7692$.

Now we can obtain the numerator of the test statistic:

$$\widehat{\sigma}^2_{between} = \frac{n_1(\overline{x}_1 - \overline{\overline{X}})^2 + n_2(\overline{x}_2 - \overline{\overline{X}})^2 + n_3(\overline{x}_3 - \overline{\overline{X}})^2}{3 - 1}$$

$$= \frac{5(64.6 - 75.7692)^2 + 4(84.0 - 75.7692)^2 + 4(81.5 - 75.7692)^2}{2} = 513.054$$

Much of the work in calculating $\widehat{\sigma}^2_{within}$ is done in the following tables.

x_1	$x_1 - \overline{x}_1$	$(x_1 - \overline{x}_1)^2$		x_2	$x_2 - \overline{x}_2$	$(x_2 - \overline{x}_2)^2$
57	−7.6	57.76		90	6	36
70	5.4	29.16		81	−3	9
79	14.4	207.36		69	−15	225
60	−4.6	21.16		96	12	144
57	−7.6	57.76				$\Sigma d_2^2 = 414$
		$\Sigma d_1^2 = 373.2$				

x_3	$x_3 - \overline{x}_3$	$(x_3 - \overline{x}_3)^2$
65	−16.5	272.25
78	−3.5	12.25
97	15.5	240.25
86	4.5	20.25
		$\Sigma d_3^2 = 545.0$

Then $\widehat{\sigma}^2_{within} = \dfrac{\Sigma d_1^2 + \Sigma d_2^2 + \Sigma d_3^2}{T - k} = \dfrac{373.2 + 414 + 545}{13 - 3} = 133.22$. Next we calculate the test statistic.

$F = \dfrac{\widehat{\sigma}^2_{between}}{\widehat{\sigma}^2_{within}} = \dfrac{513.054}{133.22} = 3.85$. Because 3.85 is not greater than 7.56, we fail to reject H_0 and conclude

that there is not enough evidence to show a significant difference in the mean ratings.

10-3 The One-Way ANOVA Table and Computers to the Rescue

Study Objectives
You should be able to:
1. Read a computer-generated ANOVA table.
2. Use Fisher's procedure to compare pairs of means.

Section Overview
   Usually the steps and results of an analysis of variance are presented in an analysis of variance table. This table contains sources of variability, degrees of freedom, sums of squares (the numerators of $\hat{\sigma}^2_{between}$ and $\hat{\sigma}^2_{within}$), mean squares ($\hat{\sigma}^2_{between}$ and $\hat{\sigma}^2_{within}$), and the computed F value, F. Packages such as MINITAB produce ANOVA tables that usually include p values. The p values can be used to indicate our decision without recourse to the F table. Additionally, MINITAB provides confidence intervals on the difference between pairs of means that indicate which means are significantly different.

Key Terms & Formulas
One-Way ANOVA Table A table that provides a summary of the one-way analysis of variance calculations.

Factor (or Treatment) Variation The variation between samples.

Error Variation The variation within samples.

Mean Squares (MS) ANOVA table designation for the $\hat{\sigma}^2$'s:

   MS factor $= \hat{\sigma}^2_{between}$, and MS error $= \hat{\sigma}^2_{within}$

Fisher's (LSD) Procedure A method of determining which pairs of means are significantly different.

Roped in Again

In the climbing rope experiment, a mountaineering shop dropped four brands of ropes and measured the increase in the length of the rope. The measurements are in the table below. Use MINITAB to do an ANOVA on this data using $\alpha = 0.05$.

Rope 1	Rope 2	Rope 3	Rope 4
13	12	7	17
17	14	9	15
18	15	11	14
15	15		
	20		
	13		

Solution
The MINITAB output:
One-way Analysis of Variance

```
Analysis of Variance for Increase
Source      DF        SS        MS        F         P
Rope         3     96.69     32.23      5.84     0.011
Error       12     66.25      5.52
Total       15    162.94
```

```
                                  Individual 95% CIs For Mean
                                  Based on Pooled StDev
Level       N      Mean     StDev  ---+---------+---------+---------+---
1           4    15.750     2.217                    (------*------)
2           6    14.833     2.787                   (-----*-----)
3           3     9.000     2.000    (--------*-------)
4           3    15.333     1.528                     (--------*-------)
                                  ---+---------+---------+---------+---
Pooled StDev =     2.350            7.0       10.5      14.0      17.5
```

```
Fisher's pairwise comparisons

    Family error rate = 0.184
Individual error rate = 0.0500

Critical value = 2.179

Intervals for (column level mean) - (row level mean)

                    1               2               3

     2          -2.388
                 4.222

     3           2.840           2.213
                10.660           9.454

     4          -3.494          -4.120         -10.514
                 4.327           3.120          -2.153
```

The p value is 0.011. Since $0.011 < 0.05$, H_0 is rejected and we conclude that there is enough evidence to indicate that there is at least one difference among the population means. Examining Fisher's pairwise comparisons, and noting which intervals have two positive or two negative endpoints, it appears that the mean of the third rope has significantly less stretch than the others.

Practice Exercises 10-3

1. In the previous exercises, we looked at the number of errors made by three different grammar-checking programs. The number of errors is in the table that follows. Do an ANOVA using a computer package with $\alpha = 0.05$. If H_0 is rejected, use Fisher's procedure to compare the means.

Grammar1	Grammar2	Grammar3
10	11	13
8	12	10
9	8	14
9	9	15

2. In the previous exercises, we looked at the batting averages of baseball players using three methods of batting instruction. The batting averages (with the decimals omitted) are in the table that follows. Use a computer package to do an ANOVA on these four methods of instruction with $\alpha = 0.05$. If H_0 is rejected, use Fisher's procedure to compare the means.

HitMeth1	HitMeth2	HitMeth3	HitMeth4
311	320	325	265
283	280	299	278
302	298	305	291
290	295	297	284
295	281	304	284

3. We previously looked at the ratings given to books involving "fuzzy logic," with the data in the table that follows. Perform an ANOVA using a computer package with $\alpha = 0.01$. If H_0 is rejected, use Fisher's procedure to compare the means.

Book 1	Book 2	Book 3
57	90	65
70	81	78
79	69	97
60	96	86
57		

Solutions to Practice Exercises 10-3

1. MINITAB output:

```
Analysis of Variance for Errors
Source       DF        SS         MS        F        P
Grammar       2     34.67      17.33     6.00    0.022
Error         9     26.00       2.89
Total        11     60.67
```

```
                                  Individual 95% CIs For Mean
                                  Based on Pooled StDev
Level     N      Mean     StDev   --+---------+---------+---------+----
1         4     9.000     0.816   (-------*-------)
2         4    10.000     1.826       (-------*-------)
3         4    13.000     2.160                       (-------*-------)
                                  --+---------+---------+---------+----
Pooled StDev =    1.700          7.5       10.0      12.5      15.0
```

Fisher's pairwise comparisons

     Family error rate = 0.113
Individual error rate = 0.0500

Critical value = 2.262

Intervals for (column level mean) - (row level mean)

```
                    1                2

        2       -3.719
                 1.719


        3       -6.719           -5.719
                -1.281           -0.281
```

Since the p-value $= 0.022 < 0.05$, we reject H_0 and conclude that these is a significant difference in the mean number of errors recognized by the grammar-checking programs. Looking at the Fisher's pairwise comparisons, it appears that program 3 finds significantly more errors than the other two.

2. MINITAB output:

One-way Analysis of Variance

```
Analysis of Variance for Bat Aver
Source      DF        SS        MS        F        P
Method       3      1670       557     3.71    0.034
Error       16      2399       150
Total       19      4069
```

```
                                  Individual 95% CIs For Mean
                                  Based on Pooled StDev
Level     N      Mean     StDev   -+---------+---------+---------+-----
1         5    296.20     10.80                (------*-------)
2         5    294.80     16.24            (-------*------)
3         5    306.00     11.14                    (-------*-------)
4         5    280.40      9.76   (-------*-------)
                                  -+---------+---------+---------+-----
Pooled StDev =   12.24           270       285       300       315
```

Fisher's pairwise comparisons

     Family error rate = 0.189
Individual error rate = 0.0500

Critical value = 2.120

Intervals for (column level mean) - (row level mean)

	1	2	3
2	-15.02 17.82		
3	-26.22 6.62	-27.62 5.22	
4	-0.62 32.22	-2.02 30.82	9.18 42.02

Since the p value $= 0.034 < 0.05$, we reject H_0 and conclude that at least one of the mean batting averages is different from the others. Looking at the Fisher pairwise comparisons, it appears that method 3 is significantly greater than method 4.

3. MINITAB output:
One-way Analysis of Variance

```
Analysis of Variance for Ratings
Source      DF        SS        MS        F        P
Book         2      1026       513     3.85    0.058
Error       10      1332       133
Total       12      2358
```

```
                                   Individual 95% CIs For Mean
                                   Based on Pooled StDev
Level     N     Mean    StDev    ------+---------+---------+---------+
1         5    64.60     9.66    (---------*--------)
2         4    84.00    11.75                   (----------*----------)
3         4    81.50    13.48               (----------*----------)
                                   ------+---------+---------+---------+
Pooled StDev =   11.54            60        72        84        96
```

Since the p value $= 0.058$ is not less than 0.01, we fail to reject H_0 and conclude that there is not a significant difference in the mean ratings. This being the case, we do not examine any Fisher pairwise comparisons.

Solutions to Odd-Numbered Exercises

1. (a) $F_{5,8,0.05} = 3.69$.
   (b) $F_{5,20,0.05} = 2.71$.
   (c) $F_{5,30,0.05} = 2.53$.

3. (a) $F_{5,25,0.01} = 3.85$.
   (b) $F_{7,25,0.01} = 3.46$.
   (c) $F_{10,25,0.01} = 3.13$.

5. (a) $F_{8,15,0.05} = 2.64$.
   (b) $F_{5,20,0.01} = 4.10$.

7. Step 1: H_0: All population means are equal, and H_1: Not all population means are equal.
   Step 2: $\alpha = 0.05$.
   Step 3: We use a F distribution.
   Step 4: $F_{3,19,0.05} = 3.13$.
   Step 5: Reject H_0 and accept H_1 if $F > 3.13$, otherwise fail to reject H_0.

Step 6: Calculate the necessary values to obtain F.

	Group 1			Group 2	
x_1	$x_1 - \overline{x}_1$	$(x_1 - \overline{x}_1)^2$	x_2	$x_2 - \overline{x}_2$	$(x_2 - \overline{x}_2)^2$
27	−10.5	110.25	39	−4.25	18.063
50	12.5	156.25	36	−7.25	52.563
43	5.5	30.25	47	3.75	14.063
31	−6.5	42.25	51	7.75	60.063
37	−0.5	0.25	173		144.75
37	−0.5	0.25			
225		339.50			

$$\overline{x}_1 = \frac{225}{6} = 37.5 \qquad\qquad \overline{x}_2 = \frac{173}{4} = 43.25$$

$$\Sigma d_1^2 = 339.50 \qquad\qquad \Sigma d_2^2 = 144.75$$

	Group 3			Group 4	
x_3	$x_3 - \overline{x}_3$	$(x_3 - \overline{x}_3)^2$	x_4	$x_4 - \overline{x}_4$	$(x_4 - \overline{x}_4)^2$
37	1.86	3.46	24	−24.17	584.03
28	−7.14	50.98	53	4.83	23.36
44	8.86	78.50	51	2.83	8.03
36	0.86	0.74	51	2.83	8.03
30	−5.14	26.42	45	−3.17	10.03
27	−8.14	66.26	65	16.83	283.36
44	8.86	78.50	289		916.84
246		304.86			

$$\overline{x}_3 = \frac{246}{7} = 35.14 \qquad\qquad \overline{x}_4 = \frac{289}{6} = 48.17$$

$$\Sigma d_3^2 = 304.86 \qquad\qquad \Sigma d_4^2 = 916.84$$

$$\overline{\overline{X}} = \frac{6(37.5) + 4(43.25) + 7(35.14) + 6(48.17)}{6 + 4 + 7 + 6} = \frac{933}{23} = 40.565. \text{ So}$$

$$\hat{\sigma}^2_{between} = \frac{6(37.5 - 40.565)^2 + 4(43.25 - 40.565)^2 + 7(35.14 - 40.565)^2 + 6(48.17 - 40.565)^2}{4 - 1}$$

$$= \frac{6(9.394) + 4(7.209) + 7(29.431) + 6(57.836)}{3} = \frac{638.233}{3} = 212.744$$

$$\hat{\sigma}^2_{within} = \frac{339.50 + 144.75 + 304.86 + 916.84}{23 - 4} = \frac{1705.95}{19} = 89.787.$$

$$F = \frac{\hat{\sigma}^2_{between}}{\hat{\sigma}^2_{within}} = \frac{212.744}{89.787} = 2.37.$$

Step 7: Since $F = 2.37$ is < 3.13, we fail to reject H_0, the mean ages are equal.

Using MINITAB to analyze the data we obtain:

```
One-way Analysis of Variance

Analysis of Variance
Source      DF        SS         MS .        F         P
Factor       3      637.7      212.6      2.37      0.103
Error       19     1705.9       89.8
Total       22     2343.7

                                     Individual 95% CIs For Mean
                                     Based on Pooled StDev
Level       N       Mean      StDev   ------+---------+---------+---------+
Group 1     6     37.500      8.240      (---------*---------)
Group 2     4     43.250      6.946            (-----------*-----------)
Group 3     7     35.143      7.128    (--------*--------)
Group 4     6     48.167     13.541                  (---------*---------)
                                     ------+---------+---------+---------+
```

```
Pooled StDev =     9.476              32.0      40.0      48.0      56.0
```

9. Step 1: H_0: All population means are equal, and H_1: Not all population means are equal.

Step 2: $\alpha = 0.01$.

Step 3: We use a F distribution.

Step 4: $F_{2,14,0.01} = 6.51$.

Step 5: Reject H_0 and accept H_1 if $F > 6.51$, otherwise fail to reject H_0.

Step 6: Calculate the values to obtain F.

	Lower			Middle			Upper	
x_1	$x_1 - \overline{x}_1$	$(x_1 - \overline{x}_1)^2$	x_2	$x_2 - \overline{x}_2$	$(x_2 - \overline{x}_2)^2$	x_3	$x_3 - \overline{x}_3$	$(x_3 - \overline{x}_3)^2$
32	−3	9	45	8	64	38	1	1
36	1	1	42	5	25	38	1	1
40	5	25	34	−3	9	31	−6	36
32	−3	9	42	5	25	41	4	16
33	−2	4	29	−8	64	148		54
37	2	4	33	−4	16			
210		52	34	−3	9			
			259		212			

$$\overline{x}_1 = \frac{210}{6} = 35 \qquad \overline{x}_2 = \frac{259}{7} = 37 \qquad \overline{x}_3 = \frac{148}{4} = 37$$

$$\Sigma d_1^2 = 52 \qquad\qquad \Sigma d_2^2 = 212 \qquad\qquad \Sigma d_3^2 = 54$$

$$\overline{\overline{X}} = \frac{6(35) + 7(37) + 4(37)}{6 + 7 + 4} = \frac{617}{17} = 36.29.$$

$$\widehat{\sigma}^2_{between} = \frac{6(35 - 36.29)^2 + 7(37 - 36.29)^2 + 4(37 - 36.29)^2}{3 - 1}$$

$$= \frac{6(1.6641) + 7(0.5041) + 4(0.5041)}{2} = \frac{15.5297}{2} = 7.765.$$

$$\widehat{\sigma}^2_{within} = \frac{52 + 212 + 54}{17 - 3} = \frac{318}{14} = 22.714.$$

$$F = \frac{\widehat{\sigma}^2_{between}}{\widehat{\sigma}^2_{within}} = \frac{7.765}{22.714} = 0.342.$$

Step 7: Since 0.342 is < 6.51, we fail to reject H_0 that there's no difference in career decision-making attitudes among the populations.

Using MINITAB to analyze the data we obtain:

```
One-way Analysis of Variance

Analysis of Variance
Source      DF         SS         MS         F          P
Factor       2       15.5        7.8       0.34      0.716
Error       14      318.0       22.7
Total       16      333.5

                                   Individual 95% CIs For Mean
                                   Based on Pooled StDev
Level       N       Mean       StDev    --+---------+---------+---------+----
Lower       6     35.000       3.225    (-----------*-----------)
Middle      7     37.000       5.944             (----------*----------)
Upper       4     37.000       4.243       (--------------*-------------)
                                          --+---------+---------+---------+----
Pooled StDev =     4.766                  31.5      35.0      38.5      42.0
```

11. Step 1: H_0: All population means are equal, and H_1: Not all the population means are equal.

Step 2: $\alpha = 0.01$.

Step 3: We use a F distribution.

Step 4: $F_{2,11,0.01} = 7.21$.

Step 5: Reject H_0 and accept H_1 if $F > 7.21$, otherwise fail to reject H_0.

Step 6: Calculate the values to obtain F.

	Moody			Lindavic			Bell	
x_1	$x_1 - \overline{x}_1$	$(x_1 - \overline{x}_1)^2$	x_2	$x_2 - \overline{x}_2$	$(x_2 - \overline{x}_2)^2$	x_3	$x_3 - \overline{x}_3$	$(x_3 - \overline{x}_3)^2$
4	−0.6	0.36	7	−1.2	1.44	6	−2	4
2	−2.6	6.76	6	−2.2	4.84	9	1	1
5	0.4	0.16	11	2.8	7.84	7	−1	1
9	4.4	19.36	11	2.8	7.84	10	2	4
3	−1.6	2.56	6	−2.2	4.84	32		10
23		29.20	41		26.80			

$$\overline{x}_1 = \frac{23}{5} = 4.6 \qquad\qquad \overline{x}_2 = \frac{41}{5} = 8.2 \qquad\qquad \overline{x}_3 = \frac{32}{4} = 8$$

$$\Sigma d_1^2 = 29.20 \qquad\qquad \Sigma d_2^2 = 26.80 \qquad\qquad \Sigma d_3^2 = 10$$

$$\overline{X} = \frac{5(4.6) + 5(8.2) + 4(8)}{5 + 5 + 4} = \frac{96}{14} = 6.857$$

$$\widehat{\sigma}^2_{\text{between}} = \frac{5(4.6 - 6.857)^2 + 5(8.2 - 6.857)^2 + 4(8 - 6.857)^2}{3 - 1}$$

$$= \frac{5(5.094) + 5(1.804) + 4(1.306)}{2} = \frac{39.714}{2} = 19.857.$$

$$\widehat{\sigma}^2_{\text{within}} = \frac{29.20 + 26.80 + 10}{14 - 3} = \frac{66}{11} = 6.00.$$

$$F = \frac{\widehat{\sigma}^2_{\text{between}}}{\widehat{\sigma}^2_{\text{within}}} = \frac{19.857}{6.00} = 3.31.$$

Step 7: Since 3.31 is less than 7.21, we fail to reject H_0, the population mean number of errors is the same for all three groups.

Using MINITAB to analyze the data we obtain:

```
One-way Analysis of Variance

Analysis of Variance
Source      DF          SS          MS          F          P
Factor       2        39.71       19.86       3.31       0.075
Error       11        66.00        6.00
Total       13       105.71
                                         Individual 95% CIs For Mean
                                         Based on Pooled StDev
Level       N         Mean        StDev   --+---------+---------+---------+----
Moody        5        4.600       2.702   (--------*---------)
Lindavic     5        8.200       2.588                (---------*--------)
Bell         4        8.000       1.826                (----------*----------)
                                         --+---------+---------+---------+----
Pooled StDev =        2.449               2.5        5.0        7.5       10.0
```

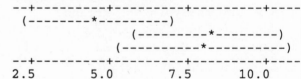

13. Step 1: H_0: All population means are equal, and H_1: Not all the population means are equal.

 Step 2: $\alpha = 0.05$.

 Step 3: We use a F distribution.

 Step 4: $F_{2, 20, 0.05} = 3.49$.

 Step 5: Reject H_0 and accept H_1 if $F > 3.49$, otherwise fail to reject H_0.

 Step 6: Calculate the values to obtain F.

Suits			Casual			Upper		
x_1	$x_1 - \overline{x}_1$	$(x_1 - \overline{x}_1)^2$	x_2	$x_2 - \overline{x}_2$	$(x_2 - \overline{x}_2)^2$	x_3	$x_3 - \overline{x}_3$	$(x_3 - \overline{x}_3)^2$
26	−8.43	71.0649	19	−7.11	50.5521	22	−5.57	31.0249
37	2.57	6.6049	24	−2.11	4.4521	33	5.43	29.4849
41	6.57	43.1649	31	4.89	23.9121	34	6.43	41.3449
35	0.57	0.3249	28	1.89	3.5721	19	−8.57	73.4449
29	−5.43	29.4849	23	−3.11	9.6721	25	−2.57	6.6049
33	−1.43	2.0449	25	−1.11	1.2321	29	1.43	2.0449
40	5.57	31.0249	24	−2.11	4.4521	31	3.43	11.7649
241		183.7143	29	2.89	8.3521	193		195.7143
			32	5.89	34.6921			
			235		140.8889			

$$\overline{x}_1 = \frac{241}{7} = 34.43 \qquad \overline{x}_2 = \frac{235}{9} = 26.11 \qquad \overline{x}_3 = \frac{193}{7} = 27.57$$

$$\Sigma d_1^2 = 183.7143 \qquad \Sigma d_2^2 = 140.8889 \qquad \Sigma d_3^2 = 195.7143$$

$$\overline{\overline{X}} = \frac{7(34.43) + 9(26.11) + 7(27.57)}{7 + 9 + 7} = \frac{669}{23} = 29.09.$$

$$\hat{\sigma}^2_{between} = \frac{7(34.43 - 29.09)^2 + 9(26.11 - 29.09)^2 + 7(27.57 - 29.09)^2}{3 - 1}$$

$$= \frac{7(28.5156) + 9(8.8804) + 7(2.3104)}{2} = \frac{295.7056}{2} = 147.85.$$

$$\hat{\sigma}^2_{within} = \frac{183.7143 + 140.8889 + 195.7143}{23 - 3} = \frac{520.3175}{20} = 26.02.$$

$$F = \frac{\hat{\sigma}^2_{between}}{\hat{\sigma}^2_{within}} = \frac{147.85}{26.02} = 5.68.$$

Step 7: Since 5.68 is greater than 3.49, we reject the H_0 that the population mean sales are equal for the three groups.

Using MINITAB to analyze the data we obtain:

```
One-way Analysis of Variance

Analysis of Variance
Source      DF          SS          MS          F          P
Factor       2       295.5       147.8       5.68      0.011
Error       20       520.3        26.0
Total       22       815.8
                                        Individual 95% CIs For Mean
                                        Based on Pooled StDev
Level        N        Mean      StDev    -----+---------+---------+---------+-
Suits        7      34.429      5.533                        (-------*-------)
Casual       9      26.111      4.197    (------*------)
Jeans        7      27.571      5.711    (-------*-------)
                                         -----+---------+---------+---------+-
Pooled StDev =       5.101               25.0      30.0      35.0      40.0
```

15. Error is the natural variation of the members within the same population.

17. The SS error is the numerator for the $\hat{\sigma}^2_{within}$ estimate of σ^2. It is found by computing the mean for each of the four groups or factors, finding the deviation between each value and its corresponding mean, squaring the deviations, and then adding these squared deviations (see the procedure spelled out in the numerator of formula 10.4).

19. The MS error is the $\dfrac{\text{SS error}}{\text{DF error}}$, which is $\hat{\sigma}^2_{within}$.

21. The $F = \dfrac{\text{MS factor}}{\text{MS error}} = \dfrac{\hat{\sigma}^2_{\text{between}}}{\hat{\sigma}^2_{\text{within}}}$. In this case, $\dfrac{0.69}{3.22} = 0.22$.

23. "LEVEL" refers to the different factors or treatments.

25. Because we failed to reject H_0, there is no need to compare means.

27. With three factors, the DF factor is $3 - 1$ or 2. With a total of 24 values within the factor groups, DF error is $24 - 3$ or 21.

29. With a p-value of 0.009, we reject H_0 since $p < 0.05$.

31. Using the DF column, since $2 + A = 19$, we must have $A = 17$.

33. Using the FACTOR row, $C = \dfrac{\text{SS factor}}{\text{DF factor}} = \dfrac{1515}{2} = 757.5$.

35. Using $C = 757.5$ from Exercise 39 and $D = 632.4$ from Exercise 40, $E = \dfrac{757.5}{632.4} = 1.20$.

37. A is the total of the DF column, so $A = 3 + 20 = 23$.

39. Using $B = 8.50$ from Exercise 44, $C = \dfrac{8.50}{3} = 2.83$.

41. Using C from Exercise 45 and D from Exercise 46, $E = \dfrac{2.83}{2.17} = 1.31$.

43. One-way Analysis of Variance

    Analysis of Variance for C10

Source	DF	SS	MS	F	P
C11	2	1652	826	1.42	0.261
Error	24	13951	581		
Total	26	15603			

45. Because we failed to reject H_0, there is no need to compare means.

47. Step 1: H_0: All population means are equal, and H_1: Not all population means are equal.
    Step 2: $\alpha = 0.05$.
    Step 3: We'll use an F distribution with $df_{\text{num}} = 3$, and $df_{\text{den}} = 60$.
    Step 4: The critical F value is $F_{3,\,60,\,0.05} = 2.76$.
    Step 5: Reject H_0 in favor of H_1 if the p-value is less than 0.05. Otherwise, fail to reject H_0.
    Step 6: From the ANOVA table, the p-value is 0.001
    Step 7: Since the p-value < 0.05, we reject H_0, there is a significant difference in the mean number of unopened buds on hibiscus plants.

49. One-way Analysis of Variance

    Analysis of Variance for Fac

Source	DF	SS	MS	F	P
Factor	5	3679438	735888	18.50	0.000
Error	12	477252	39771		
Total	17	4156691			

51. $\mu_{20\%} \neq \mu_{0\%};\ \mu_{30\%} \neq \mu_{0\%};\ \mu_{30\%} \neq \mu_{20\%};\ \mu_{30\%} \neq \mu_{60\%};\ \mu_{40\%} \neq \mu_{0\%};\ \mu_{40\%} \neq \mu_{20\%};\ \mu_{40\%} \neq \mu_{30\%};$
    $\mu_{40\%} \neq \mu_{60\%};\ \mu_{50\%} \neq \mu_{0\%};\ \mu_{50\%} \neq \mu_{20\%};\ \mu_{50\%} \neq \mu_{60\%}.$

Chapter 11 Chi-Square Test: Goodness-of-Fit and Contingency Table Methods

Study Aids and Practice Exercises

11-1 Chi-Square Testing: Purpose and Procedure

Study Objectives
You should be able to:
1. Recognize the purpose and details of a goodness-of-fit test.
2. Recognize the purpose and details of a contingency table test.

Section Overview
   Often we collect data that involve the number of people or things that fall into various categories. For example, we might want to know the number of people who are for, against, or neutral about gun control for assault weapons. Or we might count the number of people who are liberal, moderate, conservative, or apolitical. If we take a random sample and collect this type of data, it is a multinomial experiment. With a multinomial experiment, we might have a theory about the percentage we should get in each category. We might hypothesize that among college students, 35% are liberal, 25% are moderate, 30% are conservative, and 10% are apolitical. To see if these percentages for the population distribution are correct, we can perform a goodness-of-fit test. This test is based on the χ^2 distribution. From the hypothesized percentages, we calculate the expected frequency (denoted E) for each category. Also for each category, we note the observed count or frequency (denoted O). If the expected and observed counts are close, that supports the hypothesized population distribution; if they are markedly different, that tends to refute the hypothesized population distribution. The test statistic that does the comparison is:
$$\chi^2 = \Sigma \left| \frac{(O-E)^2}{E} \right|.$$
This test statistic has a χ^2 distribution with $k-1$ degrees of freedom, where k is the number of possible categories. The more the O and E values differ, the larger this test statistic becomes. So the decision rule is to reject H_0 for large values of the test statistic.
   We can also use a χ^2 test to examine the relationship between two count variables. For example, we may cross-classify people by learning both their opinion on gun control and their political leanings. We would summarize this information in a contingency table. Each cell in the table counts how many people have matching values of the two categorical variables, e.g., how many people are moderates for gun control or liberals against gun control. We test for the independence of the two variables by performing a χ^2 test with $(r-1)(c-1)$ degrees of freedom,

where r and c are the number of rows and columns, respectively, in the contingency table. While the way to calculate E is different, the test statistic is the same as before:

$$\chi^2 = \Sigma \left| \frac{(O - E)^2}{E} \right|.$$

Key Terms & Formulas

Multinomial Experiment The extension of a binomial experiment to more than two outcomes. The experiment involves a fixed number of independent, identical trials where the outcome on any trial is one of k categories or cells. The probability of falling into any cell is constant from trial to trial.

Goodness-of-Fit Test A test to decide if a hypothesized set of probabilities for the cells in a multinomial experiment is true. It involves determining the values of E, the expected cell frequency based on the hypothesized population distribution, and comparing these to the observed cell frequencies, denoted O. The test statistic is $\chi^2 = \Sigma \left| \frac{(O - E)^2}{E} \right|$. This has a χ^2 distribution with $k - 1$ degrees of freedom.

Contingency Table Test (or Test of Independence) A test to learn if the row categories and the column categories in a contingency table are independent. The test statistic is $\chi^2 = \Sigma \left| \frac{(O - E)^2}{E} \right|$. This has a χ^2 distribution with $(r - 1)(c - 1)$ degrees of freedom.

Caution: Both the goodness-of-fit test and the contingency table test require random samples and that the expected cell frequencies, the E's, to be at least 5.

11-2 The Goodness-of-Fit Test

Study Objectives

You should be able to:
1. Know when it is appropriate to do a goodness-of-fit test.
2. Do a goodness-of-fit test.

Section Overview

We use the goodness-of-fit test to decide if a multinomial population fits a specified distribution. To validly run this test, we must take a random sample that results in frequency counts for each of k categories. Additionally, the expected frequency for each of the k categories must be at least 5. If so, the test statistic is:

$$\chi^2 = \Sigma \left| \frac{(O - E)^2}{E} \right|,$$

where O represents the observed count for a cell, and E represents the expected count for a cell under the assumption that H_0 is true. This test statistic has a chi-square distribution with $k - 1$ degrees of freedom. We reject the null hypothesis for the hypothesized population distribution if the test statistic, χ^2, is greater than a χ^2 table value.

Key Terms & Formulas

Uniform Distribution A population distribution that says that each category or cell is equally likely.

Worked Examples

Floor Plans

An architect is trying to decide if the five floor plans he has designed are equally preferred by potential customers. He takes a survey of 80 people and asks each which floor plan they prefer. The responses are below. Use the goodness-of-fit test at $\alpha = 0.05$ to determine if the floor plans are equally preferred, i.e., the population distribution is uniform.

Plan	O
Cavalier	13
Esquire	22
Jolie	10
Twain	20
Hacienda	15
TOTAL	80

Solution

The hypotheses are:

H_0: The population distribution is uniform, i.e., each floor plan is equally preferred.

H_1: The population distribution is not uniform, i.e., the floor plans are not equally preferred.

The level of significance is 0.05. As long as all the values of E are 5 or more, we will use a χ^2 distribution. This χ^2 has $5 - 1 = 4$ degrees of freedom. The critical value is found in the χ^2 table in the 0.05 column and the 4-df row. This table value is 9.49. The decision rule is to reject H_0 in favor of H_1 if $\chi^2 > 9.49$. Otherwise, we fail to reject H_0. The expected frequencies are each 20% of $80 = 16$. The arithmetic to calculate the test statistic is given in the following table.

Plan	O	E	$O-E$	$(O-E)^2$	$\dfrac{(O-E)^2}{E}$
Cavalier	13	16	-3	9	0.5625
Esquire	22	16	6	36	2.25
Jolie	10	16	-6	36	2.25
Twain	20	16	4	16	1
Hacienda	15	16	-1	1	0.0625
TOTAL	80	80			6.125

Since 6.125 is less than 9.49, we fail to reject H_0 and conclude that it is reasonable to believe that the floor plans are equally preferred.

Centerism

There is a theory called "visual orientation" that says that people tend to select toward the middle of a group of items. One specific part of this theory involves selecting items from a list. The theory says that, for two adjacent items, the one closer to the middle is twice as likely to be selected as the other. A list of five nonsense words in random order is given to a sample of 200 people, and they are asked to select one word. The position of the word in the list is noted and the results given in the table that follows. Test the appropriate hypothesis at $\alpha = 0.01$.

Position	O
First	11
Second	42
Third	62
Fourth	58
Fifth	27
TOTAL	200

Solution

According to the theory, the second and fourth position should be twice as preferred as the first and fifth position. Additionally, the third position should be twice as preferred as the second or fourth position (and therefore four times as preferred as the first or fifth position). So the preferences under H_0 are in a 1:2:4:2:1 ratio. The hypotheses are:

H_0: Population percentages: First = Fifth = 10%, Second = Fourth = 20%, Third = 40%.

H_1: Population percentages are not as specified in H_0.

The level of significance is 0.01. The expected frequencies are 200 times the hypothesized percentages. The smallest E will be $200(0.10) = 20$. Since all the value of E are 5 or more, we may use a χ^2 distribution. This χ^2 will have $5 - 1 = 4$ degrees of freedom. The critical value is found in the χ^2 table in the 0.01 column and the 4-df row. This table value is 13.28. The decision rule is to reject H_0 in favor of H_1 if $\chi^2 > 13.28$. Otherwise, we fail to reject H_0. The arithmetic used to calculate the test statistic is given in the following table.

Position	O	E	$O-E$	$(O-E)^2$	$\frac{(O-E)^2}{E}$
First	11	20	−9	81	4.05
Second	42	40	2	4	0.1
Third	62	80	−18	324	4.05
Fourth	58	40	18	324	8.1
Fifth	27	20	7	49	2.45
TOTAL	200	200			18.75

Since $18.75 > 13.28$, we reject H_0 and conclude that the hypothesized population distribution is incorrect.

Practice Exercises 11-2

1. A sales manager is trying to decide if four package designs would be equally preferred by consumers. She takes a random sample of 120 people and asks them to select the most attractive package. The results are listed below. Test the appropriate hypothesis, using $\alpha = 0.05$.

Package	O
Solid red	19
Solid blue	25
Red on blue	39
Yellow on blue	37
TOTAL	120

2. A survey is done at a chain of health clubs. Members are asked their primary reason for joining their club. Their responses are given in the table below. Use this information to test at $\alpha = 0.01$ the hypothesis that the reasons are uniformly distributed throughout the population.

Reason	O
Health	45
Appearance	63
Strength	29
Weight	19
Meeting people	21
Enjoyment	3
TOTAL	180

3. A few years back, the students of UC Santa Cruz changed the name of their sports teams to the Banana Slugs. A new campus of CSU is opening on the Monterey peninsula. The college is trying to choose a name for the sports teams. They have narrowed the choices down to the Crustaceans, the Spiny Sea Urchins, and the Brain Corals. The college president, who prefers Crustaceans, claims that 50% of the students prefer Crustaceans, while the remainder would be equally split among the other two names. The student body president samples 50 students with the results in the following table. Test the appropriate hypothesis at $\alpha = 0.01$ to see if the college president is correct in his statement.

Name	O
Crustaceans	15
Sea urchins	18
Brain coral	17
TOTAL	50

4. It is suggested that one-half of New Jersey residents believe in the existence of Oklahoma, 40 percent deny that such a state exists, while the remaining residents have no opinion. A random sample of 60 New Jersey residents results in the following table. Use this data to test the appropriate hypothesis at $\alpha = 0.05$.

Oklahoma	O
Believe	25
Deny	32
No opinion	3
TOTAL	60

Solutions to Practice Exercises 11-2

1. The hypotheses are:

    H_0: The population distribution is uniform, i.e., each design is equally preferred

    H_1: The population distribution is not uniform, i.e., the designs are not equally preferred.

    The level of significance is 0.05. The expected frequencies are all 25% of 120 or 30. Since all the values of E are 5 or more, we will use a χ^2 distribution. This χ^2 will have $4 - 1 = 3$ degrees of freedom. The critical value is found in the χ^2 table in the 0.05 column and the 3-df row. The table value is 7.81, leading to the decision rule to reject H_0 in favor of H_1 if $\chi^2 > 7.81$. Otherwise, we fail to reject H_0. The arithmetic to calculate the test statistic is given in the following table.

Package	O	E	$O - E$	$(O - E)^2$	$\dfrac{(O - E)^2}{E}$
Solid red	19	30	−11	121	4.033
Solid blue	25	30	−5	25	0.833
Red on blue	39	30	9	81	2.7
Yellow on blue	37	30	7	49	1.633
TOTAL	120	120			9.2

    Because $9.2 > 7.81$, we reject H_0 and conclude that there is a difference in the preferences for the package designs.

2. The hypotheses are:

    H_0: The population distribution is uniform, i.e., each reason is selected equally often

    H_1: The population distribution is not uniform, i.e., each reason is not selected equally often.

    The level of significance is 0.01. The expected frequencies are all $1/6$ of 180 or 30. Since all the E's are more than 5, we can use a χ^2 distribution. This χ^2 will have $6 - 1 = 5$ degrees of freedom. The critical value is found in the χ^2 table in the 0.01 column and the 5-df row. This table value is 15.09. The decision rule is to reject H_0 in favor of H_1 if $\chi^2 > 15.09$. Otherwise, we fail to reject H_0. The arithmetic is given in the following table.

Reason	O	E	$O - E$	$(O - E)^2$	$\dfrac{(O - E)^2}{E}$
Health	45	30	15	225	7.5
Appearance	63	30	33	1089	36.3
Strength	29	30	−1	1	0.033
Weight	19	30	−11	121	4.033
Meeting people	21	30	−9	81	2.7
Enjoyment	3	30	−27	729	24.3
TOTAL	180	180			74.87

    Because $74.87 > 15.09$, we reject H_0 and conclude that there is not a uniform set of choices for reasons to join a health club.

3. The hypotheses are:

    H_0: 50% prefer Crustaceans, 25% prefer Spiny Sea Urchins, and 25% prefer Brain Coral

    H_1: The population distribution is not as stated.

    The level of significance is 0.01. The expected frequencies are 50%, 25%, and 25% of 50, i.e., 25, 12.5, and 12.5. Since all the E's are more than 5, we will use the χ^2 distribution. This χ^2 will have $3 - 1 = 2$ degrees of freedom. The critical value is taken from the χ^2 table in the 0.01 column and the 2-df row. This table value is 9.21. The decision rule is to reject H_0 in favor of H_1 if $\chi^2 > 9.21$; otherwise, fail to reject H_0. The arithmetic to calculate the test statistic is in the following table.

Name	O	E	$O - E$	$(O - E)^2$	$\dfrac{(O - E)^2}{E}$
Crustaceans	15	25	−10	100	4
Sea Urchins	18	12.5	5.5	30.25	2.42
Brain Coral	17	12.5	4.5	20.25	1.62
TOTAL	50	50			8.04

Because 8.04 is not greater than 9.21, we fail to reject H_0 and conclude that the given percentages are possible.

4. The hypotheses are:

H_0: 50% believe in Oklahoma, 40% deny it exists, and 10% have no opinion

H_1: The population distribution is not as stated.

The level of significance is 0.05. The expected frequencies are 50%, 40%, and 10% of 60, i.e., 30, 24, and 6. Because all the E's are more than 5, we will use a χ^2 distribution. This χ^2 will have $3 - 1 = 2$ degrees of freedom. The critical value is found in the χ^2 table in the 0.05 column and the 2 df row. This table value is 5.99. The decision rule is to reject H_0 in favor of H_1 if $\chi^2 > 5.99$; otherwise, fail to reject H_0. The arithmetic to calculate the test statistic is given in the following table.

Oklahoma	O	E	$O - E$	$(O - E)^2$	$\dfrac{(O - E)^2}{E}$
Believe	25	30	−5	25	0.833
Deny	32	24	8	64	2.667
No Opinion	3	6	−3	9	1.5
TOTAL	60	60			5

Because 5 is not greater than 5.99, we fail to reject H_0 and conclude that the given percentages are possible.

Hint: Notice that there is a 3 in the O column. This is not a problem because it is each value of E that must be at least 5, not O.

11-3 The Contingency Table Test

Study Objectives

You should be able to:

1. Calculate expected cell frequencies for independent categorical variables.
2. Perform a χ^2 contingency table test.

Section Overview

Two variables are independent if the occurrence or nonoccurrence of one variable does not change the probability of the other. For example, we might want to know whether smoking and the occurrence of cancer are independent. To do a test of independence on two categorical variables, we use the chi-square contingency table test. The assumptions for this test are similar to those of the goodness-of-fit test. We need a random sample that results in frequency counts for two categorical variables, and we need the expected frequency for each of the k categories to be at least 5. In such a case, the test statistic is again:

$$\chi^2 = \Sigma \left| \frac{(O - E)^2}{E} \right|.$$

This will have a chi-square distribution with $(r - 1)(c - 1)$ degrees of freedom, where r and c are the number of rows and columns, respectively. The values of E are calculated with:

$$E = \frac{(\text{row total})(\text{column total})}{\text{grand total}}.$$

We reject the null hypothesis of independence if the test statistic is greater than a χ^2 table value.

Key Terms & Formulas

Contingency Table Test (or Test of Independence) A test to decide if the row categories and the column categories in a contingency table are independent. The test statistic $\chi^2 = \Sigma \left| \frac{(O-E)^2}{E} \right|$ has a χ^2 distribution with $(r-1)(c-1)$ degrees of freedom.

Expected Cell Frequency for Contingency Table The value of E for a contingency table is

$$E = \frac{\text{(row total)}\text{(column total)}}{\text{grand total}}.$$

Worked Examples

Political Assault

A national survey was done asking people to indicate their political philosophy and whether they were for, against, or had no opinion on gun control for assault weapons. The results of the survey are given in the table that follows. At $\alpha = 0.05$, is this evidence that there is a relation between political philosophy and opinion on gun control?

	Liberal	Moderate	Conservative	Apolitical	TOTAL
For	53	32	16	17	118
Against	42	27	54	19	142
No opinion	23	11	3	26	63
TOTAL	118	70	73	62	323

Solution

The hypotheses are:

H_0: Opinion on gun control is independent of political philosophy.

H_1: Opinion on gun control is related to political philosophy.

> Hint: The null hypothesis does not mean that the percentage for, against, or with no opinion are the same, i.e., all $1/3$.
> Rather it means this if 40% of the liberals are in the for category, then so are 40% of the moderates, conservatives, and
> apoliticals. So the hypotheses may also be stated:

H_0: Percentage having each opinion on gun control is the same for each political philosophy.

H_1: Percentage having each opinion on gun control is not the same for each political philosophy.

The level of significance of this test is 0.05. As long as all the expected cell frequencies work out to be at least 5, we will be able to use a χ^2 test. The degrees of freedom for the test are $(3-1)(4-1) = 6$. The critical value for the test is found in the 0.05 column and the 6-df row. This value is 12.59. Therefore the decision rule is to reject H_0 in favor of H_1 if $\chi^2 > 12.59$. To calculate the test statistic, we first need to obtain the expectations. They are calculated with the formula $E = \frac{\text{(row total)}\text{(column total)}}{\text{grand total}}$. The expectations are in the table below.

	Liberal	Moderate	Conservative	Apolitical	TOTAL
For	43.1083591	25.5727554	26.6687307	22.6501548	118
Against	51.876161	30.7739938	32.0928793	27.2569659	142
No opinion	23.0154799	13.6532508	14.2383901	12.0928793	63
TOTAL	118	70	73	62	323

> Hint: The obnoxious number of decimal points in the above table are unnecessary-we allowed the word processor used
> to type the manuscript to do the arithmetic, and this is what we got. Three or four decimal points are all that would be
> necessary for the calculation of the test statistic.

The remainder of the calculations are in the following table.

Row/Column	O	E	$O-E$	$(O-E)^2$	$\dfrac{(O-E)^2}{E}$
$1-1$	53	43.108	9.892	97.851664	2.2699189
$1-2$	32	25.573	6.427	41.306329	1.61523204
$1-3$	16	26.669	−10.669	113.827561	4.26816007
$1-4$	17	22.650	−5.65	31.9225	1.4093819
$2-1$	42	51.876	−9.876	97.535376	1.88016378
$2-2$	27	30.774	−3.774	14.243076	0.46282823
$2-3$	54	32.093	21.907	479.916649	14.9539354
$2-4$	19	27.257	−8.257	68.178049	2.50130422
$3-1$	23	23.015	−0.015	0.000225	0.00000978
$3-2$	11	13.653	−2.653	7.038409	0.51552106
$3-3$	3	14.238	−11.238	126.292644	8.87011125
$3-4$	26	12.093	13.907	193.404649	15.9931075
TOTAL	323	322.999			54.7396741

Because $54.74 > 12.59$, we reject the null hypothesis and decide that there is a relation between political philosophy and opinion on gun control.

Practice Exercises 11-3

1. A sociologist is interested in examining the relation that might exist between the size of a college student's family and the student's favorite sport. She interviews 305 people with the results in the table that follows. Test, using $\alpha = 0.01$, to see if these two categorical variables are independent.

	Baseball	Football	Basketball	Hockey	Tennis	TOTAL
No sibling	23	21	44	8	11	107
1 sibling	21	26	21	9	13	90
> 1 sibling	19	33	32	12	12	108
TOTAL	63	80	97	29	36	305

2. The chancellor of Wyoming's junior college system wants to compare the languages studied at junior colleges to those taken at the University of Wyoming. He takes a sample from Caspar Junior College and a sample from U Wyo. The results are in the table that follows. Use a contingency table test to decide if there is a difference in the languages taken using a level of significance of 0.05.

	Spanish	French	German	TOTAL
Caspar	54	14	23	91
U Wyo	54	29	11	94
TOTAL	108	43	34	185

3. A judge asked the director of a women's shelter to do a survey. She wanted to investigate a potential relation between wife abuse and alcohol/drug use of the husband. Each resident was asked if the primary reason they were in the shelter was physical abuse from the husband and if he was an alcoholic or habitual drug user. The results are in the table that follows. Test for a relation using $\alpha = 0.05$.

	Physical Abuse	No Physical Abuse	TOTAL
Substance abuse	65	45	110
No Substance abuse	28	22	50
TOTAL	93	67	160

Solutions to Practice Exercises 11-3

1. The hypotheses are:
   H_0: Favorite sport is independent of family size
   H_1: Favorite sport is related to family size.
   The level of significance of this test is 0.01. As long as all the expected cell frequencies work out to be at least 5, we can use a χ^2 test. The degrees of freedom for the test are $(3-1)(5-1) = 8$. The critical value for the test is found in the 0.05 column and the 8-df row. This value is 20.1. Therefore the decision rule is to

reject H_0 in favor of H_1 if $\chi^2 > 20.1$. Next we need the expectations: $E = \dfrac{\text{(row total)(column total)}}{\text{grand total}}$. The expectations are in the table below.

	Baseball	Football	Basketball	Hockey	Tennis	TOTAL
No Sibling	22.10	28.07	34.03	10.17	12.63	107
1 Sibling	18.59	23.61	28.62	8.56	10.62	90
> 1 Sibling	22.31	28.33	34.35	10.27	12.75	108
TOTAL	63	80	97	29	36	305

The remainder of the calculations are in the following table.

Row/Column	O	E	$O - E$	$(O - E)^2$	$\dfrac{(O - E)^2}{E}$
1 − 1	23	22.10	0.9	0.81	0.03665158
1 − 2	21	28.07	−7.07	49.9849	1.78072319
1 − 3	44	34.03	9.97	99.4009	2.92097855
1 − 4	8	10.17	−2.17	4.7089	0.46301868
1 − 5	11	12.63	−1.63	2.6569	0.21036421
2 − 1	21	18.59	2.41	5.8081	0.31243141
2 − 2	26	23.61	2.39	5.7121	0.24193562
2 − 3	21	28.62	−7.62	58.0644	2.02880503
2 − 4	9	8.56	0.44	0.1936	0.02261682
2 − 5	13	10.62	2.38	5.6644	0.533371
3 − 1	19	22.31	−3.31	10.9561	0.49108472
3 − 2	33	28.33	4.67	21.8089	0.76981645
3 − 3	32	34.35	−2.35	5.5225	0.16077147
3 − 4	12	10.27	1.73	2.9929	0.29142162
3 − 5	12	12.75	−0.75	0.5625	0.04411765
TOTAL	293	292.26			10.308108

Since 10.31 is not greater than 20.1, we fail to reject H_0 and conclude that there is not significant evidence of a relation between favorite sport and the size of a family.

2. The hypotheses are:

H_0: Language studied is independent of college
H_0: Language studied is related to college.

The level of significance of this test is 0.05. If all the expected cell frequencies work out to be at least 5, we can use a χ^2 test. The degrees of freedom for the test are $(2 - 1)(3 - 1) = 2$. The critical value for the test is found in the 0.05 column and the 2-df row. This value is 5.99. Therefore the decision rule is to reject H_0 in favor of H_1 if $\chi^2 > 5.99$. Next we need the expectations: $E = \dfrac{\text{(row total)(column total)}}{\text{grand total}}$. The expectations are contained in the table that follows.

	Spanish	French	German	TOTAL
Caspar	53.12	21.15	16.72	91
U Wyo	54.88	21.85	17.28	94
TOTAL	108	43	34	185

The remainder of the calculations are in the following table.

Row/Column	O	E	$O - E$	$(O - E)^2$	$\dfrac{(O - E)^2}{E}$
1 − 1	54	53.12	0.88	0.7744	0.01457831
1 − 2	14	21.15	−7.15	51.1225	2.41713948
1 − 3	23	16.72	6.28	39.4384	2.35875598
2 − 1	54	54.88	−0.88	0.7744	0.01411079
2 − 2	29	21.85	7.15	51.1225	2.33970252
2 − 3	11	17.28	−6.28	39.4384	2.28231481
TOTAL	185	185			9.42660189

Since $9.43 > 5.99$, we reject H_0 and conclude that there is evidence of a relation between language studied and type of college.

3. The hypotheses are:

H_0: Physical and substance abuse are independent

H_0: Physical and substance abuse are related.

The level of significance of this test is 0.05. If all the expected cell frequencies work out to be at least 5, we can use a χ^2 test. The degrees of freedom for the test are $(2-1)(2-1) = 1$. The critical value for the test is found in the 0.05 column and the 1-df row. This value is 3.84. Therefore the decision rule is to reject H_0 in favor of H_1 if $\chi^2 > 3.84$. Next we need the expectations: $E = \dfrac{(\text{row total})(\text{column total})}{\text{grand total}}$. The expectations are in the table below.

	Physical abuse	No physical abuse	TOTAL
Substance abuse	63.94	46.06	110
No substance abuse	29.06	20.94	50
TOTAL	93	67	160

The remainder of the calculations are in the following table.

Row/Column	O	E	$O - E$	$(O - E)^2$	$\dfrac{(O - E)^2}{E}$
$1 - 1$	65	63.94	1.06	1.1236	0.01757272
$1 - 2$	45	46.06	-1.06	1.1236	0.02439427
$2 - 1$	28	29.06	-1.06	1.1236	0.03866483
$2 - 2$	22	20.94	1.06	1.1236	0.05365807
TOTAL	160	160			0.13428989

Since 0.13 is not in the rejection region, there is no significant evidence of a relation between physical and substance abuse.

Solutions to Odd-Numbered Exercises

1. Step 1: H_0: There was an equal preference for each brand or orange juice. H_1: The distribution was not uniform.

Step 2: $\alpha = 0.01$.

Step 3: χ^2.

Step 4: The critical χ^2 value with $4 - 1 = 3$ df is 11.34.

Step 5: Reject H_0 in favor of H_1 if $\chi^2 > 11.34$. Otherwise, fail to reject H_0.

Step 6: We make a table and compute the necessary values.

Brand	O	E	$O - E$	$(O - E)^2$	$\dfrac{(O - E)^2}{E}$
A	10	7.5	2.5	6.25	0.833
B	8	7.5	0.5	0.25	0.033
C	5	7.5	-2.5	6.25	0.833
D	7	7.5	-0.5	0.25	0.033
	30	30.0			1.732

Thus $\chi^2 = 1.73$.

Step 7: Since χ^2 does not fall in the rejection region, we fail to reject H_0. There was an equal preference for each brand or orange juice.

3. Step 1: H_0: Marital status and race are independent variables for those who have legal abortions. H_1: The variables are not independent.

    Step 2: $\alpha = 0.01$.

    Step 3: χ^2.

    Step 4: The critical χ^2 value with $(2-1)(2-1) = 1$ df is 6.63.

    Step 5: Reject H_0 in favor of H_1 if $\chi^2 > 6.63$. Otherwise, fail to reject H_0.

    Step 6: We produce the following contingency table.

Row-Column (Cell)	Number Preferring (O)	E	$O - E$	$(O - E)^2$	$\dfrac{(O-E)^2}{E}$
$1-1$	68	66.19	1.81	3.2761	0.04950
$1-2$	31	32.81	-1.81	3.2761	0.09985
$2-1$	279	280.81	-1.81	3.2761	0.01167
$2-2$	141	139.19	1.81	3.2761	0.02354
	519	519.0			0.18456

Thus, $\chi^2 = 0.18456$.

Step 7: Since χ^2 does not fall in the rejection region, we fail to reject the H_0. Marital status and race are independent variables for those who have legal abortions.

Using the chisquare command in MINITAB with the data we get:

```
Chi-Square Test

Expected counts are printed below observed counts

          C1      C2    Total
1         68      31      99
        66.19   32.81

2        279     141     420
       280.81  139.19

Total    347     172     519

Chi-Sq =  0.049 +  0.100 +
          0.012 +  0.024 = 0.184
DF = 1, P-Value = 0.668
```

5. Step 1: H_0: There's no difference in the percentage favoring gun control in the three districts. H_1: The variables are not independent.

    Step 2: $\alpha = 0.01$.

    Step 3: χ^2.

    Step 4: The critical χ^2 value with $(3-1)(2-1) = 2$ df is 9.21.

    Step 5: Reject H_0 in favor of H_1 if $\chi^2 > 11.34$. Otherwise, fail to reject H_0.

    Step 6: We produce the following contingency table.

Row-Column	O	E	$O - E$	$(O - E)^2$	$\dfrac{(O-E)^2}{E}$
$1-1$	89	81.38	7.62	58.0644	0.7135
$1-2$	46	53.62	-7.62	58.0644	1.0829
$2-1$	65	66.31	-1.31	1.7161	0.0259
$2-2$	45	43.69	1.31	1.7161	0.0393
$3-1$	60	66.31	-6.31	39.8161	0.6005
$3-2$	50	43.69	6.31	39.8161	0.9113
	355	355.0			3.3734

Thus $\chi^2 = 3.3734$.

Step 7: Since χ^2 does not fall in the rejection region, we fail to reject H_0. There's no difference in the percentage favoring gun control in the three districts.

Using the chisquare command in MINITAB with the data we get:
```
Chi-Square Test

Expected counts are printed below observed counts

            C1        C2     Total
    1       89        46       135
          81.38     53.62

    2       65        45       110
          66.31     43.69

    3       60        50       110
          66.31     43.69

Total      214       141       355

Chi-Sq =   0.713 +   1.083 +
           0.026 +   0.039 +
           0.600 +   0.911 = 3.373
DF = 2, P-Value = 0.185
```

7. Step 1: H_0: The number of defects is uniformly distributed among the shifts. H_1: The defects aren't uniformly distributed.
   Step 2: $\alpha = 0.01$.
   Step 3: χ^2.
   Step 4: The critical χ^2 value with $(3 - 1) = 2$ df is 9.21.
   Step 5: Reject H_0 in favor of H_1 if $\chi^2 > 9.21$. Otherwise, fail to reject H_0.
   Step 6: We make a table and compute the necessary values.

Shift	Number of Defects (O)	E	$O - E$	$(O - E)^2$	$\dfrac{(O - E)^2}{E}$
8 AM to 4 PM	27	37	-10	100	2.703
4 PM to Midnight	35	37	-2	4	0.108
Midnight to 8 AM	49	37	12	144	3.892
	111	111	0		6.703

   Thus $\chi^2 = 6.703$.
   Step 7: Since χ^2 does not fall in the rejection region, we fail to reject H_0. The number of defects is uniformly distributed among the shifts.

9. Step 1: H_0: The type of adverse effect and the dosage level are independent variables. H_1: The variables are not independent.
   Step 2: $\alpha = 0.01$.
   Step 3: χ^2.
   Step 4: The critical χ^2 value with $(3 - 1)(3 - 1) = 4$ df is 13.28.
   Step 5: Reject H_0 in favor of H_1 if $\chi^2 > 11.34$. Otherwise, fail to reject H_0.
   Step 6: We make a table and compute the necessary values.

Row-Column	O	E	$O-E$	$(O-E)^2$	$\dfrac{(O-E)^2}{E}$
1 – 1	9	11.27	−2.27	5.1529	0.4572
1 – 2	29	23.80	5.20	27.0400	1.1361
1 – 3	5	7.93	−2.93	8.5849	1.0826
2 – 1	10	7.34	2.66	7.0756	0.9640
2 – 2	9	15.50	−6.50	42.2500	2.7258
2 – 3	9	5.17	3.83	14.6689	2.8373
3 – 1	8	8.39	−0.39	0.1521	0.0181
3 – 2	19	17.71	1.29	1.6641	0.0940
3 – 3	5	5.90	−0.90	0.8100	0.1373
TOTAL	103	103.01	−0.01		9.4524

Thus $\chi^2 = 9.4524$.

Step 7: Since χ^2 does not fall in the rejection region, we fail to reject H_0. The type of adverse effect and the dosage level are independent variables.

Using the chisquare command in MINITAB with the data we get:

```
Chi-Square Test

Expected counts are printed below observed counts

            C1        C2        C3    Total
1            9        29         5       43
         11.27     23.80      7.93

2           10         9         9       28
          7.34     15.50      5.17

3            8        19         5       32
          8.39     17.71      5.90

Total       27        57        19      103

Chi-Sq =  0.458 +   1.138 +   1.084 +
          0.964 +   2.723 +   2.847 +
          0.018 +   0.094 +   0.138 = 9.464
DF = 4, P-Value = 0.050
```

11. Step 1: H_0: The distribution fits the one claimed by the sales manager. H_1: The distribution doesn't fit the sales manager's claim.

Step 2: $\alpha = 0.05$.

Step 3: χ^2.

Step 4: The critical χ^2 value with $5 - 1 = 4$ df is 9.49.

Step 5: Reject H_0 in favor of H_1 if $\chi^2 > 9.49$. Otherwise, fail to reject H_0.

Step 6: We make a table and compute the necessary values.

Number of Cars Sold Per Day	Number of Days (O)	E	$O-E$	$(O-E)^2$	$\dfrac{(O-E)^2}{E}$
0	8	8.20	−0.20	0.040	0.0049
1	11	12.30	−1.30	1.690	0.1374
2	10	12.30	−2.30	5.290	0.4301
3	6	6.15	−0.15	0.023	0.0037
4 or more	6	2.05	3.95	15.603	7.6110
TOTAL	41	41.00	0.00		8.18710

Thus $\chi^2 = 8.187$.

Step 7: Since χ^2 does not fall in the rejection region, we fail to reject H_0. The distribution fits the sales manager's claim.

13. Step 1: H_0: The distribution on Bougainville Island matched that of the general population. H_1: The distribution didn't match the one found in the general population.

Step 2: $\alpha = 0.01$.

Step 3: χ^2.

Step 4: The critical χ^2 value with $(4 - 1) = 3$ df is 11.34.

Step 5: Reject H_0 in favor of H_1 if $\chi^2 > 11.34$. Otherwise, fail to reject H_0.

Step 6: We make a table and compute the necessary values.

Blood Type	Number of Bougainville Islanders (O)	E	$O - E$	$(O - E)^2$	$\dfrac{(O - E)^2}{E}$
A	74	48.72	25.28	639.08	13.1174
B	12	11.60	0.40	0.16	0.0138
AB	11	3.48	7.52	56.55	16.2501
O	19	52.20	−33.20	1102.24	21.1157
TOTAL	116	116.00	0.00		50.4970

Thus $\chi^2 = 50.4970$.

Step 7: Since χ^2 falls in the rejection region, we reject H_0. The distribution of blood types doesn't match the one found in the general population.

15. Step 1: H_0: The selection of the preferred restaurant is independent of gender. H_1: The variables are not independent.

Step 2: $\alpha = 0.05$.

Step 3: χ^2.

Step 4: The critical χ^2 value with $(2 - 1)(2 - 1) = 1$ df is 9.49.

Step 5: Reject H_0 in favor of H_1 if $\chi^2 > 9.49$. Otherwise, fail to reject H_0.

Step 6: We produce the following contingency table.

Row-Column	O	E	$O - E$	$(O - E)^2$	$\dfrac{(O - E)^2}{E}$
$1 - 1$	7	13.5	−6.5	42.25	3.130
$1 - 2$	18	11.5	6.5	42.25	3.674
$2 - 1$	20	13.5	6.5	42.25	3.130
$2 - 2$	5	11.5	−6.5	42.25	3.674
	50	50.0			13.607

Thus, $\chi^2 = 13.607$.

Step 7: Since χ^2 does not fall in the rejection region, we fail to reject the H_0. Marital status and race are independent variables for those who have legal abortions.

Using the chisquare command in MINITAB with the data we get:

```
Chi-Square Test

Expected counts are printed below observed counts

            C1        C2      Total
  1          7        18         25
          13.50     11.50

  2         20         5         25
          13.50     11.50

Total       27        23         50
```

```
Chi-Sq =   3.130 +   3.674 +
           3.130 +   3.674 = 13.607
DF = 1, P-Value = 0.000
```

17. Step 1: H_0: Age is independent of the side to which the occipital rest extends. H_1: The variables are not independent.

Step 2: $\alpha = 0.05$.

Step 3: χ^2.

Step 4: The critical χ^2 value with $(2-1)(3-1) = 2$ df is 5.99.

Step 5: Reject H_0 in favor of H_1 if $\chi^2 > 5.99$. Otherwise, fail to reject H_0.

Step 6: We produce the following contingency table.

Row-Column	O	E	$O-E$	$(O-E)^2$	$\dfrac{(O-E)^2}{E}$
$1-1$	48	45.03	2.97	8.8209	0.186
$1-2$	35	40.16	-5.16	26.6256	0.663
$1-3$	28	25.81	2.19	4.7961	0.186
$2-1$	109	280.81	-2.97	8.8209	0.079
$2-2$	105	99.84	5.16	26.6256	0.267
$2-3$	62	64.19	-2.19	4.7961	0.074
	387	387.0			1.455

Thus, $\chi^2 = 1.455$.

Step 7: Since χ^2 does not fall in the rejection region, we fail to reject the H_0. Marital status and race are independent variables for those who have legal abortions.

Using the chisquare command in MINITAB with the data we get:

```
Chi-Square Test

Expected counts are printed below observed counts

            C1        C2        C3     Total
    1       48        35        28       111
         45.03     40.16     25.81

    2      109       105        62       276
        111.97     99.84     64.19

Total     157       140        90       387

Chi-Sq =   0.196 +   0.662 +   0.185 +
           0.079 +   0.266 +   0.074 = 1.462
DF = 2, P-Value = 0.481
```

19. Three in one and three in the other.

21. $\chi^2 = 28.511$.

23. No other information is need since the p-value $= 0.0000$. We will reject H_0 at any level.

25. Two in one and five in the other.

27. $\chi^2 = 21.193$.

29. 84.

31. df $= (2-1)(2-1) = 1$.

33. No other information is need since the p-value $= 0.596$. Since the p-value is greater than 50%, we will fail to reject H_0 at (almost) any level.

Chapter 12 Linear Regression and Correlation

Study Aids and Practice Exercises

12-1 Introductory Concepts

Study Objectives
You should be able to:
1. Distinguish between simple linear and multiple regression.
2. Recognize the roles of dependent and independent variables in a regression equation.
3. Use scatter diagrams to see if and how two variables are related.

Section Overview

In many different areas, it is important to be able to predict a variable. A physician prescribing a drug would want to know the relation between the dose of a drug and the decrease in a person's blood pressure. An electrical engineer would want to know the change in resistance produced by a change in the gage of a wire. A businessperson would want to know the effect on sales of real estate associated with a change in the prime interest rate. The analysis of each of these situations would fall under the heading of regression analysis. In regression analysis, we are trying to predict or estimate the value of one variable based on the value or values of one or more other variables. The usual result of a regression analysis includes a regression or estimating equation that shows how to use one variable to predict the other. The variable we are trying to predict is the response or the dependent variable, denoted y. The variable being used to provide the prediction is the explanatory or independent variable, denoted x.

Closely related to regression analysis is correlation analysis. In correlation analysis, we are examining the nature and strength of the relationships between variables. It makes sense to only use independent variables that are "correlated" with the dependent variable in the regression equation.

One type of regression analysis is simple linear regression. Simple regression uses only a single independent variable to predict the dependent variable (as opposed to multiple regression that uses two or more independent variables to predict the response). A simple regression is a simple linear regression if the form of the relationship between x and y is best expressed by the equation of a straight line.

Since relationships can happen by chance, it is better to start a regression analysis with some idea about what variables might be related to the response. That is, we should try to determine the logical relationships that might exist between the response and candidates for explanatory variables. The logical relationships that might exist can fall into two general but potentially overlapping categories. One is a cause-and-effect relationship. In that, one variable is believed to directly change the response. We can usually suggest one or more variables that could have an effect on a response. An economist can quickly provide a list of variables that can change the prime interest rate. A real estate agent can easily (and unstoppably) provide a list of characteristics of a house that will raise or lower its selling price (raise if you are buying, lower if you are selling). A biologist can name several factors that can change the density of butterflies in a Eucalyptus grove. A second type of logical relationship is a common-cause factor. In that we believe two variables are related because they are affected similarly by another variable or variables. Scores on algebra and statistics exams can be affected by an unmeasurable factor, say mathematical aptitude. While algebra does not cause statistics and vice-versa, they may change in a similar way, and then we can use one to estimate the other.

One visual way of examining the relationship between variables is to draw a scatter diagram or scatterplot. On a scatter diagram, the horizontal axis represents the values of x and the vertical axis represents the values of y. Points are drawn that represent individual data points. From this diagram, we can usually see whether the two variables are related and, if they are, the nature of the relation. For example, it is easy to spot a simple linear relation: the points would suggest a straight line. A line that is ascending, i.e., one for which y increases as x increases, indicates a positive relationship. A line that is descending, i.e., one for which y decreases as x increases, indicates a negative relationship.

Key Terms & Formulas

Regression Analysis The area of statistics that produces estimates or predictions of one variable based on the value(s) of one or more other variables. For example, we might predict future income based on level of education (or level of education combined with years of experience).

> Hint: A bit of jargon. If a statistician is using regression to estimate the mean response of y for some x, it is called an estimate; if the statistician is trying to guess a single response of y for some x, it is called a prediction.

Correlation Analysis The study of the nature and strength of the relations between two or more variables.

> Hint: Regression analysis and correlation analysis are not totally separate entities. A regression analysis only makes sense if the variables involved are "correlated." Most studies involving the relationship between variables combine aspects of regression and correlation. There is little or no point in looking at a study and saying this part is regression and that part is correlation. Rather, we use the tools from both methods to examine and use the relationships that exist.

Simple Regression & Simple Correlation The study and measurement of the relationship between two variables.

Multiple Regression & Multiple Correlation The study and measurement of the relationship between three or more variables.

Estimating (or Regression) Equation The equation that produces an estimate or prediction of one variable based on the value(s) of other variable(s).

Dependent Variable (or Response) The variable predicted by a regression equation. It is usually denoted as y.

Independent (or Explanatory) Variable The variable used to predict the dependent variable in a regression equation. It is usually denoted as x.

> Hint: In simple regression there will be one independent variable; in multiple regression there will be more than one independent variable.

Cause-and-Effect Relationship A relationship between two variables in which one variable produces a change or response in the other. An example might be temperature and amount of heating oil consumed. Or (heaven forbid) amount studied and test grade.

Common-Cause Factor When two variables change in a similar pattern because of the effect of a third factor, this third factor or variable is called "common-cause factor." An example might be the length and weight of a baby being "caused" by nutrition (and perhaps genetics). Or fading and cracking of a house's siding being effected by the house's age (and perhaps weather factors).

Spurious Correlation Purely accidental, meaningless correlations that happen by chance and are unlikely to recur in the future. They are useless in analyzing data.

Hint: A wonderful example was developed by the famous statistician Mark J. Nicolich when he noticed a strong correlation between wine consumption and SAT scores.

Logical Relationship A starting point for finding variables that are possible candidates for regression and correlation analysis. By working with variables that would have some logical reason for being related, we save time and help prevent working with spurious correlations.

Scatter Diagram (or Scatterplot) A plot of bivariate (two variables) data. The horizontal axis represents the values of x and the vertical axis represents the values of y. This scattergram has two purposes.

1. To see if the two variables are related.
2. If they are related, detect the nature of the relationship.

Linear Relationship When the regression relation between two variables can best be described by the equation of a straight line, it is called a linear relationship.

Positive Linear Relationship A linear relationship in which y increases as x increases.

Negative Linear Relationship A linear relationship in which y decreases as x increases.

Worked Examples

Possible Probabilities

The following graphs represent possible scatterplots between x and y. Indicate whether there is a relation and, if so, its nature.

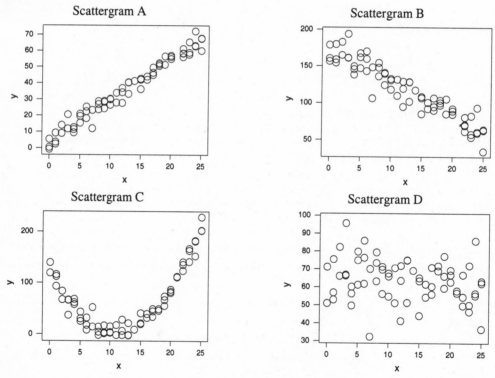

Solution

Scattergram A is typical of the appearance of a positive simple linear relationship between x and y. Scattergram B is similar in that it appears that there is a simple linear relationship between x and y, only here it is a negative relation. Scattergram C suggests that there is a relation between x and y, but it is not a linear one (it is an example of what is called a quadratic regression). Scattergram D does not show any relationship between x and y.

Home Buying

A new resident of a town collects the data in the following table. It represents the square footage of recently sold homes and their selling price in thousands of dollars. Draw a scatter diagram of the data and reach a decision about the type of relation that exists between the two variables.

Sq. Footage	Selling Price
1600	117
1600	207
3000	431
2200	312
2000	288
2200	240
1300	140
1000	90
2900	440
1600	190
1400	114
1500	156
2000	289
2400	349

Solution

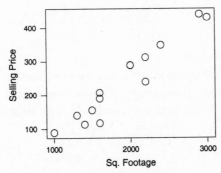

From the scatter diagram, it appears that there is a positive linear relation between the size of a house and its selling price.

Practice Exercises 12-1

Hint: In the exercises of this chapter we will use the same three data sets repeatedly. (If we changed data sets, we would need to repeat some tedious calculations over and over.) So it would be a good idea to retain a copy of your work from early sections to refer to when doing exercises in the later sections.

1-2. The dean of admissions of a college wants to evaluate a new entrance exam first given a year ago. She took a random sample of first-year students and learned their entrance exam score and first-year GPA. The data is in the table that follows.

Student	Exam	College GPA
1	16	2.64
2	22	3.40
3	23	3.15
4	20	3.13
5	18	2.77
6	10	2.20
7	22	3.56
8	22	3.31
9	13	2.44
10	20	3.17
11	18	3.14
12	20	3.03

1. Create a scatter diagram for this data.
2. What can you say about the relationship between these two variables?

3-4. James C. Daly, statistician extraordinaire and college football fan, wanted to examine the relation between pass completion percent and points scored per game. He took the top 15 rated teams in the country and collected the data that follows.

Pass %	Points
0.494	25.1
0.522	19.4
0.508	27.5
0.487	27.0
0.485	19.2
0.476	25.8
0.559	29.9
0.477	32.9
0.494	20.4
0.652	32.9
0.643	36.6
0.661	31.6
0.531	30.6
0.572	33.6
0.472	25.0

   3.   Create a scatter diagram for this data.

   4.   What can you say about the relationship between these two variables?

5-6. A kinesiologist surveyed students at a state college and learned the amount of time they spend exercising. She also measured the percent of body fat for each student. The data is in the table that follows.

Exercise	Body Fat %
80	22
40	28
0	33
60	25
120	16
240	8
0	28
30	27
60	26
90	21
180	11

   5.   Create a scatter diagram for this data.

   6.   What can you say about the relationship between these two variables?

Solutions to Practice Exercises 12-1

   1.

   2.   There appears to be a positive linear relationship between exam scores and college GPA.

3.

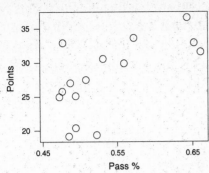

4. If there is any relation at all, it appears to be a weak positive one.

5.

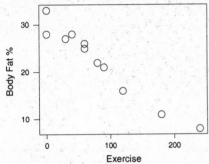

6. There appears to be a negative simple linear relationship between time exercised and percent of body fat.

12-2 Simple Linear Regression Analysis

Study Objectives

You should be able to:
1. Write the model for a simple linear regression of y on x.
2. Understand the properties of the linear regression line.
3. Calculate estimates of the slope and y-intercept of the linear regression equation.
4. Use the linear regression line to provide predictions of y for values of x.
5. Calculate the standard error of estimate.

Section Overview

We are using simple linear regression (SLR) of y on x when we use a straight line to describe the relation between x and y. In SLR, we search for the best line, the line that minimizes the amount of error we make when we use x to predict y. The criterion for selecting this line is called least squares, and the resulting line minimizes the quantity $\Sigma(y - \hat{y})^2$. This equation, often called the least squares regression line, is $\hat{y} = a + bx$
where $\hat{y} =$ is the estimated value of y, the dependent variable

$x =$ the independent variable

$a =$ the y intercept $=$ the value of $\hat{y}$ when x is equal to zero $=$ the point where the line goes through the y or vertical axis

$b =$ the slope of the regression line $=$ the change in $\hat{y}$ for a 1-unit increase in $x =$ the "rise over run" of the line

The values of the constants (or coefficients) of the regression line are computed by:

$$b = \frac{n(\Sigma xy) - (\Sigma x)(\Sigma y)}{n(\Sigma x^2) - (\Sigma x)^2} \text{ and } a = \overline{y} - b\overline{x}$$

We can calculate a least squares regression line even if x and y are totally unrelated. This means that after the calculations are done, we still do not know how well the regression line predicts y. If the points on the scatter diagram are close to the regression line, the line will do a good job of predicting y. If there is a great deal of spread around the regression line, the predictions of y will not be very good. To describe the scatter of points, we use the standard error of estimate. Denoted $s_{y.x}$, this quantity is the standard deviation of the observations around the regression line. To calculate this standard error of estimate, we use:

$$s_{y.x} = \sqrt{\frac{\Sigma(y-\hat{y})^2}{n-2}} = \sqrt{\frac{\Sigma(y^2) - a(\Sigma y) - b(\Sigma xy)}{n-2}}.$$

Key Terms & Formulas

Simple Linear Regression Often shortened to SLR, it is when the relationship between two variables is described by a straight line.

Straight-Line Equation The equation of a straight line is $\hat{y} = a + bx$. In this equation:

a = the y intercept = the value of $\hat{y}$ when x is equal to zero = the point where the line goes through the y or vertical axis;

b = the slope of the regression line = the change in $\hat{y}$ for a 1-unit increase in x = the "rise over run" of the line.

Properties of the Linear Regression Line The regression line that we will use to estimate or predict y has two important properties:

1. $\Sigma(y-\hat{y}) = 0$, i.e., the positive deviations above the line balance out the negative deviations below the line.
2. $\Sigma(y-\hat{y})^2$ = the smallest value possible for any line.

Hint: Because of the second property, the line we use is often called the least squares regression line and the method used to obtain the line is called the method of least squares.

Coefficients of the Least Squares Regression Line To calculate the estimates of the y intercept and slope of the regression line, we use: $b = \dfrac{n(\Sigma xy) - (\Sigma x)(\Sigma y)}{n(\Sigma x^2) - (\Sigma x)^2}$ and $a = \bar{y} - b\bar{x}$.

Standard Error of Estimate $s_{y.x}$ The standard deviation of the observations around the regression line. The formula for its calculation is: $s_{y.x} = \sqrt{\dfrac{\Sigma(y-\hat{y})^2}{n-2}} = \sqrt{\dfrac{\Sigma(y^2) - a(\Sigma y) - b(\Sigma xy)}{n-2}}.$

Worked Examples

Still Home Buying

We return to the new resident of a town who collected data on the size and selling price of houses. For this data, calculate the estimated regression line and the standard error of estimate. Use the regression line to estimate the selling price of a house that has 2100 square feet of space.

Sq. Footage	Selling Price
1600	117
1600	207
3000	431
2200	312
2000	288
2200	240
1300	140
1000	90
2900	440
1600	190
1400	114
1500	156
2000	289
2400	349

Solution

The calculations necessary to obtain the estimates are in and below the following table.

$x =$ Sq. Footage	$y =$ Selling Price	xy	x^2	y^2
1600	117	187200	2560000	13689
1600	207	331200	2560000	42849
3000	431	1293000	9000000	185761
2200	312	686400	4840000	97344
2000	288	576000	4000000	82944
2200	240	528000	4840000	57600
1300	140	182000	1690000	19600
1000	90	90000	1000000	8100
2900	440	1276000	8410000	193600
1600	190	304000	2560000	36100
1400	114	159600	1960000	12996
1500	156	234000	2250000	24336
2000	289	578000	4000000	83521
2400	349	837600	5760000	121801
26700	3363	7263000	55430000	980241

$$\bar{x} = \frac{26700}{14} = 1907.14 \qquad\qquad \bar{y} = \frac{3363}{14} = 240.21$$

$$b = \frac{n(\Sigma xy) - (\Sigma x)(\Sigma y)}{n(\Sigma x^2) - (\Sigma x)^2} = \frac{14(7263000) - (26700)(3363)}{14(55430000) - (26700)^2} = \frac{11889900}{63130000} = 0.18834$$

$a = \bar{y} - b\bar{x} = (240.21) - 0.18834(1907.14) = -118.98$.

Therefore, the regression line is: $\hat{y} = -118.98 + 0.18834x$. Using these regression coefficients and the sums from the prior table, we get the standard error of estimate:

$$s_{y.x} = \sqrt{\frac{\Sigma(y^2) - a(\Sigma y) - b(\Sigma xy)}{n - 2}}$$

$$= \sqrt{\frac{980241 - (-118.98)(3363) - 0.18834(7263000)}{14 - 2}}$$

$$= \sqrt{\frac{12447}{12}} = 32.21.$$

The estimate of y for $x = 2100$ is: $\hat{y} = -118.98 + 0.18834(2100) = 276.53$.

Practice Exercises 12-2

1-3. The dean of admissions of a college took a random sample of first-year students and learned their entrance exam score and their first-year GPA. The data is in the table that follows.

Student	Exam	College GPA
1	16	2.64
2	22	3.40
3	23	3.15
4	20	3.13
5	18	2.77
6	10	2.20
7	22	3.56
8	22	3.31
9	13	2.44
10	20	3.17
11	18	3.14
12	20	3.03

1. Calculate the coefficients of the regression line, b and a.
2. Estimate the GPA of a person who scores 21 on the entrance exam.
3. Calculate the standard error of estimate.

4-6. JC Daly examined the relation between pass completion percent and points scored per football game. He took the top 15 rated teams in the country and collected the data that follows.

Pass %	Points
0.494	25.1
0.522	19.4
0.508	27.5
0.487	27.0
0.485	19.2
0.476	25.8
0.559	29.9
0.477	32.9
0.494	20.4
0.652	32.9
0.643	36.6
0.661	31.6
0.531	30.6
0.572	33.6
0.472	25.0

4. Calculate the coefficients of the regression line, b and a.
5. Estimate the points scored by a team whose passing percentage is 0.500.
6. Calculate the standard error of estimate.

7-9. A kinesiologist surveyed students at a state college and determined the amount of time they spend exercising and the percent of body fat for each respondent. The data is in the table that follows.

Exercise	Body Fat %
80	22
40	28
0	33
60	25
120	16
240	8
0	28
30	27
60	26
90	21
180	11

7. Calculate the coefficients of the regression line, b and a.
8. Estimate the body fat % for a person who exercises 100 minutes per week.
9. Calculate the standard error of estimate.

Solutions to Practice Exercises 12-2

1. Most of the calculations are shown in the following table:

Student	x = Exam	y = College GPA	xy	x^2	y^2
1	16	2.64	42.24	256	6.9696
2	22	3.40	74.8	484	11.56
3	23	3.15	72.45	529	9.9225
4	20	3.13	62.6	400	9.7969
5	18	2.77	49.86	324	7.6729
6	10	2.20	22	100	4.84
7	22	3.56	78.32	484	12.6736
8	22	3.31	72.82	484	10.9561
9	13	2.44	31.72	169	5.9536
10	20	3.17	63.4	400	10.0489
11	18	3.14	56.52	324	9.8596
12	20	3.03	60.6	400	9.1809
TOTAL	224	35.94	687.33	4354	109.4346

$$\bar{x} = \frac{224}{12} = 18.6667 \qquad\qquad \bar{y} = \frac{35.94}{12} = 2.995$$

$$b = \frac{n(\Sigma xy) - (\Sigma x)(\Sigma y)}{n(\Sigma x^2) - (\Sigma x)^2} = \frac{12(687.33) - (224)(35.94)}{12(4354) - (224)^2} = \frac{197.40}{2072} = 0.09527 \approx 0.095.$$

$a = \bar{y} - b\bar{x} = (2.995) - 0.095(18.6667) = 1.12166.$

The regression line is: $\hat{y} = 1.2166 + 0.09527x.$

2. The estimate of y for $x = 21$ is $\hat{y} = 1.2166 + 0.09527(21) = 3.22.$

3. $s_{y.x} = \sqrt{\dfrac{\Sigma(y^2) - a(\Sigma y) - b(\Sigma xy)}{n - 2}} = \sqrt{\dfrac{109.4346 - (1.2166)(35.94) - (0.9527)(687.33)}{12 - 2}}$

$\qquad = \sqrt{\dfrac{0.228067}{10}} = 0.151.$

4. Most of the calculations are shown in the following table:

$x = $ Pass %	$y = $ Points	xy	x^2	y^2
0.494	25.1	12.3994	0.244036	630.01
0.522	19.4	10.1268	0.272484	376.36
0.508	27.5	13.97	0.258064	756.25
0.487	27	13.149	0.237169	729
0.485	19.2	9.312	0.235225	368.64
0.476	25.8	12.2808	0.226576	665.64
0.559	29.9	16.7141	0.312481	894.01
0.477	32.9	15.6933	0.227529	1082.41
0.494	20.4	10.0776	0.244036	416.16
0.652	32.9	21.4508	0.425104	1082.41
0.643	36.6	23.5338	0.413449	1339.56
0.661	31.6	20.8876	0.436921	998.56
0.531	30.6	16.2486	0.281961	936.36
0.572	33.6	19.2192	0.327184	1128.96
0.472	25	11.8	0.222784	625
8.033	417.5	226.863	4.365003	12029.33

$$\bar{x} = \frac{224}{12} = 18.6667 \qquad\qquad \bar{y} = \frac{35.94}{12} = 2.995$$

$$b = \frac{n(\Sigma xy) - (\Sigma x)(\Sigma y)}{n(\Sigma x^2) - (\Sigma x)^2} = \frac{15(226.863) - (8.033)(417.5)}{15(4.365003) - (8.033)^2} = \frac{49.16750}{0.945956} = 51.9765.$$

$a = \bar{y} - b\bar{x} = (27.8333) - 51.9765(0.5355) = -0.001825.$

Therefore, the regression line is: $\hat{y} = -0.001825 + 51.9765x.$

5. The estimate of y for $x = 0.500$ is $\hat{y} = -0.001825 + 51.9765(0.500) = 25.99.$

6. $s_{y.x} = \sqrt{\dfrac{\Sigma(y^2) - a(\Sigma y) - b(\Sigma xy)}{n - 2}}$

$\qquad = \sqrt{\dfrac{12029.33 - (-0.001825)(417.5) - 51.9765(226.863)}{15 - 2}}$

$\qquad = \sqrt{\dfrac{238.543}{13}} \approx 4.2836.$

7. Most of the calculations are shown in the following table:

$x =$ Exercise	$y =$ Body Fat %	xy	x^2	y^2
80	22	1760	6400	484
40	28	1120	1600	784
0	33	0	0	1089
60	25	1500	3600	625
120	16	1920	14400	256
240	8	1920	57600	64
0	28	0	0	784
30	27	810	900	729
60	26	1560	3600	676
90	21	1890	8100	441
180	11	1980	32400	121
900	245	14460	128600	6053

$$\bar{x} = \frac{900}{11} = 81.818 \qquad\qquad \bar{y} = \frac{245}{11} = 22.2727$$

$$b = \frac{n(\Sigma xy) - (\Sigma x)(\Sigma y)}{n(\Sigma x^2) - (\Sigma x)^2} = \frac{11(14460) - (900)(245)}{11(128600) - (900)^2} = \frac{-61440}{604600} = -0.1016.$$

$a = \bar{y} - b\bar{x} = (22.2727) - (-0.1016)(81.818) = 30.587.$

Therefore, the regression line is: $\hat{y} = 30.587 - 0.1016x$.

8. The estimate of y for $x = 100$ is $\hat{y} = 30.587 - 0.1016(100) = 20.4.$

9.
$$s_{y.x} = \sqrt{\frac{\Sigma(y^2) - a(\Sigma y) - b(\Sigma xy)}{n - 2}}$$

$$= \sqrt{\frac{6053 - (30.587)(245) - (-0.1016)(14460)}{11 - 2}}$$

$$= \sqrt{\frac{28.583}{9}} \approx 1.782.$$

12-3 Relationship Tests and Prediction Intervals in Simple Linear Regression Analysis

Study Objectives

You should be able to:

1. Do a t test to decide if the independent and dependent variables in a regression are related.
2. Use a computer package such as MINITAB to test to decide if the independent and dependent variables in a regression are related.
3. Calculate confidence intervals for the mean value of the dependent variable for a given value of the independent variable.
4. Calculate prediction intervals for a value of the dependent variable for a given value of the independent variable.

Section Overview

We can always calculate a least squares regression line. All we do is substitute numbers into formulas. This in no sense proves that there is a relationship between x and y. While we may think that the two variables are related, we usually need to substantiate this belief. The ability to do this is provided by a hypothesis test about the slope of the regression line. Our premise is that there is a population regression line that says that the mean value of y is $A + Bx$. Then the values of a and b are estimates of A and B in this population regression line. If the value of B is zero, this shows that a change in the value of x does not correspond to a change in y, i.e., x and y are unrelated. So by testing the hypothesis that B is equal to zero, we are testing that there is not a linear relation between x and y. This test requires a few assumptions. The first is the one already mentioned, that there is a population regression line with slope B and intercept A estimated by b and a, respectively. Second, we need to believe that, for each value of x, the distribution of the y values is normal. Third, for each value of x, the distribution of y must have the same standard deviation. Finally, each y value must be independent of the other y values. When these assumptions are satisfied, we can do either a t test or an F test to see if the slope in the population is zero.

Hint: In simple linear regression, the t and F tests test the same hypotheses and always give the same results. In multiple regression, they test different hypotheses and will have different outcomes.

The hypothesis of the t test is:

H_0: $B = 0$.

This is because if H_0: $B = 0$ is true, then x and y are not related. Since we are interested in using x to predict y whether the relation is positive or negative, we usually do a two-tailed test. The test statistic has $n - 2$ degrees of freedom, and is calculated with:

$$t = \frac{b - 0}{s_b} = \frac{b}{s_b}, \text{ where}$$

b = slope of the estimated (least squares) regression line

s_b = standard error of the sampling distribution of $b = \dfrac{s_{y.x}}{\sqrt{\Sigma(x^2) - \dfrac{(\Sigma x)^2}{n}}}$.

A rejection of H_0 implies that it is reasonable to use x to predict or estimate y.

We can also test for a relation between x and y with an analysis of variance F test. In SLR the hypotheses are the same as the t test. The test statistic is usually obtained through a computer package. While we could go into the details of the test, it is easier to work with the computer-generated p-value for the test. If it is less than the selected level of significance of the test, we reject the null hypothesis and conclude that there is a relationship between x and y.

One of the primary purposes of doing a regression is to obtain estimates or predictions for y for various values of x. Of course, since these estimates are based on sample information rather than from the entire population, they are likely to be "off" by some amount. We can show how far off they are likely to be by using interval estimates. There are two types of intervals that might be of interest for any given x. We might like to know the average value of y for a given value of x. For example, we might want to know the average growth of a crop subjected to a certain amount of fertilizer. Alternatively, we might want to know the value of y for one particular time x is a given amount. For example, we may want to know the amount of crop we get from a single field treated with a certain amount of fertilizer. This is called a prediction interval for y (as opposed to the previous interval which is a confidence interval for the mean value of y) for a particular value of x. To obtain a confidence interval for the mean value of y for $x = x_g$, we calculate:

$$\hat{y} \pm t_{\alpha/2}\left[s_{y.x}\sqrt{\frac{1}{n} + \frac{(x_g - \overline{x})^2}{\Sigma(x^2) - \dfrac{(\Sigma x)^2}{n}}}\right].$$

The prediction interval for y when $x = x_g$ is almost identical; it differs by a factor of 1 under the square root:

$$\hat{y} \pm t_{\alpha/2} \left[s_{y.x} \sqrt{1 + \frac{1}{n} + \frac{(x_g - \overline{x})^2}{\Sigma(x^2) - \dfrac{(\Sigma x)^2}{n}}} \right].$$

Key Terms & Formulas

Homoscedasticity A regression assumption, it says that the standard deviation of y around the regression line is constant for all values of the independent variable x.

Test Statistic for H_0: $B = 0$ To test the hypothesis H_0: $B = 0$, the test statistic, which has a t distribution with $n - 2$ degrees of freedom, is $t = \dfrac{b}{s_b}$, where

$b =$ slope of the estimated (least squares) regression line

$s_b =$ standard error of the sampling distribution of $b = \dfrac{s_{y.x}}{\sqrt{\Sigma(x^2) - \dfrac{(\Sigma x)^2}{n}}}$

Analysis of Variance Test As an alternative to the t test, we can also test the hypothesis H_0: $B = 0$ by using an analysis of variance F test.

Interval Estimate for the Mean Value of y Given $x = x_g$ With the regression assumptions satisfied, we can calculate a confidence interval for the mean value of y for $x = x_g$ with:

$$\hat{y} \pm t_{\alpha/2} \left[s_{y.x} \sqrt{\frac{1}{n} + \frac{(x_g - \overline{x})^2}{\Sigma(x^2) - \dfrac{(\Sigma x)^2}{n}}} \right].$$

Prediction Interval for the Value of y Given $x = x_g$ Again if the regression assumptions are satisfied, we can calculate a prediction interval for the value of y for $x = x_g$ with:

$$\hat{y} \pm t_{\alpha/2} \left[s_{y.x} \sqrt{1 + \frac{1}{n} + \frac{(x_g - \overline{x})^2}{\Sigma(x^2) - \dfrac{(\Sigma x)^2}{n}}} \right].$$

Worked Examples

Home Buying Tests

We return to the data on the size of homes and their selling price in thousands of dollars. We have calculated an estimated regression line: $\hat{y} = -118.98 + 0.18834x$. Recognizing that this regression line could happen by chance, we want to test the hypothesis H_0: $B = 0$. Test this hypothesis first by doing a t test by hand and then by using MINITAB to do the analysis using a level of significance of 0.01. The data and some quantities we have already calculated follow.

x = Sq. Footage	y = Selling Price	xy	x^2	y^2
1600	117	187200	2560000	13689
1600	207	331200	2560000	42849
3000	431	1293000	9000000	185761
2200	312	686400	4840000	97344
2000	288	576000	4000000	82944
2200	240	528000	4840000	57600
1300	140	182000	1690000	19600
1000	90	90000	1000000	8100
2900	440	1276000	8410000	193600
1600	190	304000	2560000	36100
1400	114	159600	1960000	12996
1500	156	234000	2250000	24336
2000	289	578000	4000000	83521
2400	349	837600	5760000	121801
26700	3363	7263000	55430000	980241

$$b = 0.18834 \qquad s_{y.x} = 32.21$$

Solution

The hypotheses of interest are:

H_0: $B = 0$

H_1: $B \neq 0$.

The level of significance is 0.01. We will assume that the selling prices are normal with a constant standard deviation. Then we may use the t distribution with $14 - 2 = 12$ degrees of freedom to do the test. The boundaries of the rejection region are found in the t table in the 12-df row and the $0.01/2 = 0.005$ column. This t value is 3.055. The decision rule is to reject H_0 if the test statistic is either less than -3.055 or greater than $+3.055$. To calculate the test statistic, we first need to obtain the estimated standard error of b.

$$s_b = \frac{s_{y.x}}{\sqrt{\Sigma(x^2) - \frac{(\Sigma x)^2}{n}}} = \frac{32.21}{\sqrt{55430000 - \frac{(26700)^2}{14}}} = 0.01517.$$

Then the test statistic is:

$$t = \frac{b}{s_b} = \frac{0.18834}{0.01517} = 12.42.$$

The test statistic 12.42 is larger than 3.055, so we reject H_0 and conclude that there is a significant relationship between square footage and the selling price of a house.

MINITAB output for this data:

```
The regression equation is
Selling Price = - 119 + 0.188 Sq. Footage

Predictor        Coef        StDev           T          P
Constant      -118.98        30.18       -3.94      0.002
Sq. Foot      0.18834      0.01517       12.42      0.000

S = 32.21       R-Sq = 92.8%       R-Sq(adj) = 92.2%

Analysis of Variance

Source            DF          SS          MS          F          P
Regression         1      159953      159953     154.21      0.000
Residual Error    12       12447        1037
Total             13      172400
```

We could use the upper portion of the output to perform the test. MINITAB calculated the value of the test statistic as 12.42 and gave the associated p value, which in this problem is 0 (to three decimal places). We can also use the analysis of variance table to do this test. Notice that the p value for the ANOVA table is also 0.000. The p values for the t and F tests will always match for simple linear regression, so either can be used for the test.

Home Estimates and Predictions

Suppose that the home buyer is undecided about the size of the house he wants. He might only require a house of 1800 square feet. But there is a possibility that his mother-in-law will move in with his family, and, if so, he figures they will need a 2500-square-foot house. Because of this, he wants to estimate the mean cost of an 1800- and a 2500-square-foot house with 95% confidence intervals. In addition, to make sure that the price of any individual house is reasonable, he wants 95% prediction intervals for these sizes. Obtain each interval the home buyer needs.

Solution

We will try to present the confidence and prediction intervals for 1800 and 2500 simultaneously so you can see their similarities and differences. For $x = 1800$ we have:

$$\hat{y} = -118.98 + 0.18834x = -118.98 + 0.18834(1800) = 220.03$$

while for $x = 2500$ we obtain:

$$\hat{y} = -118.98 + 0.18834x = -118.98 + 0.18834(2500) = 351.87.$$

These are the same for the confidence intervals and the prediction intervals. For both types of intervals, the degrees of freedom will be $14 - 2 = 12$, and, since we are using 95% confidence for both, the t table values will be the same. We look in the t table in the 12-df row and the $0.05/2 = 0.025$ column. There we find 2.179. The multiplier of this t value is slightly different for each interval. For the confidence interval for the mean selling price when $x = 1800$, we have:

$$s_{y.x}\sqrt{\frac{1}{n} + \frac{(x_g - \bar{x})^2}{\Sigma(x^2) - \frac{(\Sigma x)^2}{n}}} = (32.21)\sqrt{\frac{1}{14} + \frac{(1800 - 1907.14)^2}{55430000 - \frac{(26700)^2}{14}}} = (32.21)\sqrt{0.07397} \approx 8.76.$$

For the mean selling price when $x = 2500$, we calculate:

$$s_{y.x}\sqrt{\frac{1}{n} + \frac{(x_g - \bar{x})^2}{\Sigma(x^2) - \frac{(\Sigma x)^2}{n}}} = (32.21)\sqrt{\frac{1}{14} + \frac{(2500 - 1907.14)^2}{55430000 - \frac{(26700)^2}{14}}} = (32.21)\sqrt{0.14938} \approx 12.49.$$

For the prediction interval for the selling price when $x = 1800$, we have:

$$s_{y.x}\sqrt{1 + \frac{1}{n} + \frac{(x_g - \bar{x})^2}{\Sigma(x^2) - \frac{(\Sigma x)^2}{n}}} = (32.21)\sqrt{1 + \frac{1}{14} + \frac{(1800 - 1907.14)^2}{55430000 - \frac{(26700)^2}{14}}}$$

$$= (32.21)\sqrt{1.07397} \approx 33.38.$$

For the prediction interval for the selling price when $x = 2500$, we calculate:

$$s_{y.x}\sqrt{1 + \frac{1}{n} + \frac{(x_g - \bar{x})^2}{\Sigma(x^2) - \frac{(\Sigma x)^2}{n}}} = (32.21)\sqrt{1 + \frac{1}{14} + \frac{(2500 - 1907.14)^2}{55430000 - \frac{(26700)^2}{14}}}$$

$$= (32.21)\sqrt{1.14938} \approx 34.53.$$

The confidence interval for the mean selling price of an 1800-square-foot house is:

$$\hat{y} \pm t_{\alpha/2}\left[s_{y.x}\sqrt{\frac{1}{n} + \frac{(x_g - \bar{x})^2}{\Sigma(x^2) - \frac{(\Sigma x)^2}{n}}}\right] = 220.03 \pm (2.179)(8.76) = 220.03 \pm 19.09$$

$$200.9 < \mu < 239.1,$$

while the confidence interval for the mean selling price of a 2500-square-foot house is:

$$\hat{y} \pm t_{\alpha/2}\left[s_{y.x}\sqrt{\frac{1}{n} + \frac{(x_g - \bar{x})^2}{\Sigma(x^2) - \frac{(\Sigma x)^2}{n}}}\right] = 351.87 \pm (2.179)(12.49) = 351.87 \pm 27.22$$

$$324.7 < \mu < 379.1.$$

The prediction interval for the selling price of an 1800-square-foot house is:

$$\hat{y} \pm t_{\alpha/2}\left[s_{y.x}\sqrt{1 + \frac{1}{n} + \frac{(x_g - \bar{x})^2}{\Sigma(x^2) - \frac{(\Sigma x)^2}{n}}}\right] = 220.03 \pm (2.179)(33.38) = 220.03 \pm 72.74$$

$$147.3 < y < 292.8,$$

while the prediction interval for the selling price of a 2500-square-foot house is:

$$\hat{y} \pm t_{\alpha/2}\left[s_{y.x}\sqrt{1 + \frac{1}{n} + \frac{(x_g - \bar{x})^2}{\Sigma(x^2) - \frac{(\Sigma x)^2}{n}}}\right] = 351.87 \pm (2.179)(34.53) = 351.87 \pm 75.24$$

$$276.6 < y < 427.1.$$

Practice Exercises 12-3

1-4. We return to the dean of admissions of a college who took a random sample of first-year students and learned their entrance exam score and their first-year GPA. The data and some calculations are in the table that follows.

Student	x = Exam	y = College GPA	xy	x^2	y^2
1	16	2.64	42.24	256	6.9696
2	22	3.40	74.8	484	11.56
3	23	3.15	72.45	529	9.9225
4	20	3.13	62.6	400	9.7969
5	18	2.77	49.86	324	7.6729
6	10	2.20	22	100	4.84
7	22	3.56	78.32	484	12.6736
8	22	3.31	72.82	484	10.9561
9	13	2.44	31.72	169	5.9536
10	20	3.17	63.4	400	10.0489
11	18	3.14	56.52	324	9.8596
12	20	3.03	60.6	400	9.1809
TOTAL	224	35.94	687.33	4354	109.4346

Additionally, the following were previously calculated:

$$\bar{x} = \frac{224}{12} = 18.6667, \qquad \hat{y} = 1.2166 + 0.09527x, \qquad s_{y.x} = 0.151.$$

1. Test to decide if there is a relationship between the entrance exam and the GPA. Use a level of significance of 0.01.

2. Use MINITAB or a similar computer package to do the test of Exercise 1.

3. Obtain a 99% confidence interval for the mean GPA of people who score 19 on the entrance exam.

4. Calculate a 99% prediction interval for the GPA of a person who scores 22 on the entrance exam.

5-8. We return to JC Daly who examined the relation between pass completion percent and points scored per game. His data and some associated calculations are in the table that follows.

x = Pass %	y = Points	xy	x^2	y^2
0.494	25.1	12.3994	0.244036	630.01
0.522	19.4	10.1268	0.272484	376.36
0.508	27.5	13.97	0.258064	756.25
0.487	27	13.149	0.237169	729
0.485	19.2	9.312	0.235225	368.64
0.476	25.8	12.2808	0.226576	665.64
0.559	29.9	16.7141	0.312481	894.01
0.477	32.9	15.6933	0.227529	1082.41
0.494	20.4	10.0776	0.244036	416.16
0.652	32.9	21.4508	0.425104	1082.41
0.643	36.6	23.5338	0.413449	1339.56
0.661	31.6	20.8876	0.436921	998.56
0.531	30.6	16.2486	0.281961	936.36
0.572	33.6	19.2192	0.327184	1128.96
0.472	25	11.8	0.222784	625
8.033	417.5	226.863	4.365003	12029.33

Additionally, we have already calculated:
$$\bar{x} = 0.5355, \qquad \hat{y} = -0.001825 + 51.9765x, \qquad s_{y.x} = 4.2836.$$

5. Test to decide if there is a relationship between passing percentage and points scored. Use a level of significance of 0.05.

6. Use MINITAB or a similar computer package to do the test of Exercise 5.

7. Obtain a 95% confidence interval for the mean points of a team with a passing percentage of 0.55

8. Calculate a 95% prediction interval for the points of a team with a passing percentage of 0.48.

9-12. The data from the kinesiologist who determined the amount of time students spent exercising and their percent of body fat is in the table that follows.

x = Exercise	y = Body Fat %	xy	x^2	y^2
80	22	1760	6400	484
40	28	1120	1600	784
0	33	0	0	1089
60	25	1500	3600	625
120	16	1920	14400	256
240	8	1920	57600	64
0	28	0	0	784
30	27	810	900	729
60	26	1560	3600	676
90	21	1890	8100	441
180	11	1980	32400	121
900	245	14460	128600	6053

The kinesiologist has also calculated the following:
$$\bar{x} = 81.818, \qquad \hat{y} = 30.587 - 0.1016x \qquad s_{y.x} = 1.782 .$$

9. Test to decide if there is a relationship between amount of exercise and percentage of body fat. Use $\alpha = 0.05$.

10. Use MINITAB or a similar computer package to perform the test of Exercise 9.

11. Obtain a 95% confidence interval for the mean percentage of body fat for people that exercise 120 minutes in a week.

12. Calculate a 95% prediction interval for the percentage of body fat for a person who exercises 120 minutes in a week.

Solutions to Practice Exercises 12-3

1. The hypotheses of interest are:
   H_0: $B = 0$
   H_1: $B \neq 0$.

The level of significance is 0.01. We will assume that the GPA's are normal with a constant standard deviation. Then we may use the t distribution with $12 - 2 = 10$ degrees of freedom to do the test. The boundaries of the rejection region are found in the t table in the 10-df row and the $0.01/2 = 0.005$ column. This t value is 3.169. The decision rule is to reject H_0 if the test statistic is either less than -3.169 or greater than $+3.169$. To calculate the test statistic, we first need to obtain the estimated standard error of b:

$$s_b = \frac{s_{y.x}}{\sqrt{\Sigma(x^2) - \frac{(\Sigma x)^2}{n}}} = \frac{0.151}{\sqrt{4354 - \frac{(224)^2}{12}}} \approx 0.01149 .$$

Then the test statistic is $t = \dfrac{b}{s_b} = \dfrac{0.09527}{0.01149} \approx 8.29$.

The test statistic 8.29 is larger than 3.169, so we reject H_0 and conclude that there is a significant relationship between the entrance exam and GPA.

2. The MINITAB output for this data follows.

```
Regression Analysis

The regression equation is

College GPA = 1.22 + 0.0953 Exam

Predictor          Coef        StDev           T          P
Constant         1.2166       0.2185        5.57      0.000
Exam             0.09527      0.01147       8.31      0.000

S = 0.1507       R-Sq = 87.3%      R-Sq(adj) = 86.1%

Analysis of Variance

Source             DF          SS          MS          F          P
Regression          1      1.5672      1.5672      69.01      0.000
Residual Error     10      0.2271      0.0227
Total              11      1.7943
```

We could use either the upper portion of the output or the analysis of variance table to perform the test. In both cases, the p value is 0.000. Since this is less than 0.01, we again conclude that there is a significant relationship between the entrance exam and GPA.

3. To obtain a confidence interval for the mean GPA of people who score 19 on the entrance exam, we first use the regression equation to obtain an estimate of the mean GPA. For $x = 19$, we have:
$\hat{y} = 1.2166 + 0.09524x = 1.2166 + 0.09524(19) = 3.026$.
The degrees of freedom will be $12 - 2 = 10$. We find the t value for a 99% confidence interval by looking in the t table value in the 10-df row and the $0.01/2 = 0.005$ column. There we find a t value of 3.169. Then for a 99% confidence interval for the mean GPA when the entrance exam score is 19, we calculate:

$$s_{y.x}\sqrt{\frac{1}{n} + \frac{(x_g - \bar{x})^2}{\Sigma(x^2) - \frac{(\Sigma x)^2}{n}}} = (0.151)\sqrt{\frac{1}{12} + \frac{(19 - 18.667)^2}{4354 - \frac{(224)^2}{12}}} = (0.151)\sqrt{0.09398} = 0.04376.$$

The confidence interval for μ is:

$$\hat{y} \pm t_{\alpha/2}\left[s_{y.x}\sqrt{\frac{1}{n} + \frac{(x_g - \bar{x})^2}{\Sigma(x^2) - \frac{(\Sigma x)^2}{n}}}\right] = 3.026 \pm (3.169)(0.04376) = 3.026 \pm 0.139,$$

$$2.887 < \mu < 3.165.$$

4. The prediction interval for a GPA of a person with an entrance exam score of $x = 22$ begins with the prediction of the GPA:
$\hat{y} = 1.2166 + 0.09524x = 1.2166 + 0.09524(22) = 3.312$.
Then we calculate the multiplier of the t value:

$$s_{y.x}\sqrt{1+\frac{1}{n}+\frac{(x_g-\bar{x})^2}{\Sigma(x^2)-\frac{(\Sigma x)^2}{n}}} = (0.151)\sqrt{1+\frac{1}{12}+\frac{(19-18.6667)^2}{4354-\frac{(224)^2}{12}}}$$

$$= (0.151)\sqrt{1.09398} = 0.15794$$

The prediction interval is:

$$\hat{y}\pm t_{\alpha/2}\left[s_{y.x}\sqrt{1+\frac{1}{n}+\frac{(x_g-\bar{x})^2}{\Sigma(x^2)-\frac{(\Sigma x)^2}{n}}}\right] = 3.312\pm(3.169)(0.15794) = 3.312\pm0.500,$$

$$2.812 < y < 3.812.$$

5. The hypotheses are:
   H_0: $B = 0$
   H_1: $B \neq 0$.
   We want the level of significance to be 0.05. We will assume that the points scored are normal with a constant standard deviation. Then we may use the t distribution with $15 - 2 = 13$ degrees of freedom to do the test. The boundaries of the rejection region are found in the t table in the 13-df row and the $0.05/2 = 0.025$ column. This t value is 2.160. The decision rule is to reject H_0 if the test statistic is either less than -2.160 or greater than $+2.160$. To calculate the test statistic, we first need to obtain the estimated standard error of b:

$$s_b = \frac{s_{y.x}}{\sqrt{\Sigma(x^2)-\frac{(\Sigma x)^2}{n}}} = \frac{4.2836}{\sqrt{4.365003-\frac{(8.033)^2}{15}}} = 17.0576.$$

The test statistic is $t = \dfrac{b}{s_b} = \dfrac{51.9765}{17.0576} = 3.05$. Since $3.05 > 2.160$, we reject H_0 and conclude that there is a significant relationship between passing percentage and points scored.

6. The MINITAB output for this data follows:

```
Regression Analysis

The regression equation is
Points = - 0.00 + 52.0 Pass %

Predictor         Coef         StDev            T          P
Constant        -0.002         9.202        -0.00      1.000
Pass %           51.98         17.06         3.05      0.009

S = 4.284       R-Sq = 41.7%      R-Sq(adj) = 37.2%

Analysis of Variance

Source            DF           SS           MS ·         F          P
Regression         1        170.37       170.37       9.28      0.009
Residual Error    13        238.54        18.35
Total             14        408.91
```

We could use either the upper portion of the output or the analysis of variance table to perform the test. In both cases, the p-value is 0.009. Since this is less than 0.05, we again conclude that there is a significant relationship between the pass completion percentage and points scored.

7. To obtain a confidence interval for the mean points scored for teams with a pass completion percentage of 0.55, we first use the regression equation to obtain an estimate of the mean points scored. For $x = 0.55$, we have:

$$\hat{y} = -0.001825 + 51.9765x = -0.001825 + 51.9765(0.55) = 28.59.$$

The degrees of freedom will be $15 - 2 = 13$. We find the t value for a 95% confidence interval by looking in the t table value in the 13 df row and the $0.05/2 = 0.025$ column. There we find a t value of 2.160. Then we calculate:

$$s_{y.x}\sqrt{\frac{1}{n}+\frac{(x_g-\bar{x})^2}{\Sigma(x^2)-\frac{(\Sigma x)^2}{n}}}=(4.2836)\sqrt{\frac{1}{15}+\frac{(0.55-0.5355)^2}{4.365003-\frac{(8.033)^2}{15}}}$$

$$=(4.2836)\sqrt{0.07000}=1.1333$$

The confidence interval is:

$$\hat{y}\pm t_{\alpha/2}\left[s_{y.x}\sqrt{\frac{1}{n}+\frac{(x_g-\bar{x})^2}{\Sigma(x^2)-\frac{(\Sigma x)^2}{n}}}\right]=28.59\pm(2.160)(1.1333)=28.59\pm2.45$$

$$26.14<\mu<31.04.$$

8. The prediction interval for the number of points scored for a team with a pass completion percentage of 0.48 begins with an estimate of the points:

$\hat{y}=-0.001825+51.9765x=-0.001825+51.9765(0.48)=24.95.$

The degrees of freedom will be $15-2=13$. We find the t value for a 95% prediction interval by looking in the t table value in the 13 df row and the $0.05/2=0.025$ column. There we find a t value of 2.160. Then we calculate:

$$s_{y.x}\sqrt{1+\frac{1}{n}+\frac{(x_g-\bar{x})^2}{\Sigma(x^2)-\frac{(\Sigma x)^2}{n}}}=(4.2836)\sqrt{1+\frac{1}{15}+\frac{(0.55-0.5355)^2}{4.365003-\frac{(8.033)^2}{15}}}$$

$$=(4.2836)\sqrt{1.07000}=4.4310.$$

The prediction interval is:

$$\hat{y}\pm t_{\alpha/2}\left[s_{y.x}\sqrt{1+\frac{1}{n}+\frac{(x_g-\bar{x})^2}{\Sigma(x^2)-\frac{(\Sigma x)^2}{n}}}\right]=24.95\pm(2.160)(4.4310)=24.95\pm9.57$$

$$15.38<y<34.52.$$

9. The hypotheses are:

H_0: $B=0$

H_1: $B\neq0.$

We want the level of significance to be 0.05. We will assume that body fat has a normal distribution with a constant standard deviation. Then we may use the t distribution with $11-2=9$ degrees of freedom to do the test. The boundaries of the rejection region are found in the t table in the 9-df row and the $0.05/2=0.025$ column. This t value is 2.262. The decision rule is to reject H_0 if the test statistic is either less than -2.262 or greater than $+2.262$. To calculate the test statistic, we first need to obtain the estimated standard error of b:

$$s_b=\frac{s_{y.x}}{\sqrt{\Sigma(x^2)-\frac{(\Sigma x)^2}{n}}}=\frac{1.782}{\sqrt{128600-\frac{(900)^2}{11}}}=0.00760.$$

Then the test statistic is $t=\frac{b}{s_b}=\frac{-0.1016}{0.00760}=-13.37.$

Since $-13.37<-2.262$, we reject H_0 and conclude that there is a significant relationship between exercise and percentage of body fat.

10. The MINITAB output for this data follows:

```
Regression Analysis

The regression equation is
Body Fat % = 30.6 - 0.102 Exercise

Predictor        Coef        StDev            T           P
Constant       30.5872      0.8219        37.22       0.000
Exercise      -0.101621     0.007601     -13.37       0.000

S = 1.782       R-Sq = 95.2%        R-Sq(adj) = 94.7%
```

```
Analysis of Variance

Source          DF        SS          MS        F         P
Regression       1      567.60      567.60    178.72    0.000
Residual Error   9       28.58        3.18
Total           10      596.18
```

We could use either the upper portion of the output or the analysis of variance table to perform the test. In both cases, the p-value is 0.000. Since this is less than 0.05, we again conclude that there is a significant relationship between exercise and percentage of body fat.

11. To obtain a confidence interval for the mean percentage of body fat for people who exercise 120 minutes a week, we first use the regression equation to obtain an estimate of the percentage of body fat. For $x = 120$, we have:

$$\widehat{y} = 30.587 + (-0.1016)x = 30.587 - 0.1016(120) = 18.4.$$

The degrees of freedom will be $11 - 2 = 9$. We find the t value for a 95% confidence interval by looking in the t table value in the 9-df row and the $0.05/2 = 0.025$ column. There we find a t value of 2.262. Then we calculate:

$$s_{y.x}\sqrt{\frac{1}{n} + \frac{(x_g - \overline{x})^2}{\Sigma(x^2) - \frac{(\Sigma x)^2}{n}}} = (1.782)\sqrt{\frac{1}{11} + \frac{(120 - 81.818)^2}{128600 - \frac{(900)^2}{11}}}$$

$$= (1.782)\sqrt{0.11743} = 0.6106.$$

The confidence interval is:

$$\widehat{y} \pm t_{\alpha/2}\left[s_{y.x}\sqrt{\frac{1}{n} + \frac{(x_g - \overline{x})^2}{\Sigma(x^2) - \frac{(\Sigma x)^2}{n}}}\right] = 18.4 \pm (2.262)(0.61067) = 18.4 \pm 1.4$$

$$17.0 < \mu < 19.8.$$

12. The prediction interval for the percentage of body fat for a person who exercises 120 minutes a week begins with an estimate of the body fat. In 11, we found that this is 18.4. The degrees of freedom will again be $11 - 2 = 9$. We use the same t value as 11, 2.262. Then we calculate:

$$s_{y.x}\sqrt{1 + \frac{1}{n} + \frac{(x_g - \overline{x})^2}{\Sigma(x^2) - \frac{(\Sigma x)^2}{n}}} = (1.782)\sqrt{1 + \frac{1}{11} + \frac{(120 - 81.818)^2}{128600 - \frac{(900)^2}{11}}}$$

$$= (1.782)\sqrt{1.11743} = 1.88373.$$

The prediction interval is:

$$\widehat{y} \pm t_{\alpha/2}\left[s_{y.x}\sqrt{1 + \frac{1}{n} + \frac{(x_g - \overline{x})^2}{\Sigma(x^2) - \frac{(\Sigma x)^2}{n}}}\right] = 18.4 \pm (2.262)(1.88373) = 18.4 \pm 4.3$$

$$14.1 < y < 22.7.$$

12-4 Simple Linear Correlation Analysis

Study Objectives

You should be able to:

1. Understand the breakdown of total variation into a portion explained by regression and a portion unexplained by regression.
2. Calculate and understand the properties of the coefficient of determination.
3. Calculate and understand the properties of the coefficient of correlation.

Section Overview

In simple linear regression, we try to use x to predict or estimate y; in simple linear correlation analysis, we try to measure the strength of the relationship between two variables. The two most frequently used measures are the coefficient of determination and the coefficient of correlation.

The justification for using the coefficient of determination as a measure of the association between two variables comes from regression. If we do not use x to predict y, the best guess for y is $\overline{y}$. Using $\overline{y}$ as an estimator for y, our error, or total deviation, is $(y - \overline{y})$. If we use a regression equation to predict y, the best guess for y is $\widehat{y}$ and the error or unexplained deviation is $(y - \widehat{y})$. The difference between the total deviation and the unexplained deviation is called the explained deviation and equals $(\widehat{y} - \overline{y})$. It "explains" how much variation in y can be accounted for by the relationship between x and y. To measure each of these for the entire sample, we have:

Total variation = sum of squares total = SST = $\Sigma(y - \overline{y})^2$,
Explained variation = sum of squares of regression = SSR = $\Sigma(\widehat{y} - \overline{y})^2$,
Unexplained variation = sum of squares of error = SSE = $\Sigma(y - \widehat{y})^2$.
Then the coefficient of determination, r^2, is:
$$r^2 = \frac{\text{explained variation}}{\text{total variation}} = \frac{\text{SSR}}{\text{SST}} = \frac{\Sigma(\widehat{y} - \overline{y})^2}{\Sigma(y - \overline{y})^2}.$$
The calculation of r^2 is simplified with the following formula:
$$r^2 = \frac{a(\Sigma y) + b(\Sigma xy) - n(\overline{y})^2}{\Sigma(y^2) - n(\overline{y})^2}.$$

The coefficient of correlation is the square root of the coefficient of determination. It is the positive square root if b is positive; it is the negative square root if b is negative. While it probably is not as useful as the coefficient of determination, many fields have used the correlation coefficient for so long that they continue to use it. While its actual size is hard to interpret, we can at least say that the closer its value is to $+1$ or -1, the stronger the relation between x and y.

> Caution: One warning. Both the coefficient of determination and the coefficient of correlation measure the strength of the relation between x and y. However, in no way do they imply causality. While it may be true that x "causes" y to change, there is nothing in the statistics that proves this. It might be that there are one or more underlying factors that cause x and y to change similarly. We are using the fact that x and y seem to move together to provide us with the ability to estimate or predict y better than we otherwise could.

Key Terms & Formulas

Total Variation Also called the sum of squares of the total, and abbreviated SST, it is the total variation of the values of y about the sample mean: SST = $\Sigma(y - \overline{y})^2$.

Explained Variation Also called the sum of squares of regression, and abbreviated SSR, it is the portion of the total variation explained by the regression relationship between x and y. It is calculated by: SSR = $\Sigma(\widehat{y} - \overline{y})^2$.

Unexplained Variation Also called the sum of squares of error, and abbreviated SSE, it is the portion of the total variation unexplained by the regression relationship between x and y. It is calculated by: SSE = $\Sigma(y - \widehat{y})^2$.

Coefficient of Determination The coefficient of determination, r^2, is the proportion of the total variation in y that is explained or accounted for by x (or by the regression line). Its definition is:
$$r^2 = \frac{\text{explained variation}}{\text{total variation}} = \frac{\text{SSR}}{\text{SST}} = \frac{\Sigma(\widehat{y} - \overline{y})^2}{\Sigma(y - \overline{y})^2}. \text{ It can be calculated more easily with}$$
$$r^2 = \frac{a(\Sigma y) + b(\Sigma xy) - n(\overline{y})^2}{\Sigma(y^2) - n(\overline{y})^2}.$$

> Hint: This formula looks worse than the original. However, all the elements of the latter formula are available from previous work, while the elements of the former would need to be calculated from scratch.

<u>Coefficient of Correlation</u> The coefficient of correlation is the square root of the coefficient of determination. It is the positive square root if b is positive; it is the negative square root if b is negative.

Worked Examples

Home Buying Correlations

Our potential home buyer is curious about the strength of the relationship between the size of a house and its selling price. Using the data and the associated calculations presented in and below the table that follows, calculate the coefficient of determination and interpret its value. Also calculate the coefficient of correlation.

x = Sq. Footage	y = Selling Price	xy	x^2	y^2
1600	117	187200	2560000	13689
1600	207	331200	2560000	42849
3000	431	1293000	9000000	185761
2200	312	686400	4840000	97344
2000	288	576000	4000000	82944
2200	240	528000	4840000	57600
1300	140	182000	1690000	19600
1000	90	90000	1000000	8100
2900	440	1276000	8410000	193600
1600	190	304000	2560000	36100
1400	114	159600	1960000	12996
1500	156	234000	2250000	24336
2000	289	578000	4000000	83521
2400	349	837600	5760000	121801
26700	3363	7263000	55430000	980241

$$\bar{y} = \frac{3363}{14} = 240.21 \qquad \hat{y} = -118.98 + 0.18834x$$

Solution

While it is possible to calculate the coefficient of determination with:

$$r^2 = \frac{\text{explained variation}}{\text{total variation}} = \frac{\text{SSR}}{\text{SST}} = \frac{\Sigma(\hat{y} - \bar{y})^2}{\Sigma(y - \bar{y})^2},$$

it is simpler with:

$$r^2 = \frac{a(\Sigma y) + b(\Sigma xy) - n(\bar{y})^2}{\Sigma(y^2) - n(\bar{y})^2} = \frac{-118.98(3363) + 0.18834(7263000) - 14(240.21)^2}{(980241) - 14(240.21)^2} = 0.928.$$

This says that almost 93% of the variation in y can be explained or accounted for by the variation in x. The correlation coefficient for this data is:

$$r = \sqrt{0.928} = 0.963.$$

It is the positive square root because the slope b, 0.18834, is positive.

Home Buying Correlations by Computer

Use MINITAB to obtain the coefficient of determination.

Solution

The MINITAB output:

```
The regression equation is
Selling Price = - 119 + 0.188 Sq. Footage

Predictor        Coef        StDev          T          P
Constant       -118.98        30.18       -3.94      0.002
Sq. Foot       0.18834      0.01517       12.42      0.000

S = 32.21       R-Sq = 92.8%       R-Sq(adj) = 92.2%

Analysis of Variance
```

Source	DF	SS	MS	F	P
Regression	1	159953	159953	154.21	0.000
Residual Error	12	12447	1037		
Total	13	172400			

The coefficient of determination is labeled R-sq and is given in percentage form as 92.8%.

   Hint: Next to the R-sq on the output is a quantity called R-sq(adj). This is used in multiple regression and does not concern us here.

Practice Exercises 12-4

1-3. Our dean of admissions wants to know the strength of the relationship between the entrance exam and first-year GPA. The data and some calculations follow:

Student	$x =$ Exam	$y =$ College GPA	xy	x^2	y^2
1	16	2.64	42.24	256	6.9696
2	22	3.40	74.8	484	11.56
3	23	3.15	72.45	529	9.9225
4	20	3.13	62.6	400	9.7969
5	18	2.77	49.86	324	7.6729
6	10	2.20	22	100	4.84
7	22	3.56	78.32	484	12.6736
8	22	3.31	72.82	484	10.9561
9	13	2.44	31.72	169	5.9536
10	20	3.17	63.4	400	10.0489
11	18	3.14	56.52	324	9.8596
12	20	3.03	60.6	400	9.1809
TOTAL	224	35.94	687.33	4354	109.4346

$$\bar{y} = 2.995 \qquad \hat{y} = 1.2166 + 0.09527x$$

1. Calculate and interpret the coefficient of determination.
2. Calculate the coefficient of correlation.
3. Use MINITAB or a similar computer package to obtain the coefficient of determination.

4-6. Dropping back to JC Daly, we present his football data on passing percent and points scored with some associated calculations:

$x =$ Pass %	$y =$ Points	xy	x^2	y^2
0.494	25.1	12.3994	0.244036	630.01
0.522	19.4	10.1268	0.272484	376.36
0.508	27.5	13.97	0.258064	756.25
0.487	27	13.149	0.237169	729
0.485	19.2	9.312	0.235225	368.64
0.476	25.8	12.2808	0.226576	665.64
0.559	29.9	16.7141	0.312481	894.01
0.477	32.9	15.6933	0.227529	1082.41
0.494	20.4	10.0776	0.244036	416.16
0.652	32.9	21.4508	0.425104	1082.41
0.643	36.6	23.5338	0.413449	1339.56
0.661	31.6	20.8876	0.436921	998.56
0.531	30.6	16.2486	0.281961	936.36
0.572	33.6	19.2192	0.327184	1128.96
0.472	25	11.8	0.222784	625
8.033	417.5	226.863	4.365003	12029.33

$$\bar{y} = 27.8333 \qquad \hat{y} = -0.001825 + 51.9765x$$

4. Calculate and interpret the coefficient of determination.
5. Calculate the coefficient of correlation.

6. Use MINITAB or a similar computer package to obtain the coefficient of determination.

7-9. The data from the kinesiologist's study and some calculations follow.

x = Exercise	y = Body Fat %	xy	x^2	y^2
80	22	1760	6400	484
40	28	1120	1600	784
0	33	0	0	1089
60	25	1500	3600	625
120	16	1920	14400	256
240	8	1920	57600	64
0	28	0	0	784
30	27	810	900	729
60	26	1560	3600	676
90	21	1890	8100	441
180	11	1980	32400	121
900	245	14460	128600	6053

The kinesiologist has also calculated the following:

$$\bar{y} = 22.273 \qquad \hat{y} = 30.587 - 0.1016x.$$

7. Calculate and interpret the coefficient of determination.
8. Calculate the coefficient of correlation.
9. Use MINITAB or a similar computer package to obtain the coefficient of determination.

Solutions to Practice Exercises 12-4

1. Coefficient of determination:

$$r^2 = \frac{a(\Sigma y) + b(\Sigma xy) - n(\bar{y})^2}{\Sigma(y^2) - n(\bar{y})^2} = \frac{1.2166(35.94) + 0.09527(687.33) - 12(2.995)^2}{(109.4346) - 12(2.995)^2} = 0.873.$$

This says that more than 87% of the variation in y can be explained or accounted for by the variation in x.

2. The correlation coefficient for this data is $r = \sqrt{0.873} = 0.934$ It is the positive square root because the slope b, 0.09527, is positive.

3. The MINITAB output for this data follows.

```
Regression Analysis

The regression equation is
College GPA = 1.22 + 0.0953 Exam

Predictor        Coef        StDev           T          P
Constant       1.2166       0.2185        5.57      0.000
Exam          0.09527      0.01147        8.31      0.000

S = 0.1507      R-Sq = 87.3%      R-Sq(adj) = 86.1%

Analysis of Variance

Source             DF          SS           MS          F          P
Regression          1      1.5672       1.5672      69.01      0.000
Residual Error     10      0.2271       0.0227
Total              11      1.7943
```

The coefficient of determination is given as R-sq and equals 87.3%.

4. Coefficient of determination:

$$r^2 = \frac{a(\Sigma y) + b(\Sigma xy) - n(\bar{y})^2}{\Sigma(y^2) - n(\bar{y})^2} = \frac{-0.001825(417.5) + 51.9765(226.863) - 15(27.8333)^2}{(12029.33) - 15(27.8333)^2}$$
$$= 0.417.$$

This says that approximately 42% of the variation in y can be explained or accounted for by the variation in x.

5. The correlation coefficient for this data is $r = \sqrt{0.417} = 0.645$ It is the positive square root because the slope b, 51.9765, is positive.

6. The MINITAB output for this data follows.

```
Regression Analysis

The regression equation is
Points = - 0.00 + 52.0 Pass %

Predictor        Coef       StDev          T         P
Constant       -0.002       9.202      -0.00     1.000
Pass %          51.98       17.06       3.05     0.009

S = 4.284      R-Sq = 41.7%      R-Sq(adj) = 37.2%

Analysis of Variance

Source          DF         SS          MS         F         P
Regression       1      170.37      170.37      9.28     0.009
Residual Error  13      238.54       18.35
Total           14      408.91
```
The coefficient of determination is given as R-sq and is 41.7%.

7. Coefficient of determination:
$$r^2 = \frac{a(\Sigma y) + b(\Sigma xy) - n(\bar{y})^2}{\Sigma(y^2) - n(\bar{y})^2} = \frac{30.587(245) + (-0.1016)(14460) - 11(22.273)^2}{(6053) - 11(22.273)^2} = 0.952.$$
This says that more than 95% of the variation in y can be explained or accounted for by the variation in x.

8. The correlation coefficient for this data is $r = \sqrt{0.952} = -0.976$. It is the negative square root because the slope b, -0.1016, is negative.

9. The MINITAB output for this data follows:

```
Regression Analysis

The regression equation is
Body Fat % = 30.6 - 0.102 Exercise

Predictor        Coef       StDev          T         P
Constant      30.5872      0.8219      37.22     0.000
Exercise    -0.101621    0.007601     -13.37     0.000

S = 1.782      R-Sq = 95.2%      R-Sq(adj) = 94.7%

Analysis of Variance

Source          DF         SS          MS          F         P
Regression       1      567.60      567.60     178.72     0.000
Residual Error   9       28.58        3.18
Total           10      596.18
```
The coefficient of determination is given as R-sq and is 95.2%.

12-5 Multiple Linear Regression and Correlation

Study Objectives
You should be able to:
1. Understand the extension of simple linear regression to multiple regression.
2. Analyze multiple regression computer output.

Section Overview

Multiple regression involves the use of more than one independent variable to predict the dependent variable. While the mathematics of multiple regression is more esoteric than that of simple linear regression, the ideas involved are not that different. To simplify matters, we work with bivariate regression-two independent variables-but the ideas easily extend to any number of independent variables.

The multiple regression equation with two independent variables is:

$$\widehat{y} = a + b_1 x_1 + b_2 x_2.$$

In this equation, $\widehat{y}$ is again the estimated value of the dependent variable, a is the intercept, x_1 and x_2 are the independent variables, and b_1 and b_2 are the regression coefficients for the independent variables. The values of a, b_1, and b_2 are computed using the criterion of least squares. This criterion results in long formulae that can be used to calculate the individual values by hand. However, most sane people would use a computer package such as MINITAB to perform the calculations for a, b_1, and b_2. The values of b_1 and b_2 are called the estimated regression coefficients. They represent the average amount y will change for a one-unit increase in their x, assuming the other x remains constant. For example, if $b_1 = 10$, this means that we think that y will change by 10 every time x_1 goes up by one as long as x_2 does not change at all. Once the estimated multiple regression equation is calculated, it can be used to obtain estimates or predictions for y. We simply insert the values of x_1 and x_2 into the equation to compute the estimate.

One problem that can occur in multiple regression is multicollinearity. This occurs when the independent variables, x_1 and x_2, are themselves correlated with each other. This causes some difficulties with the mathematics of multiple regression and causes the estimates to be unreliable. One solution to this problem is to use only one of the two variables.

In simple linear regression we could estimate the standard deviation around the regression line, the standard error of estimate. We will again be able to calculate this standard error of estimate, though now it is a measure of the standard deviation around the regression plane rather than line. We could use the formula

$$s_{y.x_1x_2} = \sqrt{\frac{\Sigma(y - \widehat{y})^2}{n-3}},$$

though it is easier to use an equivalent formula:

$$s_{y.x_1x_2} = \sqrt{\frac{\Sigma(y^2) - a(\Sigma y) - b_1(\Sigma x_1 y) - b_2(\Sigma x_2 y)}{n-3}}.$$

Similar to simple linear regression, we have a population regression equation, but now with two regression coefficients, B_1 and B_2. We can use the t distribution to test H_0: $B_1 = 0$ and to test H_0: $B_2 = 0$. In each case the test has $n - 3$ degrees of freedom. The test statistic is similar to the test statistic in SLR. The only difference will be an additional subscript to identify the coefficient being tested. For example, the test statistic for H_0: $B_1 = 0$ is:

$$t = \frac{b_1}{\text{estimated standard deviation of } b_1}.$$

Otherwise, the test is the same. In SLR, we could also use an F test to test a hypothesis about the slope. In multiple regression, we again can do an F test. The hypothesis, however, is concerned simultaneously with both regression coefficients. It is H_0: $B_1 = B_2 = 0$. In words, this null hypothesis says that neither of the independent variables is related to y or the total regression is not significant. The alternative hypothesis is that at least one independent variable is related to y.

The F distribution involved in this test has $df_{num} = m$ and $df_{den} = n - m - 1$, where m is the number of independent variables in the regression equation. Here, where we are considering regression equations with two independent variables, $df_{num} = 2$ and $df_{den} = n - 3$. The remainder of the test is identical to the test in SLR.

Hint: At first glance, the analysis of variance test does not look particularly useful--it is not difficult to run two t tests to get more information. However, rather than working with only two variables, often a researcher will gather dozens or hundreds of variables that he or she thinks might be related to y. It is helpful to have this F test available to decide if any of a set of variables is helpful in predicting the dependent variable. Otherwise, we can spend a great deal of time running individual t tests to reach the same conclusion.

We can obtain confidence intervals for the mean of y and prediction intervals for individual y values. In either case we would supply values for the independent variables x_1 and x_2 and obtain the appropriate interval. The simplest way to do so is to use a computer package such as MINITAB and its "predict" subcommand. Another quantity that MINITAB can give us is the coefficient of multiple determination. It represents the percentage of variation in the dependent variable y explained by its relation to x_1 and x_2.

Key Terms & Formulas

Multiple Regression Multiple regression involves the use of two or more independent variables to predict the dependent variable.

Multiple Regression Equation For two independent variables, the multiple regression equation is:

$\hat{y} = a + b_1 x_1 + b_2 x_2$. In this equation

$\hat{y}$ = the estimated value of the dependent variable

x_1, x_2 = the independent variables

a = the y intercept

b_1 = the "slope" coefficient of x_1 = the change in $\hat{y}$ associated with a one-unit increase in x_1 when x_2 is held constant

b_2 = the "slope" coefficient of x_2 = the change in $\hat{y}$ associated with a one-unit increase in x_2 when x_1 is held constant.

Estimated Regression Coefficients These are the calculated values of b_1, and b_2.

Hint: Many also call the calculated value of a in the regression equation a regression coefficient.

Multicollinearity This occurs in multiple regression when two or more of the independent variables are highly correlated. This causes some problems in the estimation procedure and causes the estimates of the regression coefficients to be unreliable. One solution to the problem is to drop one or more of the variables correlated with the others.

Standard Error of Estimate for Multiple Regression A measure of the dispersion of the sample data points around the regression line. It is calculated either by:

$$s_{y.x_1x_2} = \sqrt{\frac{\Sigma(y - \hat{y})^2}{n - 3}}, \text{ or } s_{y.x_1x_2} = \sqrt{\frac{\Sigma(y^2) - a(\Sigma y) - b_1(\Sigma x_1 y) - b_2(\Sigma x_2 y)}{n - 3}}$$

t Test for Slope To test H_0: $B_1 = 0$ we use the t distribution with $n - m - 1$ degrees of freedom, where m is the number of independent variables involved. When there are two independent variables, this test has $n - 3$ degrees of freedom. The test statistic is

$$t = \frac{b_1}{\text{estimated standard deviation of } b_1}.$$

If the test statistic falls in the rejection region for a two-tailed test, we reject H_0 and conclude that there is a significant relationship between x_1 and y, i.e., x_1 does help in the estimation or prediction of y.

Hint: Of course a similar test exists for x_2.

Analysis of Variance Test Tests H_0: $B_1 = B_2 = 0$ versus H_1: At least one of B_1 and $B_2 \neq 0$. The null hypothesis says that neither of the independent variables is related to y, so the regression is pointless. The test statistic has an F distribution with $df_{num} = 2$ and $df_{den} = n - 3$. The decision rule is to reject H_0 if the test statistic exceeds an

F table value at level α with 2 and $n - 3$ degrees of freedom. The test statistic is available from computer packages.

Confidence Interval for the Mean Value of y Given the values of a set of independent variables, the confidence interval for the mean value of y for this set of x's.

Prediction Interval for the Value of y Given the values of a set of independent variables, the prediction interval for the value of y for this set of x's for any one occasion.

Coefficient of Multiple Determination The percent of the variation in y that is explained by its relation to x_1 and x_2. Its can be calculated several ways:

$$R^2 = \frac{\Sigma(\hat{y} - \bar{y})^2}{\Sigma(y - \bar{y})^2} = \frac{SSR}{SST} = \frac{n[a(\Sigma y) - b_1(\Sigma x_1 y) - b_2(\Sigma x_2 y)] - (\Sigma y)^2}{n\Sigma(y^2) - (\Sigma y)^2}$$

Worked Examples

Home Buying Multiple Regression

Our potential home buyer thinks that the age of the house combined with its size would provide a good prediction of its price. To verify this, he uses MINITAB to produce the output that follows. Use this output to:

1. Obtain the calculated multiple regression equation.
2. Interpret the regression coefficients.
3. Predict the selling price of a 20-year-old, 2500-square-foot house.
4. Determine the standard error of estimate.
5. Perform an analysis of variance test to verify that at least one of the independent variables, square footage and age, is related to the dependent variable, selling price. Use a level of significance of 0.01.
6. Do t tests to decide if each of the independent variables, square footage and age, belong in the multiple regression model. Use a level of significance of 0.01 for each test.
7. Obtain a 95% confidence interval for the mean selling price of a 10-year-old, 1800-square-foot home.
8. Obtain a 95% prediction interval for the mean price of a 10-year-old, 1800-square-foot home.
9. Find the coefficient of multiple determination.

The data is also presented in case you want to verify any of the MINITAB output by using the calculation formulae.

x_1 = Square Feet	x_2 = Age	y = Selling Price
1600	25	117
1600	5	207
3000	13	431
2200	1	312
2000	5	288
2200	23	240
1300	15	140
1000	14	90
2900	2	440
1600	3	190
1400	25	114
1500	16	156
2000	11	289
2400	8	349

```
MTB > Regress c3 2 c1 c2;
SUBC> predict 1800 10.

Regression Analysis

The regression equation is
Selling Price = - 46.9 + 0.172 Sq. Footage - 3.38 Age

Predictor        Coef        StDev           T          P
Constant        -46.86       21.21        -2.21       0.049
Sq. Foot      0.171509     0.008853       19.37       0.000
Age            -3.3753       0.6246        -5.40       0.000
```

```
S = 17.60          R-Sq = 98.0%      R-Sq(adj) = 97.7%
```

Analysis of Variance

Source	DF	SS	MS	F	P
Regression	2	168994	84497	272.88	0.000
Residual Error	11	3406	310		
Total	13	172400			

Source	DF	Seq SS
Sq. Foot	1	159953
Age	1	9041

Predicted Values

Fit	StDev Fit	95.0% CI	95.0% PI
228.11	5.01	(217.07, 239.14)	(187.84, 268.38)

Solution

1. We can obtain the multiple regression equation from two places in the top potion of the output. This portion is reproduced below, with the two sources of the equation underlined. Using the second part that gives more decimal places for the coefficients, we have:

    $\hat{y} = -46.9 + 0.171509x_1 - 3.3753x_2$.
    The regression equation is
    Selling Price = - 46.9 + 0.172 Sq. Footage - 3.38 Age

Predictor	Coef	StDev	T	P
Constant	-46.86	21.21	-2.21	0.049
Sq. Foot	0.171509	0.008853	19.37	0.000
Age	-3.3753	0.6246	-5.40	0.000

2. Because $b_1 = 0.172$, we can say that we expect the selling price to increase 0.172 (thousands of dollars or $172) for every increase of one square foot as long as the age of the house remains constant. Because $b_2 = -3.38$, we can say that we expect the selling price to decrease 3.38 (thousands of dollars or $3380) for every additional year in the age of the house for the same square footage.

3. The selling price of a 20-year-old, 2500-square-foot house is
    $\hat{y} = -46.9 + 0.171509(2500) - 3.3753(20) = 314.4$.

4. The standard error of estimate can be read directly from the output where it is designated as s:
    ```
    S = 17.60          R-Sq = 98.0%      R-Sq(adj) = 97.7%
    ```

5. We want to test H_0: $B_1 = B_2 = 0$ versus H_1: At least one of B_1 and $B_2 \neq 0$. This is possible if we may assume that the selling price is normal with constant variance. We could go through the test by determining the decision rule that would be based on the F distribution with $df_{num} = 2$ and $df_{den} = n - 3 = 14 - 3 = 11$. However, MINITAB provides the p value of the test, and it is 0.000. This implies that H_0 should be rejected, and we conclude that we have a significant regression.
    Analysis of Variance

Source	DF	SS	MS	F	P
Regression	2	168994	84497	272.88	0.000
Residual Error	11	3406	310		
Total	13	172400			

6. Again we will use the p value approach to do the tests.
    Hint: If you choose to use a non-p-value approach, you would work with the t distribution with $n - 3 = 14 - 3 = 11$ degrees of freedom. The test statistics are in the "t-ratio" portion of the MINITAB output.

    The hypotheses for square footage are:
    H_0: $B_1 = 0$
    H_1: $B_1 \neq 0$.

Examining the p value for x_1, we find 0.000. Since this is less than 0.01, we reject the null hypothesis and decide that square footage does belong in this regression equation. The hypotheses for age are:

H_0: $B_2 = 0$

H_1: $B_2 \neq 0$.

Examining the p value for x_1, we again find 0.000. Since this is less than 0.01, we reject the null hypothesis for age and decide that it also belongs in this regression equation:

```
Predictor        Coef        StDev          T        P
Constant       -46.86        21.21      -2.21    0.049
Sq. Foot      0.171509     0.008853     19.37    0.000
Age           -3.3753       0.6246      -5.40    0.000
```

7. The 95% confidence interval for the mean selling price of a 10-year-old, 1800-square-foot home is given by MINITAB because of our use of the predict subcommand. Our interval for the mean goes from $217,070 to $239,140.

```
  Fit   StDev Fit          95.0% CI                95.0% PI
228.11       5.01    ( 217.07,  239.14)   ( 187.84,  268.38)
```

8. The 95% prediction interval for the mean price of a 10-year-old, 1800-square-foot home is also given by MINITAB and is from $187,840 to $268,380.

```
  Fit   StDev Fit          95.0% CI                95.0% PI
228.11       5.01    ( 217.07,  239.14)   ( 187.84,  268.38)
```

9. The coefficient of multiple determination is again given by MINITAB in percentage form and, converted to decimals is 0.980.

```
S = 17.60        R-Sq = 98.0%       R-Sq(adj) = 97.7%
```

Practice Exercises 12-5

1-9. The dean of admissions decided to combine the college's entrance exam with high school GPA to predict students' first-year GPA. The data and MINITAB output for the data follow.

1. Obtain the calculated multiple regression equation.

2. Interpret the regression coefficients.

3. Predict the college GPA of a person who scores 22 on the exam and has a high school GPA of 3.20.

4. Determine the standard error of estimate.

5. Do an analysis of variance test to verify that at least one of the independent variables is related to the dependent variable. Use a level of significance of 0.05.

6. Perform t tests to decide if each of the independent variables belongs in the multiple regression model. Use a level of significance of 0.05 for each test.

7. Obtain a 95% confidence interval for the mean college GPA of people who score 20 on the exam and haves a high school GPA of 3.00.

8. Obtain a 95% prediction interval for the college GPA of a person who scores 20 on the exam and has a high school GPA of 3.00.

9. Determine the coefficient of multiple determination.

Student	$x_1 = $ Exam	$x_2 = $ HS GPA	$y = $ Col GPA
1	16	2.93	2.64
2	22	3.56	3.40
3	23	2.60	3.15
4	20	2.67	3.13
5	18	3.13	2.77
6	10	2.65	2.20
7	22	3.71	3.56
8	22	3.30	3.31
9	13	2.57	2.44
10	20	2.76	3.17
11	18	3.90	3.14
12	20	2.75	3.03

```
MTB > Regress c3 2 c1 c2;
SUBC> predict 20 3.00.
```

Regression Analysis

The regression equation is
College GPA = 0.675 + 0.0850 Exam + 0.241 HS GPA

Predictor	Coef	StDev	T	P
Constant	0.6753	0.2297	2.94	0.016
Exam	0.084976	0.008830	9.62	0.000
HS GPA	0.24096	0.07470	3.23	0.010

S = 0.1082   R-Sq = 94.1%   R-Sq(adj) = 92.8%

Analysis of Variance

Source	DF	SS	MS	F	P
Regression	2	1.68897	0.84448	72.16	0.000
Residual Error	9	0.10533	0.01170		
Total	11	1.79430			

Source	DF	Seq SS
Exam	1	1.56720
HS GPA	1	0.12177

Predicted Values

Fit	StDev Fit	95.0% CI	95.0% PI
3.0977	0.0340	(3.0209, 3.1745)	(2.8412, 3.3542)

10-13. A kinesiologist surveyed students at a state college and found the amount of time they spend exercising, their age, and the percent of body fat for each respondent. The data and MINITAB output follow.

10. Obtain the calculated multiple regression equation.

11. Perform an analysis of variance test to verify that at least one of the independent variables is related to the dependent variable. Use a level of significance of 0.05.

12. Do t tests to see if each of the independent variables belongs in the multiple regression model. Use a level of significance of 0.05 for each test.

13. Based on the results from the previous question, should we use this regression model to make estimates or predictions?

x_1 = TimeExer	x_2 = Age	y = Bodyfat%
80	22	22
40	22	28
0	21	33
60	25	25
120	19	16
240	23	8
0	31	28
30	25	27
60	19	26
90	22	21
180	25	11

```
MTB > Regress c1 2 c2 c3.

Regression Analysis

The regression equation is
Body Fat % = 36.0 - 0.104 TimeExer - 0.227 Age

Predictor          Coef        StDev           T          P
Constant         36.000        3.898        9.24      0.000
Exercise       -0.103574     0.007339      -14.11     0.000
Age             -0.2275       0.1605        -1.42      0.194

S = 1.690      R-Sq = 96.2%      R-Sq(adj) = 95.2%

Analysis of Variance

Source              DF           SS          MS          F          P
Regression          2        573.34      286.67     100.38     0.000
Residual Error      8         22.85        2.86
Total              10        596.18

Source        DF      Seq SS
Exercise       1      567.60
Age            1        5.74
```

14-22. JC Daly examined the relation between pass completion percent, running yards per game, and points scored per game. He took the top 15 rated teams in the country and collected data on these variables. The data and MINITAB output follows.

14. Obtain the calculated multiple regression equation.

15. Interpret the regression coefficients.

16. Predict the points scored for a team that completes 53% of their passes and runs for 300 yards per game.

17. Obtain the standard error of estimate.

18. Perform an analysis of variance test to verify that at least one of the independent variables is related to the dependent variable. Use a level of significance of 0.05.

19. Perform t tests to decide if each of the independent variables belongs in the multiple regression model. Use a level of significance of 0.05 for each test.

20. Obtain a 95% confidence interval for the mean points scored for teams that complete 50% of their passes and run for 250 yards per game.

21. Obtain a 95% prediction interval for the points scored for a team that completes 50% of their passes and runs for 250 yards per game.

22. Calculate the coefficient of multiple determination.

$x_1 = $ Passing	$x_2 = $ Running	$y = $ Points
0.494	297	25.1
0.522	265	19.4
0.508	251	27.5
0.487	310	27.0
0.485	140	19.2
0.476	259	25.8
0.559	394	29.9
0.477	455	32.9
0.494	225	20.4
0.652	271	32.9
0.643	308	36.6
0.661	257	31.6
0.531	234	30.6
0.572	366	33.6
0.472	303	25.0

```
MTB > Regress c3 2 c1 c2;
SUBC> Predict .500 250.

Regression Analysis

The regression equation is
Points = - 10.4 + 50.2 Pass % + 0.0394 Running

Predictor        Coef        StDev          T          P
Constant       -10.447       7.282       -1.43      0.177
Pass %          50.20       12.37         4.06      0.002
Running         0.03943      0.01104      3.57      0.004

S = 3.105        R-Sq = 71.7%      R-Sq(adj) = 67.0%

Analysis of Variance

Source             DF          SS          MS          F          P
Regression          2       293.25      146.62      15.21      0.001
Residual Error     12       115.66        9.64
Total              14       408.91

Source          DF      Seq SS
Pass %           1      170.37
Running          1      122.88

Predicted Values

      Fit   StDev Fit          95.0% CI              95.0% PI
   24.512      1.003     ( 22.326,   26.697)   ( 17.403,   31.620)
```

Solutions to Practice Exercises 12-5

1. The calculated multiple regression equation is $\hat{y} = 0.675 + 0.085x_1 + 0.241x_2$. This is based on the MINITAB output:

```
The regression equation is
College GPA = 0.675 + 0.0850 Exam + 0.241 HS GPA

Predictor        Coef         StDev          T          P
Constant       0.6753       0.2297       2.94      0.016
Exam           0.084976     0.008830     9.62      0.000
HS GPA         0.24096      0.07470      3.23      0.010
```

2. Because $b_1 \approx 0.0850$, we can say that we expect the college GPA to increase by 0.085 for every increase of one in the exam score as long as the high school GPA remains constant. Because $b_2 \approx 0.241$, we can say that we expect the college GPA to increase 0.241 for every increase of one in high school GPA if the exam grade is held constant.

3. $\hat{y} = 0.6753 + 0.084976(22) - 0.24096(3.20) = 1.77$.

4. The standard error of estimate is available from the MINITAB output:
   ```
   S = 0.1082
   ```

5. We will use the p value approach. The hypotheses are H_0: $B_1 = B_2 = 0$ versus H_1: At least one of B_1 and $B_2 \neq 0$. The p value of the test is 0.000. This implies that H_0 should be rejected, and we conclude that we have a significant regression.

```
Analysis of Variance

Source             DF          SS          MS          F          P
Regression          2       1.68897     0.84448     72.16      0.000
Residual Error      9       0.10533     0.01170
Total              11       1.79430
```

6. Again we will use the p value approach to do the tests. Both hypotheses H_0: $B_1 = 0$ and H_0: $B_2 = 0$ are rejected, the first because $0.000 < 0.05$ and the second because $0.010 < 0.05$. We conclude that both variables should be part of the model.

```
Predictor          Coef          StDev          T          P
Constant         0.6753         0.2297        2.94      0.016
Exam           0.084976       0.008830        9.62      0.000
HS GPA          0.24096        0.07470        3.23      0.010
```

7. The 95% confidence interval for the mean college GPA of people who score 20 on the exam and have a high school GPA of 3.00 can be taken directly from the MINITAB output:

```
Predicted Values

   Fit   StDev Fit        95.0% CI              95.0% PI
3.0977     0.0340    ( 3.0209,  3.1745)    ( 2.8412,  3.3542)
```

8. The 95% prediction interval for the college GPA of a person who scores 20 on the exam and has a high school GPA of 3.00 is also available for MINITAB:

```
Predicted Values

   Fit   StDev Fit        95.0% CI              95.0% PI
3.0977     0.0340    ( 3.0209,  3.1745)    ( 2.8412,  3.3542)
```

9. The coefficient of multiple determination can also be found on the MINITAB output:

```
S = 0.1082      R-Sq = 94.1%       R-Sq(adj) = 92.8%
```

10. The calculated multiple regression equation is $\hat{y} = 36.0 - 0.104x_1 - 0.227x_2$. This is based on the MINITAB output:

```
The regression equation is
Body Fat % = 36.0 - 0.104 TimeExer - 0.227 Age


Predictor          Coef          StDev          T          P
Constant         36.000         3.898         9.24      0.000
Exercise      -0.103574       0.007339      -14.11      0.000
Age             -0.2275        0.1605        -1.42      0.194
```

11. We will use the p value approach. The hypotheses are H_0: $B_1 = B_2 = 0$ versus H_1: At least one of B_1 and $B_2 \neq 0$. The p value of the test is 0.000. This implies that H_0 should be rejected, and we conclude that we have a significant regression.

```
Analysis of Variance

Source            DF          SS          MS          F          P
Regression         2       573.34      286.67      100.38     0.000
Residual Error     8        22.85        2.86
Total             10       596.18
```

12. Again we will use the p value approach to do the tests. The hypothesis H_0: $B_1 = 0$ is rejected because $0.000 < 0.05$, but H_0: $B_2 = 0$ is not rejected because $0.194 \not< 0.05$. We conclude that exercise time should be part of the model, but age is unnecessary.

```
Predictor          Coef          StDev          T          P
Constant         36.000         3.898         9.24      0.000
Exercise      -0.103574       0.007339      -14.11      0.000
Age             -0.2275        0.1605        -1.42      0.194
```

13. Based on the conclusions of the previous question, we should drop age and use a simple linear regression with only exercise time as the independent variable.

14. The calculated multiple regression equation is $\hat{y} = -10.4 + 50.2x_1 + 0.0394x_2$. This is based on the MINITAB output:

```
The regression equation is
Points = - 10.4 + 50.2 Pass % + 0.0394 Running


Predictor          Coef          StDev          T          P
Constant        -10.447         7.282        -1.43      0.177
```

Pass %	50.20	12.37	4.06	0.002
Running	0.03943	0.01104	3.57	0.004

15. We need to be careful in the interpretation of b_1. This is because an increase of one is nonsense-an increase of one for a passing rate of 0.50 would mean the passing rate becomes 1.50 (or using percents, 150% of the passes are completed). It would make more sense to look at an increase of 0.01 or 1%. For an increase of 1%, we expect the points to increase by 0.502 or approximately a half point, if the yardage is held constant. Because $b_2 = 0.0394$, we expect the points to increase by approximately 0.04 for every extra yard of running.

16. For a team that completes 53% of their passes and runs for 300 yards per game, we expect the following points to be scored: $\hat{y} = -10.4 + 50.2(0.530) + 0.0394(300) = 28.026$.

17. The standard error of estimate is available from the MINITAB output:
    S = 3.105   R-Sq = 71.7%   R-Sq(adj) = 67.0%

18. We will use the p value approach. The hypotheses are $H_0: B_1 = B_2 = 0$ versus H_1: At least one of B_1 and $B_2 \neq 0$. The p value of the test is 0.001. This implies that H_0 should be rejected, and we conclude that we have a significant regression.
    Analysis of Variance

Source	DF	SS	MS	F	P
Regression	2	293.25	146.62	15.21	0.001
Residual Error	12	115.66	9.64		
Total	14	408.91			

19. Again we will use the p value approach to do the tests. Both hypotheses $H_0: B_1 = 0$ and $H_0: B_2 = 0$ are rejected, the first because $0.002 < 0.05$ and the second because $0.004 < 0.05$. We conclude that both variables should be part of the model.

Predictor	Coef	StDev	T	P
Constant	-10.447	7.282	-1.43	0.177
Pass %	50.20	12.37	4.06	0.002
Running	0.03943	0.01104	3.57	0.004

20. The 95% confidence interval for the mean points scored for teams that have a 0.500 passing rate and run for 250 yards per game can be taken directly from the MINITAB output:

Fit	StDev Fit	95.0% CI	95.0% PI
24.512	1.003	(22.326, 26.697)	(17.403, 31.620)

21. The 95% prediction interval for the points scored for a team that has a 0.500 passing rate and runs for 250 yards per game can be taken directly from the MINITAB output:

Fit	StDev Fit	95.0% CI	95.0% PI
24.512	1.003	(22.326, 26.697)	(17.403, 31.620)

22. The coefficient of multiple determination can also be found on the MINITAB output:
    S = 3.105   R-Sq = 71.7%   R-Sq(adj) = 67.0%

Solutions to Odd-Numbered Exercises

1. Scatter diagram for the paired data.

3. Using the regression equation we found in Exercise 2, $\widehat{y} = 40.7 - 1.1326(12) = 27.11$ mpg.

5. Step 1. H_0: $B = 0$ and H_1: $B \neq 0$.
   Step 2. $\alpha = 0.05$.
   Step 3. t distribution.
   Step 4. With 10 df the critical t value corresponding to $t_{.025}$ is $+2.228$.
   Step 5. Reject H_0 and accept H_1 if $t < -2.228$ or if $t > +2.228$.
   Step 6. Using the values from the table we computed in Exercise 2 to find the regression equation we first

   find $s_b = \dfrac{s_{y,x}}{\sqrt{\Sigma(x^2) - \dfrac{\Sigma x^2}{n}}} = \dfrac{1.1886}{\sqrt{2,702 - \dfrac{176^2}{12}}} = \dfrac{1.1886}{\sqrt{120.6667}} = 0.1082$. Thus

   $t = \dfrac{b - B_0}{s_b} = \dfrac{-1.1326}{0.1082} = -10.467$. (This value is also found on the MINITAB output.)

   Step 7. Since $-10.467 < -2.228$, we reject H_0. There is a meaningful regression relationship--that is, the relationship is significant--between luggage capacity and gas consumption.

7. Using Formula 12.10, with the values $\widehat{y} = 27.11$, $t_{\alpha/2} = t_{.025} = 2.228$, and $s_{y,x} = 1.1886$.

$$\widehat{y} \pm t_{\alpha/2} = \left[s_{y,x} \sqrt{1 + \dfrac{1}{n} + \dfrac{(x_g - \overline{x})^2}{\Sigma(x^2) - \dfrac{(\Sigma x)^2}{n}}} \right]$$

$$= 27.11 \pm (2.228) \cdot (1.1886) \cdot \sqrt{1 + \dfrac{1}{12} + \dfrac{(12 - 14.6667)^2}{22/12}}$$

$$= 27.11 + 2.228 \cdot 1.1886 \cdot \sqrt{1.1423} = 27.11 \pm 2.83 = 24.28 \text{ to } 29.94 \text{ mpg.}$$

Using the command Regress with the subcommand Predict in MINITAB, we get the 95% P.I. This is shown underlined on the last line of the printout.

```
Regression Analysis

The regression equation is
Gas = 40.7 - 1.13 Luggage

Predictor        Coef        StDev            T          P
Constant        40.695       1.624        25.06      0.000
Luggage         -1.1326      0.1082      -10.47      0.000

S = 1.189       R-Sq = 91.6%       R-Sq(adj) = 90.8%
```

```
Analysis of Variance

Source              DF          SS          MS          F          P
Regression          1        154.79      154.79      109.56      0.000
Residual Error     10         14.13        1.41
Total              11        168.92

Predicted Values

      Fit    StDev Fit          95.0% CI              95.0% PI
   27.104       0.448    ( 26.105,  28.103)    ( 24.273,  29.934)
```

9. Using Formula 12.12 to first find the coefficient of determination we get:
$$r^2 = \frac{a(\Sigma y) + b(\Sigma xy) - n(\bar{y})^2}{\Sigma(y^2) - n(\bar{y})^2} = \frac{(40.7)(289) + (-1.1326)(4) - 12(24.0833)^2}{7.129 - 12(24.0833)^2} = 0.92527.$$
Thus the coefficient of correlation, $r = -\sqrt{0.92527} = -0.962$; the minus sign comes from b being negative.

11. Compute the linear regression equation using the table below.

Patient	Self-Reported Ht. Inches (x)	Crutch Length Inches (y)	xy	x^2	y^2
1	64	48.5	3104.0	4096	2352.25
2	65	49.0	3185.0	4225	2401.00
3	66	49.5	3267.0	4356	2450.25
4	67	50.5	3383.5	4489	2550.25
5	68	51.0	3468.0	4624	2601.00
6	69	51.5	3553.5	4761	2652.25
7	70	52.5	3675.0	4900	2756.25
8	71	53.0	3763.0	5041	2809.00
9	72	54.0	3888.0	5184	2916.00
10	73	54.5	3978.5	5329	2970.25
11	74	55.0	4070.0	5476	3025.00
12	75	56.0	4200.0	5625	3136.00
13	76	56.5	4294.0	5776	3192.25
TOTAL	910	681.5	47829.5	63882	35811.75

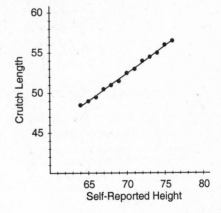

$$\bar{x} = \frac{\Sigma x}{n} = \frac{910}{13} = 70,$$

$$\bar{y} = \frac{\Sigma y}{n} = \frac{681.5}{13} = 52.4231.$$

$$b = \frac{n(\Sigma xy) - (\Sigma x)(\Sigma y)}{n(\Sigma x^2) - (\Sigma x)^2} = \frac{13(47,829.5) - (910)(681.5)}{13(63,882) - (910)^2}$$

$$= \frac{1,616.5}{2,366} = 0.68407.$$

$$a = \bar{y} - b\bar{x} = 52.4231 - (0.68407)(70) = 4.5382.$$

Thus the regression equation is: $\hat{y} = 4.5382 + 0.68407x$.

Using the regression command in MINITAB we get:
```
Regression Analysis

The regression equation is
CrtchLen = 4.54 + 0.684 Height

Predictor        Coef        StDev           T          P
Constant       4.5385       0.7940        5.72      0.000
Height        0.68407      0.01133       60.39      0.000
```

```
S = 0.1528        R-Sq = 99.7%       R-Sq(adj) = 99.7%

Analysis of Variance

Source            DF          SS          MS          F          P
Regression        1        85.166      85.166     3647.12     0.000
Residual Error    11        0.257       0.023
Total             12        85.423
```

13. Compute the standard error of the estimate using the table below.

Patient	Self-Reported Ht. Inches (x)	Crutch Length Inches (y)	$\widehat{y}$	$y - \widehat{y}$	$(y - \widehat{y})^2$
1	64	48.5	48.3090	0.191021	0.0364890
2	65	49.0	48.9930	0.006950	0.0000483
3	66	49.5	49.6771	−0.177120	0.0313716
4	67	50.5	50.3612	0.138809	0.0192680
5	68	51.0	51.0453	−0.045258	0.0020482
6	69	51.5	51.7293	−0.229328	0.0525914
7	70	52.5	52.4134	0.086601	0.0074998
8	71	53.0	53.0975	−0.097469	0.0095003
9	72	54.0	53.7815	0.218460	0.0477248
10	73	54.5	54.4656	0.034389	0.0011826
11	74	55.0	55.1497	−0.149677	0.0224033
12	75	56.0	55.8337	0.166252	0.0276398
13	76	56.5	56.5178	−0.017818	0.0003175
TOTAL	910	681.5	681.3741	0.125812	0.2580846

$$s_{y.x} = \sqrt{\frac{\Sigma(y - \widehat{y})^2}{n - 2}} = \sqrt{\frac{0.2580846}{11}} = 0.1532.$$

15. First find $\widehat{y} = 4.5382 + 0.68407(68) = 51.05496$. Then using Formula 12.9, with $t_{\alpha/2} = 2.201$, $s_{y.x} = 0.1532$, and $x_g = 68$, we get the interval:

$$\widehat{y} \pm t_{\alpha/2}\left[s_{y.x}\sqrt{\frac{1}{n} + \frac{(x_g - \overline{x})^2}{\Sigma(x^2) - \frac{(\Sigma x)^2}{n}}}\right] = 51.05496 \pm (2.201) \cdot (0.1532) \cdot \sqrt{\frac{1}{13} + \frac{(68 - 70)^2}{63,882 - \frac{(910)^2}{13}}}$$

$$= 51.05496 \pm (2.201) \cdot (0.1532) \cdot (0.3145)$$

$$= 51.05496 \pm 0.1060.$$

Thus the interval is 50.94896 to 51.16096 inches.

Using the command Regress and the subcommand Predict in MINITAB, we get the following lines added to the MINITAB output shown in Exercise 11, the desired interval is underlined.
Note: Many output lines have been omitted.

```
Predicted Values

     Fit   StDev Fit       95.0% CI              95.0% PI
  51.0549      0.0481   ( 50.9492, 51.1607)   ( 50.7024, 51.4075)
```

17. Using Formula 12.12 and the values we found in Exercise 11, we get:

$$r^2 = \frac{a(\Sigma y) + b(\Sigma xy) - n(\overline{y})^2}{\Sigma(y^2) - n(\overline{y})^2} = \frac{(4.5382)(681.5) + (0.68407)(47,829.5) - 13(52.4231)^2}{35,811.75 - 13(52.4321)^2}$$

$$= 0.997.$$

This value can also be found on the MINITAB output in Exercise 11.

19. Compute the linear regression equation using the table below.

Pre	Post	xy	x^2	y^2
5.6	7.1	39.76	31.36	50.41
8.2	7.5	61.50	67.24	56.25
12.2	12.7	154.94	148.84	161.29
10.2	10.8	110.16	104.04	116.64
6.6	8.8	58.08	43.56	77.44
7.8	8.1	63.18	60.84	65.61
7.4	8.3	61.42	54.76	68.89
6.8	7.4	50.32	46.24	54.76
10.7	8.3	88.81	114.49	68.89
6.3	5.6	35.28	39.69	31.36
9.8	8.7	85.26	96.04	75.69
11.6	11.2	129.92	134.56	125.44
11.0	11.2	123.20	121.00	125.44
9.6	10.5	100.80	92.16	110.25
7.8	8.9	69.42	60.84	79.21
11.0	12.4	136.40	121.00	153.76
7.8	9.4	73.32	60.84	88.36
11.6	9.2	106.72	134.56	84.64
9.4	7.6	71.44	88.36	57.76
6.8	6.8	46.24	46.24	46.24
7.2	7.7	55.44	51.84	59.29
10.2	10.6	108.12	104.04	112.36
195.6	198.8	1829.7	1822.5	1870.0

$$\overline{x} = \frac{\Sigma x}{n} = \frac{195.6}{22} = 8.891, \quad \overline{y} = \frac{\Sigma y}{n} = \frac{198.8}{22} = 9.036.$$

$$b = \frac{n(\Sigma xy) - (\Sigma x)(\Sigma y)}{n(\Sigma x^2) - (\Sigma x)^2} = \frac{22(1,829.7) - (195.6)(198.8)}{22(1,822.5) - (195.6)^2} = \frac{1,368.12}{1,835.64} = 0.745.$$

$a = \overline{y} - b\overline{x} = 9.036 - (0.745)(8.891) = 2.412.$

Thus the regression equation is: $\widehat{y} = 2.412 + 0.745x$.

Using MINITAB we get:
```
Regression Analysis

The regression equation is
Post = 2.41 + 0.745 Pre

Predictor        Coef        StDev           T          P
Constant        2.410        1.161        2.08      0.051
Pre            0.7453       0.1276        5.84      0.000

S = 1.166      R-Sq = 63.0%      R-Sq(adj) = 61.2%

Analysis of Variance

Source            DF           SS          MS          F          P
Regression         1       46.371      46.371      34.12      0.000
Residual Error    20       27.180       1.359
Total             21       73.551
```

21. Using Formula 12.10, with the values $\widehat{y} = 11.35$, $t_{\alpha/2} = 2.086$, and $s_{y.x} = 1.169$.

$$\widehat{y} \pm t_{\alpha/2} \left[s_{y,x} \sqrt{\frac{1}{n} + \frac{(x_g - \overline{x})^2}{\Sigma(x^2) - \frac{(\Sigma x)^2}{n}}} \right] = 11.35 \pm (2.086) \cdot (1.169) \cdot \sqrt{\frac{1}{22} + \frac{(12 - 8.891)^2}{1822.5 - \frac{(195.6)^2}{22}}}$$

$$= 11.35 \pm (2.086) \cdot (1.169) \cdot \sqrt{0.1641} = 11.35 \pm 0.979 \text{ or } 10.371 \text{ to } 12.329$$

Using MINITAB we get:
```
Predicted Values

      Fit   StDev Fit       95.0% CI           95.0% PI
   11.354      0.468   ( 10.377,  12.330)   ( 8.733,  13.974)
```

23.

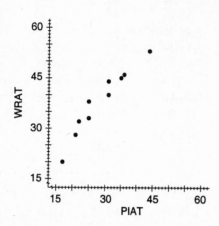

25. Using the regression equation we found in Exercise 24, $\widehat{y} = 4.91 + 1.15(40) = 50.89$ WRAT score.

27. Step 1. H_0: $B = 0$ and H_1: $B \neq 0$.
Step 2. $\alpha = 0.05$.
Step 3. t distribution.
Step 4. With 8 df the critical t value corresponding to $t_{.025}$ is 2.306.
Step 5. Reject H_0 and accept H_1 if $t < -2.306$ or if $t > +2.306$.
Step 6. Using the values from the table we computed in Exercise 24 to find the regression equation we first

find $s_b = \dfrac{s_{y.x}}{\sqrt{\Sigma(x^2) - \dfrac{(\Sigma x)^2}{n}}} = \dfrac{2,785}{\sqrt{8,843 - \dfrac{(287)^2}{10}}} = \dfrac{2,785}{\sqrt{606.1}} = 0.1131$. So

$t = \dfrac{b}{s_b} = \dfrac{1.1495}{0.1131} = 10.164.$
Step 7. Since 10.164 is greater than 2.306, we reject H_0. The relationship is meaningful.

29. Using the regression equation we found in Exercise 28, $\widehat{y} = 2.88 + 0.885(25.50) = 25.448$.

31. Using Formula 12.12 and the values we found in Exercise 28, we get:
$$r^2 = \frac{a(\Sigma y) + b(\Sigma xy) - n(\bar{y})^2}{\Sigma(y^2) - n(\bar{y})^2} = \frac{(25.629)(413.29) + (0.00802)(10,463.7) - 16(25.831)^2}{10,681.4 - 16(25.831)}$$
$$= 0.001.$$

33. Compute the linear regression equation using the table below.

	AQR Stock (x)	Amcorp Stock (y)	xy	x^2	y^2
	51.375	21.375	1098.14	2639.39	456.891
	50.625	20.250	1025.16	2562.89	410.062
	51.500	21.000	1081.50	2652.25	441.000
	50.750	20.375	1034.03	2575.56	415.141
	52.750	21.875	1153.91	2782.56	478.516
	50.125	20.500	1027.56	2512.52	420.250
Total	307.125	125.375	6420.30	15725.17	2621.860

$$\overline{x} = \frac{\Sigma x}{n} = \frac{307.125}{6} = 51.188, \qquad \overline{y} = \frac{\Sigma y}{n} = \frac{125.375}{6} = 20.896.$$

$$b = \frac{n(\Sigma xy) - (\Sigma x)(\Sigma y)}{n(\Sigma x^2) - (\Sigma x)^2} = \frac{6(6,420.30) - (307,125)(125.375)}{6(15,725.17) - (307.125)^2} = 0.6327.$$

$$a = \overline{y} - b\overline{x} = 20.896 - (0.6327)(51.188) = -11.49.$$

Thus the regression equation is: $\widehat{y} = -11.49 + 0.6327x$.

Using the regression command in MINITAB we get:

```
Regression Analysis

The regression equation is
Amcorp = - 11.5 + 0.633 AQR

Predictor          Coef          StDev            T           P
Constant         -11.488         7.472         -1.54       0.199
AQR               0.6327         0.1460         4.33       0.012

S = 0.2995       R-Sq = 82.4%      R-Sq(adj) = 78.1%

Analysis of Variance

Source             DF              SS            MS           F         P
Regression          1          1.6854        1.6854       18.79     0.012
Residual Error      4          0.3588        0.0897
Total               5          2.0443
```

35. Using Formula 12.12 and the values we found in Exercise 33, we get the coefficient of determination:

$$r^2 = \frac{a(\Sigma y) + b(\Sigma xy) - n(\overline{y})^2}{\Sigma(y^2) - n(\overline{y})^2}$$

$$= \frac{(-11.49)(125.375) + (0.6327)(6,420.30) - 6(20.896)^2}{2,621.86 - 6(20.896)^2} = 0.8487.$$

So the coefficient of correlation is $r = 0.921$.

37. Compute the linear regression equation using the table below.

Lost, y	Nodes, x	xy	x^2	y^2
0	0	0	0	0
1	3	3	9	1
12	14	168	196	144
14	19	266	361	196
5	18	90	324	25
12	19	228	361	361
0	3	0	9	0
3	3	9	9	9
3	14	42	196	9
21	18	378	324	441
10	15	150	225	100
10	12	120	144	100
7	13	91	169	49
1	6	6	36	1
1	5	5	25	1
4	17	68	289	16
26	28	728	784	676
4	15	60	225	16
6	15	90	225	36
5	11	55	121	25
145	248	2557	4032	1989

$$\bar{x} = \frac{\Sigma x}{n} = \frac{248}{20} = 12.4, \quad \bar{y} = \frac{\Sigma y}{n} = \frac{145}{20} = 7.25.$$

$$b = \frac{n(\Sigma xy) - (\Sigma x)(\Sigma y)}{n(\Sigma x^2) - (\Sigma x)^2} = \frac{20(2557) - (248)(145)}{20(4032) - (248)^2} = 0.7933.$$

$$a = \bar{y} - b\bar{x} = 7.25 - (0.7933)(12.4) = -2.587.$$

Thus the regression equation is: $\hat{y} = -2.587 + 0.7933x.$

Using the regression command in MINITAB we get:

```
Regression Analysis

The regression equation is
Lost = - 2.59 + 0.793 Nodes

Predictor           Coef        StDev           T          P
Constant          -2.587        1.982       -1.30      0.208
Nodes             0.7933        0.1396        5.68      0.000

S = 4.318       R-Sq = 64.2%      R-Sq(adj) = 62.2%

Analysis of Variance

Source            DF           SS           MS          F          P
Regression         1       602.09       602.09      32.29      0.000
Residual Error    18       335.66        18.65
Total             19       937.75
```

39. Step 1. $H_0: B = 0$ and $H_1: B \neq 0.$

Step 2. $\alpha = 0.05.$

Step 3. t distribution.

Step 4. With 18 df the critical t value corresponding to $t_{.025}$ is 2.101.

Step 5. Reject H_0 and accept H_1 if $t < -2.101$ or if $t > 2.101.$

Step 6. Using the values from the table we computed in Exercise 37 to find the regression equation we find:

$$s_{y.x} = \sqrt{\frac{\Sigma(y^2) - a(\Sigma y) - b(\Sigma xy)}{n - 2}} = \sqrt{\frac{1989 + 2.587(145) - 0.7933(2557)}{18}} = 4.318, \text{ then}$$

$$s_b = \frac{s_{y.x}}{\sqrt{\Sigma(x^2) - \frac{(\Sigma x)^2}{n}}} = \frac{4.318}{\sqrt{4032 - \frac{(145)^2}{20}}} = \frac{4.318}{\sqrt{2980.75}} = 0.0791. \text{ So } t = \frac{b}{s_b} = \frac{0.793}{0.0791} = 10.012.$$

Step 7. Since 10.012 is greater than 2.101, we reject H_0. The relationship is meaningful.

41. Using the point estimate from Exercise 38, $\hat{y} = 6.93$ when $x_g = 12$. Then with $t_{\alpha/2} = 2.101$, $s_{y.x} = 4.318$, and $x_g = 8$, we get the interval:

$$\hat{y} \pm t_{\alpha/2} \left[s_{y.x} \sqrt{1 + \frac{1}{n} + \frac{(x_g - \bar{x})^2}{\Sigma(x^2) - \frac{(\Sigma x)^2}{n}}} \right] = 6.933 \pm (2.101) \cdot (4.318) \cdot \sqrt{1 + \frac{1}{20} + \frac{(12 - 12.4)^2}{4,032 - \frac{(248)^2}{20}}}$$

$$= 6.933 \pm (2.101) \cdot (4.318) \cdot (1.025)$$
$$= 6.933 \pm 9.32.$$

Thus the interval is -2.39 to 16.25 datagrams lost.

Using the command Regress and the subcommand Predict in MINITAB, we get the following lines added to the MINITAB output shown in Exercise 11, the desired interval is underlined.

Note: Many output lines have been omitted.

```
Predicted Values

    Fit  StDev Fit        95.0% CI            95.0% PI
  6.933      0.967   (   4.901,   8.965)   (  -2.365,  16.230)
```

43. Using Formula 12.12 and the values we found in Exercise 37, we get the coefficient of determination:

$$r^2 = \frac{a(\Sigma y) + b(\Sigma xy) - n(\bar{y})^2}{\Sigma(y^2) - n(\bar{y})^2} = \frac{(-2.587)(145) + (0.7933)(2557) - 20(7.25)^2}{1989 - 20(7.25)^2}$$

$$= 0.642.$$

So the coefficient of correlation is $r = 0.801$.

45. behavior score.

47. The standard error $= 10.22$.

49. p-value for development is 0.000. Since $0.000 < 0.05$, we reject H_0. The coefficient for the development score is significant at the 0.05 level. (The t-value method confirms this. Here the t-value is 5.37, while with 18 df $t_{.025}$ is ± 2.101, so since $2.101 < 5.37$, we reject H_0.)

51. *behavior* $= 6.3 + 9.08(3) + 0.119(65) = 41.275$.

53. There are 51 degrees of freedom are available in any t test on an independent variable

```
Analysis of Variance
```

Source	DF	SS	MS	F	P
Regression	14	8483.6	606.0	2.27	0.017
Residual Error	51	13604.9	226.8		
Total	65	22088.5			

55.

Variable	Coefficient	Variable	Coefficient	Variable	Coefficient
Constant	76.68	NO_2	-29.26	TDS	-73.75
Vel	-25.16	NO_3	0.3457	Lev	-0.5426
Vol	16.04	PO_4	-1.525	Root	-0.3299
Rain	1.087	DO	-13.389	Stem	1.6846
NH_4	1.631	SAT	1.0212	Lt	0.0098

57. $H_0: B_{Rain} = 0$, $H_1: B_{Rain} \neq 0$. The corresponding t-ratio is 0.037, with a p-value of 0.714. Since p-value > 0.05, we fail to reject H_0. This predictor is not significant in the model when all the others are present.

59. The coefficient of determination is $r^2 = 0.384$.

61. $H_0: B_{pH} = B_{Conducty} = B_{COD} = 0$, H_1: At least one of B_{pH} or $B_{Conducty}$ or $B_{COD} \neq 0$. The F statistic is 9.15, with a p-value of 0.000. Since the p-value < 0.05, we reject H_0 and conclude that at least one of the predictors are useful in this model.

63. $H_0: B_{COD} = 0$, $H_1: B_{COD} \neq 0$. The corresponding t-ratio is 2.93, with a p-value of 0.006. Since p-value > 0.05, we reject H_0. This predictor is significant in the model when all the others are present.

Chapter 13 Nonparametric Statistical Methods

Study Aids and Practice Exercises

13-1 Nonparametric Methods: Uses, Benefits, and Limitations

Study Objectives
You should be able to:
1. Recognize the situations in which nonparametric methods are appropriate.
2. Understand the advantages and disadvantages of nonparametric methods.

Section Overview
Until now, we have considered methods called parametric methods. These are the methods of choice if we have large samples or if we believe the populations involved are normal. However, sometimes our samples are small and not from normal populations. For these we turn to nonparametric (or distribution-free) methods

It is convenient for us to divide our nonnormal data into three categories:
1. Data that comes from quantitative but nonnormal populations.
2. Ordinal (or rank) data.
3. Nominal data.

Each of these can be analyzed using a nonparametric technique. The method for a particular situation depends on the nature of the data and the purpose of the inference.

The advantages of nonparametric techniques include the following:
1. They can be used to analyze ordinal or nominal data.
2. They can be used with small sample sizes without requiring the assumption that the sampled population is normal.
3. The calculations usually are easier and the results more straightforward than the corresponding parametric methods.

There are some disadvantages:
1. If quantitative data is collected ordinally or nominally (e.g., ages are collected as $18 - 25$, $26 - 40$, etc.), information is lost.
2. Given the same sample size, if the parametric conditions or assumptions are met, parametric methods are more effective (powerful) than nonparametric methods.

Hint: The last disadvantage above almost makes nonparametric methods sound inferior. This is not so. The key phrase is "if the parametric conditions or assumptions are met." While it is true that nonparametric methods are weaker when the parametric assumptions are met, the power of nonparametric methods hits when the assumptions of the parametric procedures are violated. Then only nonparametric methods are valid, and using parametric techniques yields spurious results. Unfortunately, many people only know parametric methods and use them whatever the situation.

Key Terms & Formulas

<u>Nonparametric or Distribution-free Methods</u> Statistical procedures that do not require restrictive assumptions about the distribution of the population.

<u>Ordinal Data</u> Data for which an ordering exists but for which arithmetic does not make sense. For example, if we rank our favorite television shows (or college courses or foods or professors) from most favorite to least favorite, we would have an ordering. To do arithmetic, like taking an average, would be silly.

> Hint: Trying to take averages for ordinal data is a commonly committed statistical sin. Imagine a community that is equally divided on some emotional issue such as abortion. Soliciting people's opinions, we find that half the community favors abolishing abortion, while the other half favors keeping abortion legal. If we take the mean of the opinions, we would end up saying that the average opinion of the community is neutral--hardly indicative of the true situation in the community.

<u>Nominal Data</u> Data that is attribute or qualitative data. We could code nominal data into numbers for convenience or computer storage, but this does not imply that the data has any numerical qualities.

13-2 Sign Tests

Study Objectives

You should be able to:

1. Recognize when the sign test is appropriate.
2. Do a sign test based on binomial tables.
3. Perform a sign test for large samples using the normal table.

Section Overview

A sign test is a nonparametric test that compares two dependent samples. As such, it is a nonparametric alternative to the paired t test. It takes the differences of matched pairs and examines the number of positive and negative differences. If there is no difference between the two populations, the number of positive and negative differences should be approximately the same.

The null hypothesis of the sign test is:

H_0: $p = 0.5$,

where p is the number of positive differences. We will designate the test statistic as r. The sample characteristic to use as r depends on the alternative hypothesis. If the alternative hypothesis is:

H_1: $p > 0.5$,

i.e., we have a right-tailed test, r is the number of negative differences. If the alternative hypothesis is:

H_1: $p < 0.5$,

i.e., we have a left-tailed test, r is the number of positive differences. If the alternative hypothesis is

H_1: $p \neq 0.5$,

i.e., we have a two-tailed test, r is the lesser of the number of positive and negative differences. To run the test, we use the binomial tables with $n =$ the number of untied pairs to compute the p-value of the test. The p-value for this test is the probability of getting a binomial random variable less than or equal to r, i.e., we obtain $P(x \leq r)$. If we are doing a two-tailed test, we double this amount. Then the decision rule is to reject H_0 if the p-value is less than α.

It is possible to do a sign test with a large sample, $n > 30$. We let $R =$ the number of plus signs. Then the test statistic

$$z = \frac{2R - n}{\sqrt{n}}$$

has a normal distribution. The test procedure is identical in form to any other test based on the normal distribution.

Hint: For the small or large sample test, if there are any differences equal to zero, these pairs are dropped from the sample. This causes a corresponding decrease in the sample size n. The test is done on the remaining sample.

Key Terms & Formulas

Sign Test A test procedure applicable to dependent or paired samples. The test involves taking the differences between the matched pairs and basing the test on the number of positive and negative differences.

Large-Sample Sign Test With $n > 30$, we can do a sign test based on the normal distribution. In the large sample case, we can use as our test statistic $z = \dfrac{2R - n}{\sqrt{n}}$, where $R =$ the number of plus signs. This test statistic has a normal distribution.

Worked Examples

Stress Test

Eight people have had a heart valve replacement using a new procedure. Before and after the procedure, each underwent a stress test. Their pulse rates were taken immediately after each stress test. If the new procedure is effective, their pulse rates should be lower after the procedure. The pulse rates are in the table that follows. Do a sign test to see if the procedure is effective, using a level of significance of 0.05.

Patient	Before	After
1	144	140
2	160	140
3	180	156
4	132	132
5	148	140
6	156	136
7	156	160
8	136	120

Solution

We will calculate our differences by taking Before − After. Then the hypotheses are:

H_0: $p = 0.5$
H_1: $p > 0.5$,

so we are doing a right-tail test.

Hint: We could have done the subtraction as After − Before. Then we would have a left-tailed rather than a right-tailed test. But the relevant test items would also reverse, and the same decision would be reached.

The level of significance has been chosen to be 0.05. The next step is to subtract and note the signs of the differences:

Patient	Before	After	Difference	Sign
1	144	140	4	+
2	160	140	20	+
3	180	156	24	+
4	132	132	0	0
5	148	140	8	+
6	156	136	20	+
7	156	160	−4	−
8	136	120	16	+

Due to the one tie, the sample size decreases from eight to seven. Of the seven nontied data values, there are 6 positive differences and 1 negative difference. Since this is a right-tailed test, the test statistic is the number of minus signs, so $r = 1$. To perform this test, we will use the binomial distribution. The decision rule is to reject the null hypothesis if the probability we find in the binomial tables is less than $\alpha = 0.05$. We use the binomial table with $n = 7$ to find the probability of 1 or fewer minus signs by adding $P(x = 0) + P(x = 1)$ from the $p = 0.50$ column. This gives $0.0078 + 0.0547 = 0.0625$. Because 0.0625 is not less than 0.05, we fail to reject H_0 and decide that there is not significant evidence that the procedure reduces pulse rates.

Wine Tasting

The Valley of the Bears Winery wants to see how its red table wine compares to that of Cuyama Cellars. They obtain 16 volunteers willing to do a blind taste test. Each volunteer drinks one glass of wine from each winery and rates them on a scale of 0 (spit it out quick) to 10 (give me the bottle). The results are in the table that follows. Use this information to do a sign test to decide if there is any difference in the taste of the two wines. Use $\alpha = 0.10$.

VBW	CC
9	8
5	10
6	5
7	5
9	9
10	10
10	8
5	2
3	9
7	5
3	2
7	5
8	7
6	5
5	1
4	0

Solution

Because the winery wants to know if there is any difference, we will do a two-tail test. The hypotheses are:
H_0: $p = 0.5$
H_1: $p \neq 0.5$.
The level of significance has been chosen to be 0.10. Next we subtract and note the signs of the differences.

VBW	CC	Diff.	Sign
9	8	1	+
5	10	−5	−
6	5	1	+
7	5	2	+
9	9	0	0
10	10	0	0
10	8	2	+
5	2	3	+
3	9	−6	−
7	5	2	+
3	2	1	+
7	5	2	+
8	7	1	+
6	5	1	+
5	1	4	+
4	0	4	+

Due to the two ties, the sample size decreases from 16 to 14. Of these 14 there are 12 plus signs and 2 minus signs. Since this is a two-tailed test, the test statistic is the lesser of these two numbers, the number of minus signs, so $r = 2$. The decision rule is to reject the null hypothesis if the p-value we find in the binomial tables is less than

$\alpha = 0.10$. We look in the binomial table in the $n = 14$ section, the $r = 2$ row and the $p = 0.50$ column. We find 0.0 $01 + 0.0009 + 0.0056 = 0.0066$. Because this is a two-tailed test, the p value is $2(0.0066) = 0.0132$. Since $0.0132 < 0.10$, we reject the null hypothesis and conclude that there is a difference in the taste of the two wines.

News
A journalist asked 40 people to rate the reliability of information obtained from newspapers on a scale of 1 to 5. She also asked them to rate network television news programs on the same scale. Subtracting network news' ratings from those of newspapers, she found 19 positive differences, 13 negative differences, and 8 ties. Use this information to see if people perceive newspapers as more reliable at $\alpha = 0.01$.

Solution
To decide if newspapers seem more reliable, the hypotheses are:
H_0: $p = 0.5$
H_1: $p > 0.5$.
So we are doing a right-tailed test. The level of significance has been chosen as 0.01, and the data has already been summarized. Due to the eight ties, the sample size goes from 40 to 32. We will use the normal (z) distribution to do the test. The decision rule for this right-tailed test is to reject H_0 if $z > 2.33$. The test statistics is:
$$z = \frac{2R - n}{\sqrt{n}} = \frac{2(19) - 32}{\sqrt{32}} = 1.06.$$
Since this is not in the rejection region, we conclude that there is not enough evidence to show that newspapers are perceived as more reliable than network television news.

Practice Exercises 13-2
1. An instructor of statistics wanted to compare the attitudes students had about statistics before and after the class. He asked the 25 members of the class to rate the usefulness of statistics at the first class meeting and at the last class meeting. He wanted to see if the opinions on statistics had gone up. The data follows. Perform the appropriate test with $\alpha = 0.01$.

First Day	Last Day
3	5
4	4
3	5
4	8
6	9
2	7
4	4
2	2
6	9
5	5
1	3
3	2
4	10
1	5
3	8
2	4
4	4
4	8
4	2
2	6
3	7
3	4
0	5
2	6
3	2

2. Sufferers from chronic arthritis ($n = 18$) are asked to take two different pain-relief drugs and report the times for each to bring relief. The data (in minutes) is presented in the table that follows. Test to see if there is a difference in the time-to-relief for the two drugs, using $\alpha = 0.05$.

Drug A	Drug B
53	35
3	5
10	14
8	8
4	28
0	64
17	68
6	41
14	16
11	60
6	42
28	48
14	38
6	8
2	15
21	35
41	58
10	7

3. In a study on the effects of visual images on memory, volunteers are presented with paragraphs of nonsense words and are asked to memorize as many words as they can. The difference between the two groups is that they are printed using a different font or print style. Subtracting the number of words remembered with font A from font B, there were 27 positive differences, 19 negative differences, and 4 ties. Test to see if there is a significant difference in the number of words recalled, using a level of significance of 0.10.

Solutions to Practice Exercises 13-2

1. We take the first-day scores and subtract the last-day scores. If a score has gone up, the difference is negative. So we will do a left-tailed test with hypotheses:
H_0: $p = 0.5$
H_1: $p < 0.5$.
The level of significance has been chosen as 0.01. The next step is to take the differences and note the signs of the differences:

First Day	Last Day	Difference	Sign
3	5	−2	−
4	4	0	0
3	5	−2	−
4	8	−4	−
6	9	−3	−
2	7	−5	−
4	4	0	0
2	2	0	0
6	9	−3	−
5	5	0	0
1	3	−2	−
3	2	1	+
4	10	−6	−
1	5	−4	−
3	8	−5	−
2	4	−2	−
4	4	0	0
4	8	−4	−
4	2	2	+
2	6	−4	−
3	7	−4	−
3	4	−1	−
0	5	−5	−
2	6	−4	−
3	2	1	+

Due to the five ties, the sample size reduces from 25 to 20. Of these 20, there are 3 plus signs and 17 minus signs. Since this is a left-tailed test, the test statistic is the number of plus signs, so $r = 3$. The decision rule is to reject the null hypothesis if the p-value we find in the binomial tables is less than $\alpha = 0.01$. We look in the binomial table in the $n = 20$ section, the $r = 3$ row, and the $p = 0.50$ column. There we find $0.0000 + 0.0000 + 0.0002 + 0.0011 = 0.0013$. Because $0.0013 < 0.01$, we reject the null hypothesis and conclude that there is an increase in the perceived usefulness of statistics.

2. Because we are looking for any difference, we do a two-tailed test. The hypotheses are:
H_0: $p = 0.5$
H_1: $p \neq 0.5$.
The level of significance has been chosen to be 0.05. We do the subtractions and note the signs of the differences.

Drug A	Drug B	Difference	Sign
53	35	18	+
3	5	−2	−
10	14	−4	−
8	8	0	0
4	28	−24	−
0	64	−64	−
17	68	−51	−
6	41	−35	−
14	16	−2	−
11	60	−49	−
6	42	−36	−
28	48	−20	−
14	38	−24	−
6	8	−2	−
2	15	−13	−
21	35	−14	−
41	58	−17	−
10	7	3	+

Due to the one tie, the sample size decreases from 18 to 17. Of these 17, there are 2 plus signs and 15 minus signs. Since this is a two-tailed test, the test statistic is the minimum of the number of plus and minus signs, so $r = 2$. The decision rule is to reject the null hypothesis if the p-value we find in the binomial tables is less than $\alpha = 0.05$. We look in the binomial table in the $n = 17$ section, the $r = 2$ row, and the $p = 0.50$ column. There we find $0.0000 + 0.0001 + 0.0010 = 0.0011$. Because this is a two-tailed test, the p-value is twice 0.0011 or 0.0022. Since $0.0022 < 0.05$, we reject the null hypothesis and conclude that there is a difference in the time-to-relief for the two drugs.

3. Because we are looking for any difference, this is a two-tailed test. The hypotheses are:
H_0: $p = 0.5$
H_1: $p \neq 0.5$.
We have chosen the level of significance to be 0.10. The number of nontied differences is 46, so we will use the normal distribution to do the test. The decision rule for this two-tailed test is to reject H_0 if $z < -1.645$ or $z > 1.645$. The test statistic is:
$$z = \frac{2R - n}{\sqrt{n}} = \frac{2(27) - 46}{\sqrt{46}} = 1.18.$$
Since this is not in the rejection region, we conclude that there is not enough evidence to show a difference associated with the two fonts.

13-3 The Wilcoxon Signed Rank Test

Study Objectives
You should be able to:
1. Recognize when the Wilcoxon signed rank test is appropriate.
2. Find critical values from the T table
3. Perform a Wilcoxon signed rank test.

Section Overview

A sign test is a nonparametric alternative to the paired t test. A second alternative is the Wilcoxon signed rank test. This test makes more use of the data than the sign test. The test statistic involves not only the sign of the differences but also the size of their ranks. Because of this, it can produce better results than the sign test. However, it is appropriate only if the population of paired differences is continuous and symmetric.

We again have dependent samples. To simplify the explanation of the hypotheses and the test, we will assume that we subtract each observation in the second sample from its mate in the first. The null hypothesis of the Wilcoxon signed rank test is:

H_0: There is no difference between the two populations.

If the alternative hypothesis is:

H_1: The first population is greater than the second,

we have a right-tailed test as positive differences would substantiate the alternative. If the alternative hypothesis is:

H_1: The first population is less than the second,

we have a left-tailed test as negative differences would substantiate the alternative. If the alternative hypothesis is:

H_1: The first population is different from the second,

we have a two-tailed test.

In running the test, we first compute the sign and size of the differences. Any differences of zero are dropped. Then the differences are ranked from 1 to n, where the smallest difference is given a rank of 1, the second smallest 2, etc. We do this ranking ignoring the sign of the differences.

Hint: What we are really ranking is the absolute value of the differences.

The test distribution to use is the T distribution (Capital T, not Student's t). This new T table provides us with the critical values to do one- and two-tailed tests with levels of significance of 0.01 and 0.05. For a right-tailed test, the test statistic T is the sum of the ranks of the negative differences. For a left-tailed test, the test statistic T is the sum of the ranks of the positive differences. For a two-tailed test, the test statistic T is the smaller of the two sums. The null hypothesis is rejected if the computed T value is less than or equal to the T table value.

Key Terms & Formulas

Wilcoxon Signed Rank Test An alternative (as the sign test is) to the paired t test. Its test statistic involves both the signs and magnitudes of the differences between pairs. Because it uses more information than the sign test, it is a more powerful test. However, it has the more restrictive assumptions that the population of differences should be continuous and symmetric.

The Signed Rank of Differences The basis of the Wilcoxon signed rank test, they are calculated by first taking the differences of the paired data. Any zero differences are dropped. Then, ignoring the signs, the differences are ranked from smallest (given a rank of 1) to largest (given a rank of n). When the magnitudes of some differences are the same, we assign an average rank to each. Then after the ranking, the signs of the differences are attached to their ranks.

T Statistic The test statistic for the Wilcoxon signed rank test. If we are doing a right-tailed test, $T = $ the sum of the negative ranks; if we are doing a left-tailed test, $T = $ the sum of the positive ranks, if we are doing a two-tailed test, $T = $ the smaller of the two sums.

Worked Examples

Puzzles

The creator of manipulative puzzles for children wants to know if there is a difference in the time required to solve two puzzles, Catacomb and Spider. He takes a sample of 12 children and finds out how much time each takes to solve these puzzles. The data is in the table that follows. Use Wilcoxon's signed rank test to reach a decision using a level of significance of 0.05.

Catacomb	Spider
24	9
33	36
15	16
14	9
27	12
8	3
23	24
19	18
30	16
13	9
31	20
18	15

Solution

The hypotheses are:

H_0: The two puzzles take equal amounts of time

H_1: The two puzzles take different amounts of time,

so this is a two-tailed test. The level of significance has been chosen to be 0.05. The calculations necessary to do the test are in the following table:

Catacomb	Spider	Differences	Ranks	Positive Ranks	Negative Ranks
24	9	15	11.5	11.5	*
33	36	−3	4.5	*	4.5
15	16	−1	2.0	*	2.0
14	9	5	7.5	7.5	*
27	12	15	11.5	11.5	*
8	3	5	7.5	7.5	*
23	24	−1	2.0	*	2.0
19	18	1	2.0	2.0	*
30	16	14	10.0	10.0	*
13	9	4	6.0	6.0	*
31	20	11	9.0	9.0	*
18	15	3	4.5	4.5	*
TOTALS				69.5	8.5

We will use the T statistic to perform the test. We look in Appendix 7 for the critical value of the test. It is located in the $n = 12$ row and the two-tailed test, 0.05 column. There we find 13. The decision rule is to reject H_0 if $T \leq 13$. Because this is a two-tailed test, $T = $ the smaller of 69.5 and 8.5 = 8.5. Since $8.5 \leq 13$, we reject H_0 and conclude that there is a difference in the amounts of time necessary to solve the two puzzles.

Practice Exercises 13-3

1. An educational psychologist suspects that a new IQ test has a cultural bias against minorities. She takes a sample of minorities and gives them two IQ tests. One is a standard IQ test supposed to be low in its bias, and one is the test in question. The results are in the table that follows. Is there evidence that the psychologist's suspicion is correct? Use $\alpha = 0.01$.

Standard IQ Test	New IQ Test
121	115
103	95
105	105
98	95
125	105
98	97
88	90
123	121

2. Sufferers from chronic arthritis ($n = 18$) are asked to take two different pain-relief drugs and report the times required for each to bring relief. The data (in minutes) is presented in the table that follows. Use the Wilcoxon signed rank test to see if there is a difference in the time-to-relief for the two drugs using $\alpha = 0.05$.

Drug A	Drug B
53	35
3	5
10	14
8	8
4	28
0	64
17	68
6	41
14	16
11	60
6	42
28	48
14	38
6	8
2	15
21	35
41	58
10	7

Solutions to Practice Exercises 13-3

1. The hypotheses are:

    H_0: The two IQ tests have equal scores
    H_1: The standard IQ test has higher scores for minorities

    So this is a right-tailed test. The level of significance has been chosen to be 0.01. The calculations necessary for the test are in the following table:

Standard IQ Test	New IQ Test	Difference	Ranks	Positive Ranks	Negative Ranks
121	115	6	5.0	5.0	*
103	95	8	6.0	6.0	*
105	105	0	*	*	*
98	95	3	4.0	4.0	*
125	105	20	7.0	7.0	*
98	97	1	1.0	1.0	*
88	90	−2	2.5	*	2.5
123	121	2	2.5	2.5	*
TOTALS				25.5	2.5

We will use the T statistic to do the test. We look in Appendix 7 for the critical value of the test. Because of the one zero difference, we look in the $n = 7$ row and the one-tailed test, 0.01 column. There we find 0. The decision rule is to reject H_0 if $T \leq 0$. Because this is a right-tailed test, $T =$ the sum of the negative ranks $= 2.5$. Since $2.5 \not< 0$, we fail to reject H_0 and decide that there is not enough evidence to show a bias against minorities.

2. Because we are looking for a difference, we do a two-tailed test. The hypotheses are:
    H_0: The time-to-relief is the same for the two drugs
    H_1: The time-to-relief is different for the two drugs.
    The level of significance has been chosen to be 0.05. The calculations are in the following table:

Drug A	Drug B	Difference	Ranks	Positive Ranks	Negative Ranks
53	35	18	9.0	9	*
3	5	−2	2.0	*	2.0
10	14	−4	5.0	*	5.0
8	8	0	*	*	*
4	28	−24	11.5	*	11.5
0	64	−64	17.0	*	17.0
17	68	−51	16.0	*	16.0
6	41	−35	13.0	*	13.0
14	16	−2	2.0	*	2.0
11	60	−49	15.0	*	15.0
6	42	−36	14.0	*	14.0
28	48	−20	10.0	*	10.0
14	38	−24	11.5	*	11.5
6	8	−2	2.0	*	2.0
2	15	−13	6.0	*	6.0
21	35	−14	7.0	*	7.0
41	58	−17	8.0	*	8.0
10	7	3	4.0	4	*
TOTALS				13.0	140.0

Due to the one tie, the sample size decreases from 18 to 17. Of these 17 there are 2 plus signs and 15 minus signs. We will use the T statistic to perform the test. We look in Appendix 7 for the critical value of the test. It is located in the $n = 17$ row and the two-tailed test, 0.05 column. There we find 34. The decision rule is to reject H_0 if $T \leq 34$. Because this is a two-tailed test, $T =$ the smaller of 140 and 13 $= 13$. Since $13 \leq 34$, we reject H_0 and conclude that there is a difference in the time-to-relief for the two drugs.

13-4 The Mann-Whitney Test

Study Objectives

You should be able to:
1. Recognize the situations in which the Mann-Whitney test should be used.
2. Do the Mann-Whitney test.

Section Overview

So far, the nonparametric methods we have considered dealt with comparing two populations based on dependent samples. To compare two populations based on independent samples, we can use the nonparametric Mann-Whitney test. It is an alternative to the two sample t tests we considered in previous chapters. The Mann-Whitney test does not require the assumption that the samples are taken from normal populations, as the t test does. The only assumption required is that the populations involved have the same shapes and equal variances. In addition, the Mann-Whitney test can be used with ordinal data.

The test procedure involves taking independent random sample of sizes n_1 and n_2 from the two populations. These are temporarily combined into one group and each observation is ranked (as in the Wilcoxon signed rank test). Once we have assigned ranks, we can calculate:

R_1 = the sum of the ranks for the first group
R_2 = the sum of the ranks for the second group.

From these we could calculate:

$$U_1 = n_1 n_2 + \frac{n_1(n_1 + 1)}{2} - R_1 \text{ and/or } U_2 = n_1 n_2 + \frac{n_2(n_2 + 1)}{2} - R_2.$$

The test statistic is denoted by U and critical values of U are given in Appendix 8 of the text. For a right-tailed test, $U = U_1$, while for a two-tailed test, $U = U_2$. U is the smaller of U_1 and U_2 for a two-tailed test. The decision rule of the test is to reject H_0 if the value of U is less than the critical U value found in the table.

Key Terms & Formulas

Mann-Whitney Test Also known as the U test, it is used to compare populations based on two independent, small samples taken from nonnormal populations. It involves combining the two samples, ranking the combined observations, and calculating the sum of the ranks for each group.

U Test Statistic For a right-tailed test, $U = U_1 = n_1 n_2 + \frac{n_1(n_1 + 1)}{2} - R_1$; for a left-tailed test,

$U = U_2 = n_1 n_2 + \frac{n_2(n_2 + 1)}{2} - R_2$; for a two-tailed test, U is the minimum of U_1 and U_2.

Worked Examples

The Return of Electric Life

In Chapter 9, we looked at the following data on the lifetimes of electrical components from two different suppliers. We tested the hypothesis that the mean lifetimes of the components are equal using a t test based on the assumption that the lifetimes are normally distributed. Test the same idea without assuming that the lifetimes are normal. Use $\alpha = 0.10$.

Supplier 1						Supplier 2			
60	53	53	69	56	54	57	58	51	55
	64	65	47	38	40	47	57	55	58

Solution

The hypotheses are:

H_0: The distributions of the lifetimes are the same
H_1: The distributions of the lifetimes are different.

We are doing a two-tailed test. The level of significance is 0.10. The ranks and their sums are in the following table:

Supplier 1	Supplier 1 Ranks	Supplier 2	Supplier 2 Ranks
60	16.0	57	12.5
53	6.5	58	14.5
53	6.5	51	5.0
69	19.0	55	9.5
56	11.0	47	3.5
54	8.0	57	12.5
64	17.0	55	9.5
65	18.0	58	14.5
47	3.5		
38	1.0		
40	2.0		
$n_1 = 11$	$R_1 = 08.5$	$n_2 = 8$	$R_2 = 81.5$

The critical value of the test is found in the U table. We look in the $\alpha = 0.10$ portion of the table, in the $n_1 = 11$ row and the $n_2 = 8$ column. There we find 23. The decision rule is to reject the null hypothesis if $U < 23$. The value of U is the minimum of U_1 and U_2:

$$U_1 = n_1 n_2 + \frac{n_1(n_1+1)}{2} - R_1 = (11)(8) + \frac{11(12)}{2} - 108.5 = 45.5$$

$$U_2 = n_1 n_2 + \frac{n_2(n_2+1)}{2} - R_2 = (11)(8) + \frac{8(9)}{2} - 81.5 = 42.5.$$

So $U = 42.5$. Since 42.5 is not less than 23, the null hypothesis is not rejected, and we fail to see any difference between the lifetimes of the components from the two suppliers.

Practice Exercises 13-4

1. A forestry major wants to compare the number of mites found in trees exposed to sunlight versus trees that are shaded. He theorizes that there would be more in the shaded trees than in the trees in the sun. He samples five of each type of tree and counts the number of mites on a total of 20 leaves per tree. The data follows. Use the Mann-Whitney test to check the theory, using a level of significance of 0.05.

Sun	Shade
27	42
33	58
43	48
29	44
37	41

2. A criminologist wanted to compare the frequency with which police officers in New York and Los Angeles discharged their weapons. She took independent samples of twenty-year veterans from both cities and detected how many times each officer had discharged his or her weapon, with the results below. Use the Mann-Whitney test to decide if New York officers fired their weapons more often then LA officers. Use a level of significance of 0.01.

NYC	LA
0	0
2	2
3	1
1	2
1	0
1	0
5	1
4	0
1	0
2	
2	
2	
3	
4	

Solutions to Practice Exercises 13-4

1. The hypotheses are:

    H_0: The number of mites in the sun trees and the shade trees is the same

    H_1: There are fewer mites in the sun trees then in the shade trees.

    This is a left-tailed test. The level of significance has been chosen to be 0.05. The following table contains the ranks and their sums:

Sun	Ranks Sun	Shade	Ranks Shade
27	1	42	6
33	3	58	10
43	7	48	9
29	2	44	8
37	4	41	5
$n_1 = 5$	$R_1 = 17$	$n_2 = 5$	$R_2 = 38$

We find the critical value of the test by looking in the U table in the 0.05 for one-tailed test section, in the $n_1 = 5$ row and the $n_2 = 5$ column. There we find 4. The decision rule is to reject H_0 if $U < 4$. Because this a left-tailed test:

$$U = U_2 = n_1 n_2 + \frac{n_2(n_2 + 1)}{2} - R_2 = (5)(5) + \frac{5(6)}{2} - 38 = 2.$$

Because $2 < 4$, we reject the null hypothesis and conclude that there are fewer mites in the sun trees.

2. The hypotheses are:

H_0: The frequency of weapons being discharged is the same in NYC and LA

H_1: The frequency of weapons being discharged is greater in NYC than in LA.

This is a right-tailed test. The level of significance has been chosen to be 0.01. The following table contains the ranks and their sums:

NYC	Rank NYC	LA	Rank LA
0	3.5	0	3.5
2	15.5	2	15.5
3	19.5	1	9.5
1	9.5	2	15.5
1	9.5	0	3.5
1	9.5	0	3.5
5	23.0	1	9.5
4	21.5	0	3.5
1	9.5	0	3.5
2	15.5		
2	15.5		
2	15.5		
3	19.5		
4	21.5		
$n_1 = 14$	$R_1 = 208.5$	$n_2 = 9$	$R_2 = 67.5$

We find the critical value of the test by looking in the U table in the 0.01 for one-tailed test section, in the $n_1 = 14$ row and the $n_2 = 9$ column. There we find 36. The decision rule is to reject H_0 if $U < 36$. Because this a right-tailed test:

$$U = U_1 = n_1 n_2 + \frac{n_1(n_1 + 1)}{2} - R_1 = (14)99) + \frac{14(15)}{2} - 208.5 = 22.5.$$

Because $22.5 < 36$, we reject the null hypothesis and conclude that officers in NYC discharge their weapons more often than the officers in LA.

13-5 The Kruskal-Wallis Test

Study Objectives

You should be able to:

1. Know when the Kruskal-Wallis test is appropriate.
2. Run a Kruskal-Wallis test.

Section Overview

We learned how one-way analysis of variance can examine three or more independent random samples to see if they are from populations with equal means. ANOVA requires that the samples come from normal populations with equal variances. When the assumption of normality is unreasonable, we can use the Kruskal-Wallis test for the one-way analysis of variance test. The hypotheses for the Kruskal-Wallis test are the same as those of ANOVA. We rank the observations from the samples, identically to the Mann-Whitney test. The test statistic is:

$$H = \frac{12}{N(N+1)} \left(\frac{R_1^2}{n_1} + \frac{R_2^2}{n_2} + \cdots + \frac{R_k^2}{n_k} \right) - 3(N+1),$$

where N = total sample size, k = the number of samples, The R's are the rank sums, and the n's are the sample sizes. The decision rule is to reject H_0 if the value of H exceeds a χ^2 table value with $k - 1$ degrees of freedom.

Key Terms & Formulas

Kruskal-Wallis Test Also known as the H test, it is the nonparametric alternative to a one-way analysis of variance. It is used when the assumption of equal variances is reasonable, but the assumption of normality is not.

H Test Statistic This is the test statistic for the Kruskal-Wallis test. It has a χ^2 distribution with $k - 1$ degrees of freedom and is calculated with:

$$H = \frac{12}{N(N+1)} \left(\frac{R_1^2}{n_1} + \frac{R_2^2}{n_2} + \cdots + \frac{R_k^2}{n_k} \right) - 3(N+1).$$

Worked Examples

Reroped

In an experiment to evaluate the resiliency of climbing ropes, a mountaineering shop attached 250-pound weights to four brands of ropes and dropped them 150 feet. After the drops, they measured the increase in the length of the rope. The measurements are in the table below. Use the Kruskal-Wallis test to compare the ropes using $\alpha = 0.01$.

Rope 1	Rope 2	Rope 3	Rope 4
13	12	7	17
17	14	9	15
18	15	11	14
15	15		
	20		
	13		

Solution

The hypotheses of interest are:

H_0: The mean increase is the same for the ropes
H_1: At least one rope differs from the rest.

The level of significance was given as 0.01. We produced the ranks in the following table:

Rope 1	Rank 1	Rope 2	Rank 2	Rope 3	Rank 3	Rope 4	Rank 4
13	5.5	12	4.0	7	1	17	13.5
17	13.5	14	7.5	9	2	15	10.5
18	15.0	15	10.5	11	3	14	7.5
15	10.5	15	10.5				
		20	16.0				
		13	5.5				
$n_1 = 4$	$R_1 = 44.5$	$n_2 = 6$	$R_2 = 54.0$	$n_3 = 3$	$R_3 = 6.0$	$n_4 = 3$	$R_4 = 31.5$

The distribution of the test statistic is χ^2 with $4 - 1 = 3$ degrees of freedom. The critical value of the test is found in the 3-df row and the 0.01 column of the χ^2 table. We find 11.34. The decision rule is to reject H_0 if $H > 11.34$; otherwise, fail to reject H_0. Then we compute H:

$$H = \frac{12}{N(N+1)}\left(\frac{R_1^2}{n_1} + \frac{R_2^2}{n_2} + \frac{R_3^2}{n_3} + \frac{R_4^2}{n_4}\right) - 3(N+1)$$

$$= \frac{12}{16(16+1)}\left(\frac{44.5^2}{4} + \frac{54^2}{6} + \frac{6^2}{3} + \frac{31.5^2}{3}\right) - 3(16+1) = 7.40.$$

Because 7.40 is not greater than 11.34, we fail to reject H_0. There is not enough evidence at $\alpha = 0.01$ to conclude that there is any difference in the mean increase in the lengths of the ropes.

Practice Exercises 13-5

1. Three different books are on the market that discuss "fuzzy logic," a concept emerging in engineering and computer systems. A publisher assigns 13 editors to each read one book and rate the book on 10 characteristics on a scale of 0 to 10. Then each editor totals his or her points and gives the book a total score. The publisher is interested in knowing if there is a significant difference in the ratings. Use the Kruskal Wallis test at $\alpha = 0.01$ to compare the ratings.

Book 1	Book 2	Book 3
57	90	65
70	81	78
79	69	97
60	96	86
57		

2. An office manager is evaluating three grammar-checking programs. She wants to compare the number of errors each would find on a sample of the manager's writing. Running each program four times, they detect the number of errors listed in the following table. Perform the Kruskal-Wallis test at $\alpha = 0.05$.

Grammar1	Grammar2	Grammar3
10	11	13
8	12	10
9	8	14
9	9	15

Solutions to Practice Exercises 13-5

1. The hypotheses of interest are:
    H_0: The mean rating is the same for the three books
    H_1: At least one rating differs from the rest.
    The level of significance was given as 0.01. We produced the ranks in the following table:

Book 1	Rank 1	Book 2	Rank 2	Book 3	Rank 3
57	1.5	90	11	65	4
70	6.0	81	9	78	7
79	8.0	69	5	97	13
60	3.0	96	12	86	10
57	1.5				
$n_1 = 5$	$R_1 = 20$	$n_2 = 4$	$R_2 = 37$	$n_3 = 4$	$R_3 = 34$

    The distribution of the test statistic is χ^2 with $3 - 1 = 2$ degrees of freedom. The critical value of the test is found in the 2-df row and the 0.01 column of the χ^2 table. It is 9.21. The decision rule is to reject H_0 if $H > 9.21$; otherwise, fail to reject H_0. Then we compute H:

    $$H = \frac{12}{N(N+1)}\left(\frac{R_1^2}{n_1} + \frac{R_2^2}{n_2} + \frac{R_3^2}{n_3}\right) - 3(N+1)$$

    $$= \frac{12}{13(13+1)}\left(\frac{20^2}{5} + \frac{37^2}{4} + \frac{34^2}{4}\right) - 3(13+1) = 4.90.$$

    Because 4.90 is not greater than 9.21, we fail to reject H_0 and conclude that there is not enough evidence at $\alpha = 0.01$ to conclude that there is any difference in the mean ratings given to the books.

2. The hypotheses are:
    H_0: The number of errors is the same for the three grammar-checking programs

H_1: There is at least one difference in the three grammar-checking programs.

The level of significance has been chosen to be 0.05. The calculations are in the following table:

Program 1	Ranks 1	Program 2	Ranks 2	Program 3	Ranks 3
10	6.5	11	8.0	13	10.0
8	1.5	12	9.0	10	6.5
9	4.0	8	1.5	14	11.0
9	4.0	9	4.0	15	12.0
$n_1 = 4$	$R_1 = 16$	$n_2 = 4$	$R_2 = 22.5$	$n_3 = 4$	$R_3 = 39.5$

The distribution of the test statistic is χ^2 with $3 - 1 = 2$ degrees of freedom. The critical value of the test is found in the 2-df row and the 0.05 column of the χ^2 table. We find 5.99. The decision rule is to reject H_0 if $H > 5.99$; otherwise, fail to reject H_0.

$$H = \frac{12}{N(N+1)}\left(\frac{R_1^2}{n_1} + \frac{R_2^2}{n_2} + \frac{R_3^2}{n_3}\right) - 3(N+1)$$

$$= \frac{12}{12(12+1)}\left(\frac{16^2}{4} + \frac{22.5^2}{4} + \frac{39.5^2}{4}\right) - 3(12+1) = 5.66.$$

Because 5.66 is not greater than 5.99 we fail to reject H_0. There is not enough evidence at $\alpha = 0.01$ to conclude that there is any difference in the mean number of errors caught by the grammar-checking programs.

13-6 Runs Test for Randomness

Study Objectives

You should be able to:

1. Define a run.
2. Do the runs test for randomness of data.

Section Overview

Data is often collected in sequence. For example, we might observe the class GPA given by a statistics instructor over the last 10 quarters or the genders of the next 20 births at General Hospital. In such cases, we would like to be able to figure out if there is a pattern over time. This can be accomplished by using a runs test for randomness.

A run is a sequence of similar observations in a row. For example, a run might be three quarters in a row in which your stats' instructor gave above average grades, or a run could be six female births in a row at General Hospital. The runs test counts the number of runs that occur in a sequence of data. The null hypothesis says that the data is random, the alternative says that the data shows a pattern. If there are many short runs or just a few long runs, that is an indication of a lack of randomness. Appendix 9 in the text contains the unusually small and the unusually large number of runs that lead to a rejection of H_0.

Key Terms & Formulas

Runs Test for Randomness A test that decides if a sequence of data is random or if there is an underlying pattern to the data.

Run A run is a sequence of similar observations in a row.

Worked Examples

Hits

A baseball player believes that multiple home run games go in streaks, i.e., nonrandomly. To test this hypothesis, the player examines the number of homers hit by his team in June:

2, 3, 0, 1, 0, 1, 2, 3, 6, 3, 3, 1, 0, 1, 1, 2, 5, 1, 0, 2, 4, 3, 3, 2, 3, 4, 1, 0, 0, 0.

Test to see if the player's contention is true at $\alpha = 0.05$.

Solution
The hypotheses for this test are:
H_0: The multiple home run games occur randomly
H_1: The multiple home run games occur in a pattern (streaks).
The only significance level possible with our table is 0.05. We count the number of runs by first listing the observations. Below this list we place a plus sign for multiple homer games and a minus sign for nonmultiple homer games. There are $r = 8$ runs.

```
2 3 0 1 0 1 2 3 6 3 3 1 0 1 1 2 5 1 0 2 4 3 3 2 3 4 1 0 0 0
+ + − − − − + + + + + − − − − + + − − + + + + + + + − − − −
```

The next step is to identify the n_1 = the number of plus signs and n_2 = the number of minus signs. We find these are 16 and 14. Entering the tables in Appendix 9, we look for the critical values of the test in the $n_1 = 16$ row and the $n_2 = 14$ column. We find 10 and 22. The decision rule is to reject H_0 in favor of H_1 if $r \geq 22$ or $r \leq 10$. Since $8 \leq 10$, we conclude that multiple home run games do occur in streaks.

Practice Exercises 13-6

1. The manager of a movie theater believes that attendance at a movie usually follows a pattern. To test this, the manager notes the two-week attendance for a two-week revival of The Ipcress File:

    121   133   140   124   109   111   142   143   216   233   233   187   180   151

    Test to see if the manager's contention is true at $\alpha = 0.05$.

2. An inspector examines computer chips coming off an assembly line. She classifies each chip as S = satisfactory or D = defective. Inspecting 27 chips in sequence, she observes the following sequence:

    S S S S S S S D D D D S S S D D D S S S S S S S S S

    Does this suggest a lack of randomness?

Solutions to Practice Exercises 13-6

1. The hypotheses for this test are:
    H_0: The attendance is randomly
    H_1: The attendance has a pattern.
    The significance level is 0.05. We count the number of runs by first listing the observations. Below this list we place a plus sign for an increase in attendance and a minus sign for a decrease. There are $r = 4$ runs. Because of the tie, the number of signs is 12 instead of 13:

    ```
    121   133   140   124   109   111   142   143   216   233   233   187   180   151
          +     +     −     −     +     +     +     +     +     0     −     −     −
    ```

    The number of plus signs is $n_1 = 7$, and the number of minus signs is $n_2 = 5$. The critical values are found in the $n_1 = 7$ row and the $n_2 = 5$ column of Appendix 9. We find 3 and 11. The decision rule is to reject H_0 in favor of H_1 if $r \geq 11$ or $r \leq 3$. Since 4 is between these two values, we conclude that attendance could be random.

2. The hypotheses for this test are:
    H_0: There is randomness in the sequence of satisfactory and defective chips
    H_1: There is a pattern to the sequence of satisfactory and defective chips.
    Because of our tables, the significance level is 0.05. There are $r = 5$ runs. The number of satisfactory chips is $n_1 = 20$, and the number of defective chips is $n_2 = 7$. The critical values are found in the $n_1 = 20$ row and the $n_2 = 7$ column of Appendix 9. We find 6 and 16. The decision rule is to reject H_0 in favor of H_1 if $r \geq 16$ or $r \leq 6$. Since $5 < 6$, we conclude that there is a pattern to the sequence of satisfactory and defective chips.

13-7 Spearman Rank Correlation Coefficient

Study Objectives

You should be able to:

1. Calculate the Spearman rank correlation coefficient.
2. Test for a significant rank correlation.

Section Overview

In Chapter 12 we learned about the coefficient of correlation (r). It measures the strength of the relationship for two numerical variables. As an alternative for ranked data, we have the Spearman rank correlation coefficient (r_s). It measures the strength of the relationship between two ordinal variables.

To calculate r_s we first rank the data. Next, we compute the differences between ranks (D). Then:

$$r_s = 1 - \left[\frac{6 \Sigma D^2}{n(n^2 - 1)} \right].$$

We interpret the value of r_s similarly to r. Values close to $+1$ or -1 suggest a strong relationship, values close to 0 suggest a weak or no relationship.

We can test to see if r_s is significantly different from 0. The test statistic is

$$t = r_s \sqrt{\frac{n - 2}{1 - r_s^2}},$$

which has a t distribution with $n - 2$ degrees of freedom. The test is done similarly to any other t test.

Key Terms & Formulas

Spearman Rank Correlation Coefficient (r_s) The correlation of ranked or ordinal data. It can be calculated with $r_s = 1 - \left[\frac{6 \Sigma D^2}{n(n^2 - 1)} \right]$, where D represents the differences for the ranked data.

Test Statistic for Correlation of Ranked Data To test for a significant correlation between ranked data, the test statistic is $t = r_s \sqrt{\frac{n - 2}{1 - r_s^2}}$, which has a t distribution with $n - 2$ degrees of freedom.

Worked Examples

Car Rankings

An automotive magazine listed characteristics of cars and asked readers to rate their importance. The results were broken down by gender and the results (1 = most important, 10 = least important) are presented in the following table. Calculate r_s for this table, and use it to test for a significant correlation between the ranks. Let $\alpha = 0.10$ for the test.

Aspect	Females	Males
Style	4	1
Size	7	5
Comfort	2	6
Power	6	3
Mileage	9	8
Storage	5	9
Color	8	10
Speed	10	2
Price	1	4
Other	3	7

Solution

The calculations are summarized in the following table:

Aspect	Females	Males	D	D^2
Style	4	1	3	9
Size	7	5	2	4
Comfort	2	6	−4	16
Power	6	3	3	9
Mileage	9	8	1	1
Storage	5	9	−4	16
Color	8	10	−2	4
Speed	10	2	8	64
Price	1	4	−3	9
Other	3	7	−4	16
				$\Sigma D^2 = 148$

Then $r_s = 1 - \left[\dfrac{6\Sigma D^2}{n(n^2-1)} \right] = 1 - \left[\dfrac{6(148)}{10(10^2-1)} \right] = 1 - 0.897 = 0.103.$

For the test, the hypotheses of interest are:

H_0: The rankings of males and females are independent (or $\rho_s = 0$)

H_1: The rankings of males and females are related (or $\rho_s \neq 0$).

The level of significance is given as 0.10. The test ratio has a t distribution. Because there are 10 pairs of observations, the degrees of freedom are $10 - 2 = 8$. Looking in the table in the 8-df row and the $0.10/2 = 0.05$ column, we find the critical value of 1.860. The decision rule for this two-tailed test is to reject H_0 in favor of H_1 if $t > 1.860$ or if $t < -1.860$. The test ratio is

$$t = r_s \sqrt{\frac{n-2}{1-r_s^2}} = 0.103 \sqrt{\frac{10-2}{1-0.103^2}} = 0.293.$$

Since the test statistic is not in the critical region, we conclude that the rankings are independent.

Practice Exercises 13-7

1-2. A sociologist has a theory that says that the more enjoyable people perceive a career to be, the less well paid they believe it is. Listing 12 careers, she asks people to rank (1 = high, 12 = low) them on how they think they would enjoy each and how well paid they believe each career to be. The consensus is in the following table:

Career	Enjoyment	Salary
Actuary	10	2
Cleric	4	12
Clown	2	11
Cook	6	10
Duck hunter	9	8
Fire fighter	5	9
Nanny	7	7
Postal clerk	12	5
Pro football	8	3
Rodeo rider	11	6
Statistician	1	1
Thespian	3	4

For this data:

1. Calculate r_s.

2. Test for independence of perceived enjoyment and salary with $\alpha = 0.05$.

3. Employers and employees are asked to rank the most important characteristics of a good employee. The results are in the table that follows. Use this table to calculate r_s.

Characteristic	Employers	Employees
Hard worker	2	3
Willing to do overtime	3	8
Efficient	5	5
Team player	4	7
Friendly	8	6
Smart	6	2
Dedicated	1	4
Educated	7	1

Solutions to Practice Exercises 13-7

1. The calculations are summarized in the following table:

Career	Enjoyment	Salary	D	D^2
Actuary	10	2	8	64
Cleric	4	12	−8	64
Clown	2	11	−9	81
Cook	6	10	−4	16
Duck hunter	9	8	1	1
Fire fighter	5	9	−4	16
Nanny	7	7	0	0
Postal clerk	12	5	7	49
Pro football	8	3	5	25
Rodeo rider	11	6	5	25
Statistician	1	1	0	0
Thespian	3	4	−1	1
				$\Sigma D^2 = 342$

Then $r_s = 1 - \left[\dfrac{6\Sigma D^2}{n(n^2-1)}\right] = 1 - \left[\dfrac{6(342)}{12(12^2-1)}\right] = 1 - 1.195 = -0.195$.

2. The test hypotheses are:

H_0: The rankings for enjoyment and salary are independent

H_1: There is a negative relation between rankings of enjoyment and salary.

The level of significance is given as 0.05. The test ratio has a t distribution with $12 - 2 = 10$ degrees of freedom. Looking in the table in the 10-df row and the 0.05 column, we find the table value of 1.812. The decision rule for this left-tailed test is to reject H_0 in favor of H_1 if $t < -1.812$. The test statistic is

$t = r_s\sqrt{\dfrac{n-2}{1-r_s^2}} = -0.195\sqrt{\dfrac{12-2}{1-(-0.195)^2}} = -0.629$. Since the test statistic is not in the critical region, we

conclude that the rankings are independent.

3. The calculations are summarized in the following table.

Characteristic	Employers	Employees	D	D^2
Hard worker	2	3	−1	1
Willing to do overtime	3	8	−5	25
Efficient	5	5	0	0
Team player	4	7	−3	9
Friendly	8	6	2	4
Smart	6	2	4	16
Dedicated	1	4	−3	9
Educated	7	1	6	36
				$\Sigma D^2 = 100$

Then $r_s = 1 - \left[\dfrac{6\Sigma D^2}{n(n^2-1)}\right] = 1 - \left[\dfrac{6(100)}{8(8^2-1)}\right] = 1 - 1.190 = -0.190$.

Solutions to Odd-Numbered Exercises

1. Step 1. H_0: $p = 0.5$ and H_1: $p \neq 0.5$.
   Step 2. $\alpha = 0.05$.
   Step 3. Tally the sign difference.

Person	Level Before Use	Level After Use	Sign
A	263	214	−
B	194	188	−
C	273	284	+
D	185	185	0
E	238	264	+
F	212	190	−
G	189	185	−
H	164	153	−
I	248	248	0
J	261	229	−

   6 minus, 2 pluses, 2 zeroes.
   Step 4. Use the binomial probability table with $n = 8$, $r = 2$, and $p = 0.5$.
   Step 5. Reject H_0 in favor of H_1 if $0.05 >$ the probability of the sample results.
   Step 6. $P(0) + P(1) + P(2) = 0.0039 + 0.0312 + 0.1094 = 0.1445$.
   Step 7. Since $0.1445 > 0.05$, we fail to reject H_0.

3. Step 1. H_0: The data are random, H_1: the data are not random.
   Step 2. $\alpha = 0.05$.
   Step 3. Count the number of runs.

   ```
   +  +  +  −  −  −  −  +  −  0  0  +  +  +  +  +  +  −  −  −  −
         1              2  3  4                          5
   +  −  0  0  +  +  +  −  0  −
   6  7  8           9     10
   ```

   Hence r (the number of runs) is 10.
   Step 4. Calculate the number of '+', $n_1 = 14$, and the number of '−', $n_2 = 12$.
   Step 5. Fail to reject H_0 if $a <$ the sample r value $< b$. Reject H_0 in favor of H_1 if the sample r value is not between a and b.
   Step 6. Using Appendix 9 with $n_1 = 14$ and $n_2 = 12$, we get the values $a = 8$ and $b = 20$.
   Step 7. Since $a < 10 < b$, we fail to reject H_0.

5. Step 1. H_0: $p_s = 0$ and H_1: $p_s > 0$.
   Step 2. $\alpha = 0.01$.
   Step 3. Use the t distribution.
   Step 4. With $36 - 2 = 34$ df the critical t value is about 2.437 (between 30 df and 40 df).
   Step 5. Reject H_0 in favor of H_1 if $t > 2.437$. Otherwise, fail to reject H_0.
   Step 6. Compute r_s and t.

City	Watches Ranking	CHD Rank	Diff. between Ranks, D	D^2
Boston	2.5	10	−7.5	56.25
Buffalo	4	2	2.0	4.00
New York City	1	1	0.0	0.00
Salt Lake City	11	31	−20.0	400.00
Columbus, OH	19.5	26	−6.5	42.25
Worcester, MA	6	4.5	1.5	2.25
Providence, RI	19.5	3	16.5	272.25
Springfield, MA	22.5	7	15.5	240.25
Rochester, NY	7.5	14.5	−7.0	49.00
Kansas City, MO	32	21	11.0	121.00
St. Louis	15	8	7.0	49.00
Houston	19.5	36	−16.5	272.25
Paterson, NJ	31	4.5	26.5	702.25
Bakersfield, CA	17	20	−3.0	9.00
Atlanta	36	33	3.0	9.00
Detroit	2.5	11	−8.5	72.25
Youngstown, OH	30	6	24.0	576.00
Indianapolis	24.5	22	2.5	6.25
Chicago	27	13	14.0	196.00
Philadelphia	11	16	−5.0	25.00
Louisville, KY	15	18	−3.0	9.00
Canton, OH	15	9	6.0	36.00
Knoxville, TN	11	17	−6.0	36.00
San Francisco	5	27	−22.0	484.00
Chattanooga, TN	24.5	12	12.5	156.25
Dallas	28.5	32	−3.5	12.25
Oxnard, CA	7.5	34	−26.5	702.25
Nashville	33	14.5	18.5	342.25
San Diego	9	24	−15.0	225.00
East Lansing, MI	34.5	29	5.5	30.25
Fresno, CA	19.5	25	−5.5	30.25
Memphis, TN	34.5	30	4.5	20.25
San Jose, CA	22.5	35	−12.5	156.25
Shreveport, LA	28.5	19	9.5	90.25
Sacramento, CA	26	23	3.0	9.00
Los Angeles	13	28	−15.0	225.00
			0.0	5668.50

$$r_s = 1 - \frac{6 \Sigma D^2}{n(n^2 - 1)} = 1 - \frac{6(5668.5)}{36(1296 - 1)} = 1 - 0.7295 = 0.2705.$$

$$t = r_s \cdot \sqrt{\frac{n-2}{1-r_s^2}} = 0.2705 \cdot \sqrt{\frac{34}{1 - 0.0732}} = 1.638.$$

Step 7. Since 1.638 is less than the critical t value 2.437, we fail to reject H_0.

Using MINITAB we get:
```
MTB > Correlation c1-c2.

Correlations (Pearson)

Correlation of Watch and CHD = 0.269, P-Value = 0.112
```

7. Step 1. H_0: $p = 0.5$ and H_1: $p > 0.5$.
   Step 2. $\alpha = 0.05$.
   Step 3. Tally the sign difference.

Worker	Output Before	Output After	Sign
Harris Tweed	80	85	+
Stitch N. Thyme	75	75	0
Les Brown	65	71	+
Mary Taylor	82	79	−
Chuck Moore	56	68	+
Tex Tyle	70	86	+
Ray Ohn	73	71	−
Terry Clothe	62	59	−

3 minus, 4 pluses, 1 zero.
   Step 4. Use the binomial probability table with $n = 7$, $r = 3$, and $p = 0.5$.
   Step 5. Reject H_0 in favor of H_1 if $0.05 >$ the probability of the sample results. Otherwise, fail to reject H_0.
   Step 6. $\sum_{j=0}^{3} P(j) = P(0) + P(1) + P(2) + P(3) = 0.0078 + 0.0547 + 0.1641 + 0.2734 = 0.5000$.
   Step 7. Since $0.5000 > 0.05$, we fail to reject H_0.

9. Step 1. H_0: There is randomness in the data sequence, and H_1: The data are not random.
   Step 2. $\alpha = 0.05$.
   Step 3. Count the number of runs.

o   o   o   e   e   o   e   o   o   e   e   e   o   o   o   e   e   o   e   o
    1   2 3 4   5   6   7 8   9 10 11

Hence r (the number of runs) is 11.
   Step 4. Calculate the number of odds, $n_1 = 11$, and the number of evens, $n_2 = 9$.
   Step 5. Reject H_0 in favor of H_1 if the sample r value is not between a and b. Otherwise, fail to reject H_0.
   Step 6. Using Appendix 9 with $n_1 = 11$ and $n_2 = 9$, we get the values $a = 6$ and $b = 16$.
   Step 7. Since $a < 11 < b$, we fail to reject H_0.

11. Step 1. H_0: $p = 0.5$ and H_1: $p \neq 0.5$.
    Step 2. $\alpha = 0.05$.
    Step 3. Tally the sign difference.

Test 1	Test 2	Sign
12	19	+
12	11	−
10	10	0
16	16	0
14	16	+
14	14	0
12	11	−
17	14	−
16	18	+
13	14	+
16	14	−

4 minus, 4 pluses, 3 zeroes.
    Step 4. Use the binomial probability table with $n = 8$, $r = 4$, and $p = 0.5$.
    Step 5. Reject H_0 in favor of H_1 if $0.05 >$ the probability of the sample results.
    Step 6. $P(0) + P(1) + P(2) + P(3) + P(4) = 0.0039 + 0.0312 + 0.1094 + 0.2188 + 0.2734$
    $$= 0.6367$$

Step 7. Since $0.6367 > 0.05$, we fail to reject H_0.

13. Compute r_s.

Manager	Ranking By VPA	Ranking By VPB	Difference between Ranks, D	D^2
1	2	4	-2	4
2	3	1	2	4
3	6	5	1	1
4	1	2	-1	1
5	5	6	-1	1
6	4	3	1	1
			0	12

$$r_s = 1 - \frac{6\Sigma D^2}{n(n^2 - 1)} = 1 - \frac{6(12)}{6(36 - 1)} = 1 - 0.343 = 0.657.$$

Using MINITAB we get:
```
MTB > Correlation 'VPA' 'VPB'.

Correlations (Pearson)


Correlation of VPA and VPB = 0.657, P-Value = 0.156
```

15. Step 1. H_0: There is randomness in the data sequence, and H_1: The data are not random.
Step 2. $\alpha = 0.05$.
Step 3. Count the number of runs, ignore the on-times.

L L O O E L E E E O E O L L L L L L O L L E E
  1 2 3   4   5
E E E O L L L L E E L L L O O L L
  6   7   8   9

Hence r (the number of runs) is 9.
Step 4. Calculate the number of late arrivals, $n_1 = 20$, and the number of early arrivals, $n_2 = 12$.
Step 5. Reject H_0 in favor of H_1 if the sample r value is not between a and b. Otherwise, fail to reject H_0.
Step 6. Using Appendix 9 with $n_1 = 20$ and $n_2 = 12$, we get the values $a = 10$ and $b = 22$.
Step 7. Since $9 < a = 10$, we reject H_0.

17. Step 1. H_0: The measurements from the devices are the same. H_1: The measurements are lower from device 2.
Step 2. $\alpha = 0.05$.
Step 3. Compute the difference and count the number of differences $\neq 0$.

Device 1	Device 2	Difference
10.51	11.97	1.46
20.30	19.47	-0.83
27.88	30.19	2.31
40.56	38.43	-2.13
47.57	46.23	-1.34
58.35	57.15	-1.20
66.63	66.27	-0.36

The number of difference is $n = 7$.
Step 4. Use the one-tailed T statistic test. With $n = 7$, the table value of T is 3.
Step 5. Reject H_0 in favor of H_1 if the computed T value $\leq$ the table T value. Otherwise fail to reject H_0.

Step 6. Complete the table and compute T.

Device 1	Device 2	Difference	Rank	Sign Positive	Rank Negative
10.51	11.97	1.46	5	5	
20.30	19.47	−0.83	2		−2
27.88	30.19	2.31	7	7	
40.56	38.43	−2.13	6		−6
47.57	46.23	−1.34	4		−4
58.35	57.15	−1.20	3		−3
66.63	66.27	−0.36	1		−1
				12	−16

For this lower tail test, T is the sum of the positive ranks, thus $T = 12$.

Step 7. Since the computed T value 12 is greater than the table value of 3, we fail to reject H_0.

19. Step 1. H_0: The means for the two groups are equal, and H_1: The means are not equal.

Step 2. $\alpha = 0.02$.

Step 3. Rank the data irrespective of sample category.

Smokers	Rank		Nonsmokers	Rank
58	6		86	15
50	3		68	9
44	2		72	10.5
97	16		63	7.5
80	14		79	13
63	7.5		73	12
55	5		72	10.5
$n_1 = 7$	$R_1 = 53.5$		52	4
			43	1
			$n_2 = 9$	$R_2 = 82.5$

Step 4. Use the two-tailed U statistic test. With $n_1 = 7$, $n_2 = 9$, the table value of U is 9.

Step 5. Reject H_0 in favor of H_1 if the computed U value $\leq$ the table U value. Otherwise fail to reject H_0.

Step 6. Compute U_1 and U_2.

$$U_1 = n_1 n_2 + \frac{n_1(n_1 + 1)}{2} - R_1 = 7 \cdot 9 + \frac{7 \cdot 8}{2} - 53.5 = 37.5$$

$$U_2 = n_1 n_2 + \frac{n_2(n_2 + 1)}{2} - R_2 = 7 \cdot 9 + \frac{9 \cdot 10}{2} - 82.5 = 25.5$$

The U statistic is 25.5 (the lesser of 25.5 and 37.5).

Step 7. Since the computed U value 25.5 is greater than the table value of 9, we fail to reject H_0.

21. Step 1. H_0: $p = 0.5$ and H_1: $p > 0.5$.

Step 2. $\alpha = 0.05$.

Step 3. Candidate A is the choice of 43, and 51 prefer candidate B.

Step 4. Since $n = 94$, we use the z distribution. This is a one-tail test, with $\alpha = 0.05$ the critical z value is 1.645.

Step 5. Reject H_0 in favor of H_1 if $z > 1.645$.

Step 6. With $n = 94$, $p = 0.5$, $\mu = 94 \cdot 0.5 = 47$, and $\sigma = \sqrt{npq} = \sqrt{(94)(0.5)(0.5)} = 4.85$. When

$x = 51$, $z = \dfrac{51 - 47}{4.85} = 0.8251$.

Step 7. Since $0.8251 > 1.645$, we fail to reject H_0.

23. Step 1. H_0: $p = 0.5$ and H_1: $p \neq 0.5$.

Step 2. $\alpha = 0.05$.

Step 3. Tally the sign difference.

Subject	Before	After	Sign
1	135	115	−
2	167	165	−
3	205	163	−
4	115	121	+
5	175	148	−
6	134	141	+
7	110	110	0

4 minus, 2 pluses, 1 zero.

Step 4. Use the binomial probability table with $n = 6$, $r = 2$, and $p = 0.5$.

Step 5. Reject H_0 in favor of H_1 if $0.05 >$ the probability of the sample results.

Step 6. $P(0) + P(1) + P(2) = 0.0156 + 0.0938 + 0.2344 = 0.3438$.

Step 7. Since $0.3438 > 0.05$, we fail to reject H_0.

25. Step 1. H_0: $p = 0.5$ and H_1: $p > 0.5$.

Step 2. $\alpha = 0.05$.

Step 3. 22 locations had an increase in growth, 18 locations had a decrease in growth, and no change in the rest.

Step 4. Since $n = 40$, we use the z distribution. This is a one-tail test, with $\alpha = 0.05$ the critical z value is 1.645.

Step 5. Reject H_0 in favor of H_1 if $z > 1.645$.

Step 6. With $n = 40$, $p = 0.5$, $\mu = 40 \cdot 0.5 = 20$, and $\sigma = \sqrt{npq} = \sqrt{(40)(0.5)(0.5)} = 3.16$. When $x = 22$, $z = \dfrac{22 - 20}{3.16} = 0.6329$.

Step 7. Since $0.6329 < 1.645$, we fail to reject H_0. Growmoor doesn't grow more.

27. Step 1. H_0: The means for the two groups are equal, and H_1: The means are not equal.

Step 2. $\alpha = 0.02$.

Step 3. Rank the data irrespective of sample category.

Urban	Rank		Rural	Rank
32	2.5		45	13
36	8		42	11.5
40	10		34	6.5
32	2.5		42	11.5
33	4.5		29	1
37	9		33	4.5
$n_1 = 6$	$R_1 = 36.5$		34	6.5
			$n_2 = 7$	$R_2 = 54.5$

Step 4. Use the two-tailed U statistic test. With $n_1 = 6$, $n_2 = 7$, the table value of U is 4.

Step 5. Reject H_0 in favor of H_1 if the computed U value $\leq$ the table U value. Otherwise fail to reject H_0.

Step 6. Compute U_1 and U_2.

$$U_1 = n_1 n_2 + \frac{n_1(n_1 + 1)}{2} - R_1 = 6 \cdot 7 + \frac{6 \cdot 7}{2} - 36.5 = 26.5$$

$$U_2 = n_1 n_2 + \frac{n_2(n_2 + 1)}{2} - R_2 = 6 \cdot 7 + \frac{7 \cdot 8}{2} - 54.5 = 15.5$$

The U statistic is 15.5, the lesser of 26.5 and 15.5.

Step 7. Since the computed U value 15.5 is greater than the table value of 4, we fail to reject H_0.

29. Step 1. H_0: $p_s = 0$ and H_1: $p_s \neq 0$.

Step 2. $\alpha = 0.01$.

Step 3. Use the t distribution.

Step 4. With $25 - 2 = 23$ df and a two-tailed test, the critical t values are ± 2.807.

Step 5. Reject H_0 in favor of H_1 if $t > 2.807$ or if $t < -2.807$. Otherwise, fail to reject H_0.

Step 6. Compute r_s and t.

School	Academic Rank	GMAT Rank	Diff. Between Ranks, D	D^2
Stanford	1	1	0.0	0.00
Harvard	4	4.5	−0.5	0.25
U. Pennsylvania	4	4.5	−0.5	0.25
Northwestern	4	7	−3.0	9.00
M.I.T	4	3	1.0	1.00
U. of Chicago	4	11	−7.0	49.00
U. of Michigan	7	14.5	−7.5	56.25
Columbia	9	11	−2.0	4.00
Duke	13	14.5	−1.5	2.25
Dartmouth	9	6	3.0	9.00
U. of Virginia	13	18	−5.0	25.00
Cornell	13	8.5	4.5	20.25
U. Calif. Berkeley	9	11	−2.0	4.00
U.C.L.A.	13	8.5	4.5	20.25
Carnegie Mellon	13	16.5	−3.5	12.25
Yale	22	2	20.0	400.00
U. North Carolina	16.5	16.5	0.0	0.00
New York U.	16.5	20	−3.5	12.25
Indiana U.	18.5	19	−0.5	0.25
U. of Texas	18.5	13	5.5	30.25
U.S.C.	22	22	0.0	0.00
U. of Rochester	22	21	1.0	1.00
Purdue	20	24	−4.0	16.00
U. of Pittsburgh	24.5	25	−0.5	0.25
Vanderbilt	24.5	23	1.5	2.25
			0.0	675.00

$$r_s = 1 - \frac{6\sum D^2}{n(n^2 - 1)} = 1 - \frac{6(675)}{25(625 - 1)} = 1 - 0.260 = 0.740.$$

$$t = r_s \cdot \sqrt{\frac{n-2}{1-r_s^2}} = 0.740 \cdot \sqrt{\frac{23}{1 - 0.5476}} = 5.276.$$

Step 7. Since 5.276 is greater than the critical t value 2.807, we reject H_0. There is a significant correlation.

31. Step 1. H_0: $p_s = 0$ and H_1: $p_s > 0$.

Step 2. $\alpha = 0.05$.

Step 3. Use the t distribution.

Step 4. With $15 - 2 = 13$ df and a two-tailed test, the critical t values are ± 2.160.

Step 5. Reject H_0 in favor of H_1 if $t > 2.160$ or if $t < -2.160$. Otherwise, fail to reject H_0.

Step 6. Compute r_s and t.

Resident	Age Rank	Rutabaga Consumption Rank	Difference between Ranks, D	D^2
1	6	12	−6	36
2	13	15	−2	4
3	5	10	−5	25
4	10	11	−1	1
5	8	8	0	0
6	15	14	1	1
7	2	7	−5	25
8	1	4	−3	9
9	14	13	1	1
10	12	3	9	81
11	3	1	2	4
12	7	6	1	1
13	11	9	2	4
14	9	5	4	16
15	4	2	2	4
			0.0	212

$$r_s = 1 - \frac{6\Sigma D^2}{n(n^2 - 1)} = 1 - \frac{6(212)}{15(225 - 1)} = 1 - 0.379 = 0.621.$$

$$t = r_s \cdot \sqrt{\frac{n-2}{1-r_s^2}} = 0.621 \cdot \sqrt{\frac{13}{1 - 0.3856}} = 2.857.$$

Step 7. Since 2.857 is greater than the critical t value 2.160, we reject H_0. People who eat more of Placebo's rutabagas live longer.

33. Step 1. H_0: $p_s = 0$ and H_1: $p_s \neq 0$.

Step 2. $\alpha = 0.05$.

Step 3. Use the t distribution.

Step 4. With $51 - 2 = 49$ df and a two-tailed test, the critical t value is about ± 2.010 (using the values for 40 df and 60 df).

Step 5. Reject H_0 in favor of H_1 if $t > 1.771$. Otherwise, fail to reject H_0.

Step 6. Compute r_s and t.

State	Tax Rank 1 = lowest	New Cancer Cases 1 = lowest	Difference between Ranks, D	D^2
Alabama	12	32	−20.0	400.00
Alaska	1	1	0.0	0.00
Arizona	19	28	−9.0	81.00
Arkansas	27	22	5.0	25.00
California	35	51	−16.0	256.00
Colorado	31	19	12.0	144.00
Connecticut	44	27	17.0	289.00
Delaware	11	6	5.0	25.00
Dist. of Columbia	50	9.5	40.5	1640.25
Florida	4	49	−45.0	2025.00

Georgia	32	40	−8.0	64.00
Hawaii	41	8	33.0	1089.00
Idaho	36	9.5	26.5	702.25
Illinois	28	45	−17.0	289.00
Indiana	20	38	−18.0	324.00
Iowa	26	23	3.0	9.00
Kansas	24	20	4.0	16.00
Kentucky	21	30	−9.0	81.00
Louisiana	13	31	−18.0	324.00
Maine	47	16	31.0	961.00
Maryland	46	34	12.0	144.00
Massachusetts	48	41	7.0	49.00
Michigan	34	44	−10.0	100.00
Minnesota	42	29	13.0	169.00
Mississippi	14	21	−7.0	49.00
Missouri	17	36	−19.0	361.00
Montana	22	7	15.0	225.00
Nebraska	37	17	20.0	400.00
Nevada	3	14.5	−11.5	132.25
New Hampshire	7	12	−5.0	25.00
New Jersey	33	43	−10.0	100.00
New Mexico	15	14.5	0.5	0.25
New York	51	50	1.0	1.00
North Carolina	30	42	−12.0	144.00
North Dakota	10	4	6.0	36.00
Ohio	38	46	−8.0	64.00
Oklahoma	23	25	−2.0	4.00
Oregon	45	24	21.0	441.00
Pennsylvania	25	48	−23.0	529.00
Rhode Island	43	13	30.0	900.00
South Carolina	18	26	−8.0	64.00
South Dakota	6	5	1.0	1.00
Tennessee	5	37	−32.0	1024.00
Texas	8	47	−39.0	1521.00
Utah	39	11	28.0	784.00
Vermont	40	3	37.0	1369.00
Virginia	29	39	−10.0	100.00
Washington	9	33	−24.0	576.00
West Virginia	16	18	−2.0	4.00
Wisconsin	49	35	14.0	196.00
Wyoming	2	2	0.0	0.00
			0.0	18257

$$r_s = 1 - \frac{6\Sigma D^2}{n(n^2-1)} = 1 - \frac{6(18,257)}{51(2601-1)} = 1 - 0.826 = 0.174.$$

$$t = r_s \cdot \sqrt{\frac{n-2}{1-r_s^2}} = 0.174 \cdot \sqrt{\frac{49}{1-0.0303}} = 1.237.$$

Step 7. Since 1.237 is between the critical t values ± 2.010, we fail to reject H_0. There is a strong relationship between daily temperature and sales.

35. Step 1. H_0: The mean prices at her store are comparable. H_1: The mean prices at her store are not comparable.

Step 2. $\alpha = 0.05$.

Step 3. Rank the data irrespective of sample category.

Her Store	Rank	Competitor A	Rank	Competitor B	Rank
144	5	168	10.5	184	14
136	2.5	150	7.5	172	12
146	6	142	4	168	10.5
134	1	166	9	187	15
150	7.5	136	2.5	176	13
$n = 5$	22.0	$n = 5$	33.5	$n = 5$	64.5

Step 4. Use the one-tailed χ^2 test with 2 df the H statistic is 5.99.

Step 5. Reject H_0 in favor of H_1 if the computed H value > 5.99. Otherwise fail to reject H_0.

Step 6. Compute the H value using the following formula:

$$H = \frac{12}{N(N+1)} + \left(\frac{R_1^2}{n_1} + \frac{R_2^2}{n_2} + \frac{R_3^2}{n_3} \right) - 3(N+1)$$

$$= \frac{12}{15(15+1)} + \left(\frac{22^2}{5} + \frac{33.5^2}{5} + \frac{64.5^2}{5} \right) - 3(15+1) = 0.05(1153.3) - 48$$

$$= 57.665 - 48 = 9.665.$$

Step 7. Since the computed H value 9.665 is greater than the table value 5.99, we reject H_0. At least one of the stores has different prices.

37. Step 1. H_0: The means for the three plans are equal, and H_1: The means are not equal.

Step 2. $\alpha = 0.05$.

Step 3. Rank the data irrespective of sample category.

City	FFS	Rank	HMO	Rank	PPO	Rank
Atlanta	$3,425	24	$3,259	20	$3,159	16
Chicago	3,746	30	3,133	14	2,800	6
Cleveland	3,408	22	3,465	25	4,171	32
Dallas/Fort Worth	3,538	28	2,963	10	2,946	9
Houston	3,180	17	3,295	21	3,606	29
Los Angeles	3,964	31	3,025	12	4,335	34
Minn. /St. Paul	3,003	11	2,673	4	3,108	13
New York	4,336	35	3,254	19	4,216	33
Philadelphia	3,528	27	2,882	7	2,715	5
Richmond	3,227	18	2,448	1	3,418	23
San Francisco	3,527	26	2,939	8	4,617	36
Seattle	2,659	3	2,624	2	3,148	15
	$n = 12$	272	$n = 12$	143	$n = 12$	251

Step 4. Use the one-tailed χ^2 test with 2 df the H statistic is 5.99.

Step 5. Reject H_0 in favor of H_1 if the computed H value > 5.99. Otherwise fail to reject H_0.

Step 6. Compute the H value using the following formula:

$$H = \frac{12}{N(N+1)} + \left(\frac{R_1^2}{n_1} + \frac{R_2^2}{n_2} + \frac{R_3^2}{n_3} \right) - 3(N+1)$$

$$= \frac{12}{36(36+1)} + \left(\frac{272^2}{12} + \frac{143^2}{12} + \frac{251^2}{12} \right) - 3(36+1) = 0.009009(13119.5) - 111$$

$$= 118.18 - 111 = 7.19.$$

Step 7. Since the computed H value 7.19 is greater than the table value 5.99, we reject H_0.

39. Step 1. H_0: There is randomness in the data sequence, and H_1: The data are not random.

Step 2. $\alpha = 0.05$.

Step 3. Count the number of runs.

+	+	−	+	−	+	−	−	−	−	−	+	+	−	−	−
1	2	3	4	5					6			7			

−	+	+	+	+	+	−	+	+
8					9	10		11

Hence r (the number of runs) is 11.

Step 4. Calculate the number of +'s, $n_1 = 13$, and the number of −'s, $n_2 = 12$.

Step 5. Reject H_0 in favor of H_1 if the sample r value is not between a and b. Otherwise, fail to reject H_0.

Step 6. Using Appendix 9 with $n_1 = 13$ and $n_2 = 12$, we get the values $a = 8$ and $b = 19$.

Step 7. Since $a < 11 < b$, we fail to reject H_0.

41. Step 1. H_0: The mean time to complete the puzzle is the same for both groups, and H_1: The subjects using alcohol take longer.

Step 2. $\alpha = 0.05$.

Step 3. Rank the data irrespective of sample category.

Control Group	Rank	Experimental Group	Rank
63	10	78	18
57	8	77	17
44	4	75	16
70	13	74	15
50	5	80	19
42	2	55	6
64	11	62	9
56	7	72	14
41	1	66	12
$n_1 = 9$	$R_1 = 61$	43	3
		$n_2 = 10$	$R_2 = 129$

Step 4. Use the one-tailed U statistic test. With $n_1 = 9$, $n_2 = 10$, the table value of U is 24.

Step 5. Reject H_0 in favor of H_1 if the computed U value $\leq$ the table U value. Otherwise fail to reject H_0.

Step 6. Compute U_1 and U_2.

$$U_1 = n_1 n_2 + \frac{n_1(n_1 + 1)}{2} - R_1 = 9 \cdot 10 + \frac{9 \cdot 10}{2} - 61 = 74$$

$$U_2 = n_1 n_2 + \frac{n_2(n_2 + 1)}{2} - R_2 = 9 \cdot 10 + \frac{10 \cdot 11}{2} - 129 = 16$$

Because this is a left-tailed test, the computed U is $U_2 = 16$.

Step 7. Since the computed U value 16 is less than the table value of 24, we reject H_0.

43. Step 1. H_0: $p_s = 0$ and H_1: $p_s \neq 0$.

Step 2. $\alpha = 0.05$.

Step 3. Use the t distribution.

Step 4. With $20 - 2 = 18$ df and a two-tailed test, the critical t values are ± 2.101.

Step 5. Reject H_0 in favor of H_1 if $t > 2.101$ or if $t < -2.101$. Otherwise, fail to reject H_0.

Step 6. Compute r_s and t.

Value	Male Rank	Female Rank	Difference between Ranks, D	D^2
Ability utilization	1	2.5	−1.5	2.25
Achievement	2	2.5	−0.5	0.25
Advancement	3	13	−10	100
Aesthetics	4	17	−13	169
Altruism	5	5	0	0
Authority	6	11	−5	25
Autonomy	7	6	1	1
Creativity	8	10	−2	4
Economics	9	7	2	4
Life style	10	12	−2	4
Personal development	11	1	10	100
Physical activity	12	14	−2	4
Prestige	13	8	5	25
Risk	14	19	−5	25
Social interaction	15	16	−1	1
Social relations	16	4	12	144
Variety	17	15	2	4
Working conditions	18	9	9	81
Cultural identity	19	18	1	1
Physical prowess	20	20	0	0
			0.0	694.5

$$r_s = 1 - \frac{6\Sigma D^2}{n(n^2 - 1)} = 1 - \frac{6(694.5)}{20(400 - 1)} = 1 - 0.522 = 0.478.$$

$$t = r_s \cdot \sqrt{\frac{n-2}{1-r_s^2}} = 0.478 \cdot \sqrt{\frac{18}{1 - 0.2285}} = 2.31.$$

Step 7. Since 2.31 is greater than the critical t value 2.101, we reject H_0. The correlation is significant.